254

Encyclopedia of Chemical Processing and Design

12

Encyclopedia of Chemical Processing and Design

EXECUTIVE EDITOR   **John J. McKetta**
ASSOCIATE EDITOR   **William A. Cunningham**

12

**Corrosion to
Cottonseed**

MARCEL DEKKER, INC.   NEW YORK AND BASEL

Library of Congress Cataloging in Publication Data   (Revised)
Main entry under title:

Encyclopedia of chemical processing and design.

Includes bibliographical references.
1. Chemical engineering—Dictionaries. 2. Chemistry,
Technical—Dictionaries. I. McKetta, John J.
II. Cunningham, William Aaron.
TP9.E66   660.2'8'003   75-40646
ISBN   0-8247-2451-8 (v. 1)

MARCEL DEKKER, INC.
270 Madison Avenue, New York, New York, 10016

LIBRARY OF CONGRESS CATALOG CARD NUMBER: 75-40646
ISBN: 0-8247-2462-3

Current printing (last digit):
10 9 8 7 6 5 4 3 2 1

PRINTED IN THE UNITED STATES OF AMERICA

International Advisory Board

Contributors to Volume 12

Darsh K. Aggarwal   Process Engineer, Fluor Engineers and Constructors, Inc., Irvine, California: *Cost, Plant Utility*

Takeji Baba   Managing Director, Japan Cosmetic Industry Association, Tora-nomon, Minato-Ku, Tokyo, Japan: *Cosmetics, Japan*

K. D. Chandrasekaran   Chemical Engineering Education Development Centre, Madras, India: *Cost, References*

James R. Couper   Professor, Department of Chemical Engineering, University of Arkansas, Fayetteville, Arkansas: *Cost, Operating Expenses Estimation*

O. L. Culberson   Professor, Department of Chemical, Metallurgical, and Polymer Engineering, The University of Tennessee, Knoxville, Tennessee: *Cost Engineering*

William A. Cunningham   Department of Chemical Engineering, The University of Texas, Austin, Texas: *Corrosion*

Dennis E. Drayer   Senior Research Engineer, Marathon Oil Company, Littleton, Colorado: *Cost-Capacity Relationships*

John E. Eggleston   Manager, Regional Procurement Operations, Bechtel Corporation, San Francisco, California: *Cost, Equipment and Materials*

W. D. Harris   Department of Chemical Engineering, Texas A & M University, College Station, Texas: *Cottonseed*

A. W. Hawkins   Consultant, Crosslands, Kennett Square, Pennsylvania: *Cost, Cash Flow Concepts*

George M. Hoerner, Jr.   Professor and Executive Officer, Department of Chemical Engineering, Lafayette College, Easton, Pennsylvania: *Cost—Process Equipment (Updating)*

Harry Isacoff   Corporate Director, Technical Services, International Flavors and Fragrances (U.S.), New York, New York: *Cosmetics*

Candace C. Johnnie   Process Engineer, Fluor Engineers and Constructors, Inc., Irvine, California: *Cost, Plant Utility*

Carl E. Locke   Associate Professor, School of Chemical Engineering and Materials Science, The University of Oklahoma, Norman, Oklahoma: *Corrosion: Cathodic and Anodic Protection*

E. E. Ludwig   President, Ludwig Consulting Engineers, Inc., Baton Rouge, Louisiana: *Cost, Capital Estimation*

W. L. Nelson   Technical Editor, Oil and Gas Journal, Tulsa, Oklahoma: *Cost Indexes*

Ivy M. Parker   Former Editor of Publications for the National Association of Corrosion Engineers, Austin, Texas: *Corrosion*

Walter L. Sutor   Manager, Process Design, Hooker Chemical Corp., Niagara Falls, New York: *Cost, Startup for New Plants*

D. Venkateswarlu   Chemical Engineering Education Development Center, Madras, India: *Cost, References*

Contents of Volume 12

Conversion to SI Units

To convert from	To	Multiply by
acre	square meter (m²)	4.046×10^3
angstrom	meter (m)	1.0×10^{-10}
are	square meter (m²)	1.0×10^2
atmosphere	newton/square meter (N/m²)	1.013×10^5
bar	newton/square meter (N/m²)	1.0×10^5
barrel (42 gallon)	cubic meter (m³)	0.159
Btu (International Steam Table)	joule (J)	1.055×10^3
Btu (mean)	joule (J)	1.056×10^3
Btu (thermochemical)	joule (J)	1.054×10^3
bushel	cubic meter (m³)	3.52×10^{-2}
calorie (International Steam Table)	joule (J)	4.187
calorie (mean)	joule (J)	4.190
calorie (thermochemical)	joule (J)	4.184
centimeter of mercury	newton/square meter (N/m²)	1.333×10^3
centimeter of water	newton/square meter (N/m²)	98.06
cubit	meter (m)	0.457
degree (angle)	radian (rad)	1.745×10^{-2}
denier (international)	kilogram/meter (kg/m)	1.111×10^{-7}
dram (avoirdupois)	kilogram (kg)	1.772×10^{-3}
dram (troy)	kilogram (kg)	3.888×10^{-3}
dram (U.S. fluid)	cubic meter (m³)	3.697×10^{-6}
dyne	newton (N)	1.0×10^{-5}
electron volt	joule (J)	1.60×10^{-19}
erg	joule (J)	1.0×10^{-7}
fluid ounce (U.S.)	cubic meter (m³)	2.96×10^{-5}
foot	meter (m)	0.305
furlong	meter (m)	2.01×10^2
gallon (U.S. dry)	cubic meter (m³)	4.404×10^{-3}
gallon (U.S. liquid)	cubic meter (m³)	3.785×10^{-3}
gill (U.S.)	cubic meter (m³)	1.183×10^{-4}
grain	kilogram (kg)	6.48×10^{-5}
gram	kilogram (kg)	1.0×10^{-3}
horsepower	watt (W)	7.457×10^2
horsepower (boiler)	watt (W)	9.81×10^3
horsepower (electric)	watt (W)	7.46×10^2
hundred weight (long)	kilogram (kg)	50.80
hundred weight (short)	kilogram (kg)	45.36
inch	meter (m)	2.54×10^{-2}
inch mercury	newton/square meter (N/m²)	3.386×10^3
inch water	newton/square meter (N/m²)	2.49×10^2
kilogram force	newton (N)	9.806

To convert from	To	Multiply by
kip	newton (N)	4.45×10^3
knot (international)	meter/second (m/s)	0.5144
league (British nautical)	meter (m)	5.559×10^3
league (statute)	meter (m)	4.83×10^3
light year	meter (m)	9.46×10^{15}
liter	cubic meter (m³)	0.001
micron	meter (m)	1.0×10^{-6}
mil	meter (m)	2.54×10^{-6}
mile (U.S. nautical)	meter (m)	1.852×10^3
mile (U.S. statute)	meter (m)	1.609×10^3
millibar	newton/square meter (N/m²)	100.0
millimeter mercury	newton/square meter (N/m²)	1.333×10^2
oersted	ampere/meter (A/m)	79.58
ounce force (avoirdupois)	newton (N)	0.278
ounce mass (avoirdupois)	kilogram (kg)	2.835×10^{-2}
ounce mass (troy)	kilogram (kg)	3.11×10^{-2}
ounce (U.S. fluid)	cubic meter (m³)	2.96×10^{-5}
pascal	newton/square meter (N/m²)	1.0
peck (U.S.)	cubic meter (m³)	8.81×10^{-3}
pennyweight	kilogram (kg)	1.555×10^{-3}
pint (U.S. dry)	cubic meter (m³)	5.506×10^{-4}
pint (U.S. liquid)	cubic meter (m³)	4.732×10^{-4}
poise	newton second/square meter (N · s/m²)	0.10
pound force (avoirdupois)	newton (N)	4.448
pound mass (avoirdupois)	kilogram (kg)	0.4536
pound mass (troy)	kilogram (kg)	0.373
poundal	newton (N)	0.138
quart (U.S. dry)	cubic meter (m³)	1.10×10^{-3}
quart (U.S. liquid)	cubic meter (m³)	9.46×10^{-4}
rod	meter (m)	5.03
roentgen	coulomb/kilogram (c/kg)	2.579×10^{-4}
second (angle)	radian (rad)	4.85×10^{-6}
section	square meter (m²)	2.59×10^6
slug	kilogram (kg)	14.59
span	meter (m)	0.229
stoke	square meter/second (m²/s)	1.0×10^{-4}
ton (long)	kilogram (kg)	1.016×10^3
ton (metric)	kilogram (kg)	1.0×10^3
ton (short, 2000 pounds)	kilogram (kg)	9.072×10^2
torr	newton/square meter (N/m²)	1.333×10^2
yard	meter (m)	0.914

Encyclopedia of Chemical Processing and Design

12

Corrosion

There are numerous definitions of corrosion, the simplest of which is often used when ferrous metals are involved: "just ordinary rusting." A definition of much greater scientific significance is that "corrosion is destruction of a metal by chemical or electrochemical reaction with its environment."

Metals exposed to the elements tend to revert to a thermodynamically more stable form or native state. Iron rusts to produce iron oxide. Noble metals such as gold, platinum, and silver are least reactive while the alkalis and alkaline earth metals are the most reactive. The change from a metallic state to a compound is a form of corrosion. This involves formation of ions which are electrically charged. For example, iron (Fe^0) becomes a positively charged ion (Fe^{2+}) by losing two electrons. These released ions must be picked up by less-active metal ions or by a nonmetal. In acid solutions, hydrogen ions (H^+) pick up electrons to become hydrogen gas (H_2^0). When iron is immersed in a copper sulfate solution, the iron goes into solution as ferrous ion and metallic copper are plated out.

Corrosion phenomena involve a basic electrochemical process in which an anodic material corrodes by transferral of one or more electrons to its surroundings. The recipient of the discharged electrons at the cathode is usually a metal ion or atmospheric oxygen. The complete process may be depicted by the following equations:

At the anode—corrosion or oxidation:

$$M_1^0 \rightarrow M_1^+ + e$$

At the cathode—plating out or reduction:

$$M_2^+ + e \rightarrow M_2^0 \qquad \text{(for a metal)}$$

or

$$\tfrac{1}{2}O_2^0 + H_2O + e \rightarrow 2OH^- \qquad \text{(for oxygen)}$$

Thus, if two electrodes, one of iron and the other of copper, are inserted into a solution of copper sulfate and connected through an outside circuit, not only will the above half-cell reactions take place, but also an electric current will flow. The potential forcing the current to flow is a function of numerous factors, some of which are discussed later.

It is important to note that there are *three* parts involved in the corrosion battery:

1. The *anode* which is corroded or wasted away and from which electrons are released through the external circuit.

2. The *cathode*, where reduction, or electron absorption, takes place.
3. The *external circuit* (wire and solution) through which electrons pass.

All three must be present if current flow or corrosion is to occur. Each half cell, consisting of the element and its ion, contributes to the measurable potential of the battery but, obviously, the potential of one half-cell alone cannot be measured directly. However, it is feasible to measure the potential of a complete battery consisting of any reproducible half-cell and some standard reference half-cell.

By convention, the hydrogen gas–hydrogen ion half-cell, has been adopted as the standard against which all other half-cells will be compared. Unfortunately, maintenance of the conditions required for the hydrogen half-cell is a technically difficult operation. As a consequence, other much more stable half-cells have been used as secondary standards for measurement of potentials. The potentials of the half-cells in question are then easily calculated relative to the hydrogen standard. For additional information, see the *Electrochemical Engineering* article.

By arbitrarily setting the potential of the $H_2^0 \rightarrow 2H^+ + e$ half-cell at 0.00, the metals can be arranged in order of activity or electrode potential. One such arrangement, called the Electromotive Force Series (or emf series), is presented in Table 1. By definition, this is "A list of elements arranged according to their standard electrode potentials, the sign being positive for elements whose potentials are cathodic to hydrogen and negative for those anodic to hydrogen."

There are two widely accepted ways of arranging the half-cells in an orderly manner. They can be listed in descending order from most negative to most positive as in Table 1. This is the one historically and currently used in European literature and which has been adopted by the National Bureau of Standards (U.S.), and by The Electrochemical Society. The reverse arrangement, with the more positive or noble metals at the top, known as the IUPAC (International Union of Pure and Applied Chemistry) system, is used in the *Corrosion: Cathodic and Anodic Protection* article.

In each case the half-cell potential is a fixed value as defined for 25°C, the pure element (metal), and the unit activity of the corresponding ions. It is this uniformity of conditions which permits the orderly arrangement of the metals in the emf series.

The more active element with the higher negative potential has a tendency to displace the less negative element. For example:

$$Fe^0 = Fe^{2+} + 2e, \qquad -0.440 \text{ V}$$
$$Cu^0 = Cu^{2+} + 2e, \qquad +0.345 \text{ V}$$

The potential between two standard electrodes, such as, $Fe = Fe^{2+} + 2e^-$ (-0.440 V) and $Mg = Mg^{2+} + 2e^-$ (-2.34 V) is the absolute difference or 1.90 V. The position of the metal in the emf series gives a clue to its reactivity in acidic media. Those metals listed above hydrogen in Table 1 react with the acidic media to produce hydrogen and those below do not. Other factors, such as

TABLE 1 Electromotive Force Series[a]

Electrode Reaction	Standard Electrode Potential, E^0 (V), 25°C
$K = K^+ + e^-$	-2.922
$Ca = Ca^{2+} + 2e^-$	-2.87
$Na = Na^+ + e^-$	-2.712
$Mg = Mg^{2+} + 2e^-$	-2.34
$Be = Be^{2+} + 2e^-$	-1.70
$Al = Al^{3+} + 3e^-$	-1.67
$Mn = Mn^{2+} + 2e^-$	-1.05
$Zn = Zn^{2+} + 2e^-$	-0.762
$Cr = Cr^{3+} + 3e^-$	-0.71
$Ga = Ga^{3+} + 3e^-$	-0.52
$Fe = Fe^{2+} + 2e^-$	-0.440
$Cd = Cd^{2+} + 2e^-$	-0.402
$In = In^{3+} + 3e^-$	-0.340
$Tl = Tl^+ + e^-$	-0.336
$Co = Co^{2+} + 2e^-$	-0.277
$Ni = Ni^{2+} + 2e^-$	-0.250
$Sn = Sn^{2+} + 2e^-$	-0.136
$Pb = Pb^{2+} + 2e^-$	-0.126
$H_2 = 2H^+ + 2e^-$	0.000
$Cu = Cu^{2+} + 2e^-$	0.345
$Cu = Cu^+ + e^-$	0.522
$2Hg = Hg_2^{2+} + 2e^-$	0.799
$Ag = Ag^+ + e^-$	0.800
$Pd = Pd^{2+} + 2e^-$	0.83
$Hg = Hg^{2+} + 2e^-$	0.854
$Pt = Pt^{2+} + 2e^-$	~ 1.2
$Au = Au^{3+} + 3e^-$	1.42
$Au = Au^+ + e^-$	1.68

[a]H. H. Uhlig (ed.), *Corrosion Handbook*, Wiley, New York, 1948, p. 1134.

dissolved air, may promote reaction, but it is not simple displacement. The standard electrode potentials are useful in understanding the basic behavior of metals, but one does not encounter clear-cut parameters in practice.

In actual corrosion work, potential differences result in galvanic or bimetallic corrosion where two metals are in electrical contact through a conducting medium. A corrosion current is generated and the more anodic material is consumed. For example, steel rivets in copper or Monel sheathing will be consumed. Metals and alloys can be arranged in order of activity in a particular medium. Detailed studies of behavior of metals and alloys in seawater have been carried out by international Nickel Co. at the Francis L. LaQue Laboratory. Table 2, Galvanic Series of Metals and Alloys in Seawater, is very useful in practical corrosion work. Such a series can be developed for any specific medium.

TABLE 2   Galvanic Series of Metals and Alloys in Seawater[a]

Corroded end (anodic—least noble)	
Magnesium	
Magnesium alloys	
	Hastelloy A
Zinc	Hastelloy B
Aluminum (commerially pure)	Brasses
	Copper
Cadmium	Bronzes
	Copper-nickel alloys
Aluminum 2024 (4.5% Cu + 0.6% Mn + 1.5% Mg)	Titanium
	Monel
Steel or iron	Silver solder
Cast iron	
	Nickel (passive)
Chromium-iron (active)	Inconel (passive)
Ni-Resist	Chromium-iron (passive)
	18–8 Cr-Ni-Fe (passive)
18–8 Cr-Ni-Fe (active)	18–8–3 Cr-Ni-Mo-Fe (passive)
18–8–3 Cr-Ni-Mo-Fe (active)	
	Silver
Hastelloy C	
	Graphite
Lead-tin solders	
Lead	Protected end
Tin	(cathodic—most noble)
Nickel (active)	
Inconel (active)	

[a]*Corrosion Testing Bulletin*, International Nickel Co., 1944. p. 17. Reported by F. N. Speller, *Corrosion Causes and Prevention*, 3rd ed., McGraw-Hill, New York, 1951, p. 20.

    The recognition of groupings or families of metals with small potential differences is beneficial in design work. Note that as shown in the galvanic series, certain metals and alloys, e.g., nickel, chrome steels, and 18–8 stainless steels, exist in both active and passive states. The active forms are in the upper or anodic portion of Table 2 while the passive forms are toward the bottom in the cathodic range. This is a very important distinction.

    In view of the numerous names attached to corrosion phenomena, e.g., stress corrosion, stray current corrosion, bimetallic corrosion, and crevice corrosion, the foregoing presentation of corrosion phenomena in terms of electrochemical action may appear to some as being too simple. However, the various names generally indicate the special locations or conditions under which the corrosion takes place. They all involve recognized electrochemical phenomena.

Corrosion Prevention or Retardation

In the strictest sense, one does not *prevent* corrosion in any given system whose electrochemical potential indicates its probability. However, much can be done to decrease the *rate* at which corrosion occurs. More correctly, then, one should refer to it as "corrosion retardation" of the metal whose function is critical.

If the resistance to electron flow to or from any part of the battery cell can be increased to the point where the current is slowed down, or perhaps reversed, corrosion will be retarded. If the resistance is at the cathode, the cell is said to be under cathodic control; if at the anode, it is under anodic control.

Before giving consideration to the several control methods, it is essential that one recognize that most metals are heterogeneous. The granular interfaces themselves provide irregularity over the entire surface of even high purity specimens. Steels and cast irons have carbon content ranging from about 0.025 to 4.5%, usually in the form of cementite, Fe_3C, crystals of which are more or less uniformly dispersed in the ferrous metal itself. These crystals not only have a major influence on the properties of the metal, but also on the electrochemical nature of the surface exposed to a potentially corroding environment.

The physical shaping and fabrication of ferrous metals and structures result in production of areas more susceptible to corrosion. Areas subjected to deformation such as bending or stamping, drilling, threading, and rolling are basically anodic to other areas of the same metal. Hence it is these areas which tend to show the earliest and most destructive corrosion. Some decrease of these effects of physical stress can often be achieved by heat treatment.

While there is substantial overlap, the various ways in which corrosion may be controlled generally will fall within one of the following categories:

1. Control of environment
2. Use of protective coatings
3. Cathodic protection
4. Selective use of ferrous and nonferrous alloys

Control of Environment

The term "environmental control," as used herein, covers not merely the metal and the aqueous and vapor phases in contact with it, but also other factors and conditions within the system and its surroundings. In several instances, they are closely interconnected.

Effect of pH Value and Oxygen Concentration

In most aqueous systems the corrosion rate of ferrous metals is reasonably constant within the pH range of 4 to 9. A sharp increase in corrosion rate occurs as pH is decreased from 4 to 3, and a gradual decrease as pH is raised from 10 to 13. Any corrosion control method dependent on pH of the medium should provide buffer characteristics such that once the desired pH is reached, it may be maintained.

Within the range in which the corrosion rate is generally independent of pH (4 to 9), the controlling factor usually is the rate at which oxygen reaches the surface. The half-cells involved here are:

Anode:

$$2Fe \rightarrow 2Fe^{2+} \rightarrow 4e$$

Cathode:

$$O_2 + 2H_2O \rightarrow 4e \rightarrow 4OH^-$$

Since these reactions may and do occur at finite points on the metal surface, the ions formed interact to form ferrous hydroxide (hydrated ferrous oxide). At higher pH there is a surplus of hydroxyl ions (OH^-), and the precipitation of ferrous hydroxide takes place quite near the anode. This gelatinous film (which, under some circumstances, may change to ferric hydroxide or oxide) influences the corrosion rate by controlling the access of oxygen. If conditions are such that oxygen and air are excluded, the corrosion ceases.

Partial exclusion of air can promote pitting as a side effect by setting up new battery or corrosion cells based on differential oxygen concentration. Complete deaeration will eliminate oxygen corrosion. This is practical for closed systems, such as boiler operation, but impractical for general conditions where other techniques for control must be used.

CAUTION: Use of extremes of pH values above 10 or 11 for corrosion control should be avoided since some metals, e.g., aluminum and zinc, and some woods used in cooling towers may be adversely affected.

Serious problems of pitting may result from oxygen concentration cells previously mentioned. In general, this results from some form of differential aeration, such as in crevice corrosion. Small volumes of medium trapped in holes, under washers or gaskets, in lap joints, under sand, dirt, scale, etc., often result in severe localized corrosion. Alloy steels in which the metals exist in both active and passive states are particularly vulnerable. The maintenance of the passive state depends on the existence of an oxidizing atmosphere. The small amounts of metal in the shielded areas are in contact with stagnant solutions of reduced oxygen content and hence are anodic to the large outside areas exposed to the higher oxygen content. The relative sizes of the two areas and the differential oxygen concentration combine to induce intensified corrosion in the shielded area.

It is of interest to note that although oxygen is essential, the actual corrosion takes place where the oxygen concentration is lowest.

Relative Sizes of Anode and Cathode

Although potential is not a function of electrode size, the relative sizes of anodic and cathodic areas are very important. In most aqueous environments, if the anode is very large in comparison to the cathode, the corrosion effect will be spread over a larger area and hence be of relatively little concern. Conversely, if the cathodic area is large and the anode is small, the results are reversed. For example, the use of steel rivets in copper sheathing in seawater has proven to be catastrophic. Another example is that of crevice corrosion described above. The very large cathodic area in contact with oxygen, affording little or no resistance to electron flow, will accelerate the corrosion at the tiny anodic area at the base of the crevice. This might well result in pitting or perhaps a small hole in the vessel wall even though the total mass of metal corroded away is very small (See also the *Corrosion: Cathodic and Anodic Protection* article.)

Stray Currents

Another environmental effect is the ability of a structure to pick up extraneous or stray currents. The point of pick-up presents no problem, but the current follows the path of least resistance in leaving the structure and may cause serious damage when it does so. At the point of departure from, for example, a pipeline passing through an area of low resistance, the current leaves the pipe, removing an electrochemical equivalent of metal as it does so. Based on Faraday's law, the metal loss will be 20 lb of iron (or 45 lb of lead) per ampere per year. If the low resistance contact is a small area, the structure may be perforated without significant overall loss of metal.

The control is obvious; the system must be fully grounded, with care being taken to assure that the grounding cable is firmly welded to the structure to be protected.

Long line currents which develop when a pipeline passes through non-homogeneous soils may build up large corrosion cells. That is, the anode area may extend over several hundred feet of pipe in low resistance soils, resulting in severe corrosion along the entire length.

Control of this type of corrosion is effected by good coating and by application of adequate cathodic protection.

Protective Coatings

Various types of coatings have long been in use in attempts to protect structures against corrosion. Coatings used range from the near unimolecular film thickness of organic and inorganic inhibitors to mechanical coatings such as the deliberately precipitated scales and complex wrappings and coatings used on pipelines.

Effectiveness of the coatings depends on complete and uninterrupted coverage of the area subject to corrosion. If this is not done, the resultant "hot spots" caused by breaks or "holidays" in the protective covering may result in heavy localized corrosion. Severity of the corrosion may well be enhanced by the relatively small anodic area as compared with the much larger cathodic area.

Inhibitors or Passivators

Inhibitors are materials—organic or inorganic—which change or cover the metallic surface in such a way that corrosion is either greatly reduced or essentially prevented. Both anodic and cathodic inhibitors find wide application. Each produces a very thin film to stifle the corrosion reaction. Some inhibitors, such as chromates, phosphates, and nitrites, retard the anodic reaction and, quite naturally, are known as "anodic inhibitors." Likewise, cathodic inhibitors are those whose effectiveness is obtained through coverage of the cathodic area. The coverage results from reaction with the hydroxide ions (OH^-) or other anions in neutral aqueous solutions so that access to oxygen is limited. Examples are controlled calcium carbonate precipitation and precipitated metal hydroxides such as nickel and zinc.

Other inhibitors, notably polar organic compounds, whose effectiveness results from a chemisorbed layer over the entire metal surface, are more like barrier coatings. Complete coverage is an equilibrium condition so adequate concentration of the inhibitor must be maintained in the contact solution.

A full discussion of characteristics and use of inhibitors is given in this Encyclopedia under the heading *Inhibitors: Corrosion*.

Chemical and Mechanical Coating

Under some conditions substantial retardation of corrosion may result through accumulation of corrosion products generated at the anode area. Such protection depends on the physical nature of the deposit. Porous corrosion products offer little protection. The porosity of the products depends on many factors. Under atmospheric conditions the addition of a small amount of copper to steel results in a tightly adhering protective scale. This effect does not carry over to aqueous solutions.

Hard water scales of $CaCO_3$ and $Mg(OH)_2$ exhibit protective action when formed under controlled conditions. If these are laid down on top of or are mixed with hydrated ferrous–ferric oxides, they are not effective.

On some occasions special scales have been deposited deliberately as a corrosion retardant. Engineers with one of the major sulfur-producing companies added soluble barium salts to the hot seawater being pumped underground in the Frasch operation. The resultant tightly-adherent hard scale of $BaSO_4$ was quite effective in holding corrosion to an acceptable minimum. The treatment had to be repeated from time-to-time in order to cover areas exposed by mechanical erosion.

Paints and organic coatings deposit a protecting film with evaporation of solvent and/or catalyst-induced reactions to produce protective barrier coatings for both atmospheric protection and tank linings for immersion exposure. Metal coatings created by galvanizing, aluminizing, etc. have a long history of atmospheric protection. In addition, metallizing is an important technique used for lining special vessels.

The term "mechanical coatings" covers such things as wrappings, tar coatings, paints, sophisticated chemical resistant coatings, and similar linings. Pipelines have long been protected by layers of multicomponent wrappings, e.g., primer, hot or cold bitumastic reinforced with fiber glass or asbestos fibers, and kraft paper. Recent practices call for use of a large number of thin plastic tapes. In order to minimize "hot spot" corrosion at breaks in the covering, often caused by pipe expansion and contraction after it is buried, the entire line is placed under cathodic protection.

Mechanical coatings cover paints, sophisticated chemical-resistant organic coatings and linings, metallic coatings, wrappings, and tar coatings. Paints and organic coatings provide a protective film or layer by evaporation of solvent and/or catalyst-induced reaction to provide protective barrier coatings for atmospheric protection and linings for immersion service. Hot spray application (metallizing) is an important technique used in lining special vessels.

Cathodic Protection

Corrosion is effected by passage of electrons from the metal (anode) into the surrounding environment. It is obvious that a reversal of electron flow will retard or even totally stop corrosion of the anode. This may be done by use of a sacrificial anode (e. g., zinc or magnesium) or by an impressed current in conjunction with an external electrode.

A detailed discussion of both cathodic and anodic protection is given in the article on *Corrosion: Cathodic and Anodic Protection.*

Corrosion-Resistant Alloys

One obvious and widely used way to reduce the problems of corrosion is to use a metal or alloy not so susceptible to deterioration in the particular medium involved. It is likewise obvious that such metals or alloys are more expensive than iron, so their use will involve an economic balance between added cost and longer life or product improvement.

Alloying elements effective in combating corrosion of ferrous metals are relatively few. Copper, silicon, nickel, and chromium are the ones generally considered to be most useful.

Carbon steels containing from 0.1 to 0.25% copper have been found to be quite effective against atmospheric corrosion. Unfortunately, the same beneficial effect is not observed in fresh or salt water exposure.

Silicon alloyed with iron results in a material highly resistant to corrosion in acid solutions but its resistance is sharply decreased in alkaline media. Silicon is relatively ineffective in concentration below 13% but from that point to about 14.5% the corrosion rate drops very sharply. The commercial iron-silicon alloys have that ratio of iron to silicon; it also contains about 0.85% carbon. As would be expected, it is brittle, subject to both thermal and mechanical shock, and suitable for casting only.

The silicon alloys find their greatest use as sewer and drainage lines and in processes involving strong mineral acids such as H_2SO_4 and H_3PO_4 as well as pure organic acids such as acetic acid. This resistance is decreased sharply by the presence of chlorides.

Nickel and Chromium Alloys

The two most widely used alloying metals are nickel and chromium. Either, or a combination of the two, when alloyed with low carbon steel develops increasing resistance to corrosion in aqueous or atmospheric environment as the concentration increases.

Basically, the effectiveness of nickel is directly related to its concentration in the alloy. Chromium exhibits maximum effectiveness per unit at lower concentrations (up to about 3.5%); some increase in benefit is achieved up to about 20% but above that there is little change in corrosion rate with added chromium.

The Cr-Ni ferrous alloys are generally referred to as stainless steels and, along with high nickel alloys and so-called superalloys, exhibit high corrosion resistance, high oxidation resistance at elevated temperatures, and outstanding mechanical properties. Machining often requires special techniques because many show substantial work-hardening properties.

Stainless steels are grouped into four classes, in each of which are several types of differing but related composition, physical, magnetic, and corrosion-resistant properties. (See the *Stainless Steels* article.)

Low carbon steel containing small additions of copper, nickel, chromium, phosphorus, and silicon performs quite favorably in atmospheric exposures. Nickel or chromium alloyed with low carbon steel shows increasing resistance to corrosion in aqueous and atmospheric conditions as concentration increases, but chromium is more beneficial at a lower concentration.

The highly corrosion-resistant metals with outstanding mechanical properties and oxidation resistance at elevated temperatures range from stainless steels to high nickel alloys to superalloys. Stainless steels are classified in four groups:

1. Martensitic: iron-chromium alloys (400 Series based on Type 420) which are magnetic and heat treatable or hardenable, containing 15 to 17%

chromium. They are widely used in chemical plants and give best performance when used in quenched and tempered or quenched and stress-relieved condition.

2. Ferritic: iron-chromium alloys (400 Series based on Type 430) which are magnetic but not heat treatable (nonhardenable). Type 430, containing 17% chromium, is the basic alloy. These alloys develop maximum softness, ductility, and corrosion resistance in annealed conditions.

3. Austenitic: chrome-nickel alloys (300 Series based on Type 302) with 18% chromium and 8% nickel which are nonmagnetic and nonhardenable by heat treatment. They have the widest range of useful properties— extreme ductility and toughness, superior corrosion resistance, and high strength developed through cold work.

4. Precipitation hardening stainless steels may be either martensitic, semi-austenitic, or austenitic. They are high strength alloys which can be fabricated easily and then hardened. They develop strength through precipitation hardening rather than through phase transformations as in the case of iron-chromium martensitic alloys.

The chromium steels and stainless steels have poor corrosion resistance in reducing atmospheres but have superior resistance in oxidizing atmospheres. They are very susceptible to chloride contamination.

Nickel, high-nickel alloys, and the superalloys complement the stainless steels.

Nickel has outstanding resistance to caustics, high-temperature halogens and hydrogen halide, salts (beware of oxidizing halide salts such as hypohalites), and foods. Nickel-copper (Monels) exceed pure nickel in resistance to sulfuric acid, hydrofluoric acid, and brine. They are free of chloride-ion stress-corrosion cracking and find wide use in handling seawater and brackish water.

More sophisticated alloys of nickel with molybdenum and with or without chromium have given good service in the chemical industry. With proper design and engineering their high costs are balanced by superior performance in both oxidizing and reducing atmosphere.

The subject of corrosion crosses many scientific and engineering fields. The theoretical aspects of corrosion have received brief coverage as have procedures and approaches available for controlling corrosion in the chemical process industries. Further and more detailed information can be found in the many books and current literature on the subject, some of which are listed in the Bibliography.

Bibliography

Gavert, R. B., Moore, R. L., and Westbrook, J. H., *Critical Surveys of Data Sources: Mechanical Properties of Metals*, NBS Special Publication 396–1, U.S. Department of Commerce, National Bureau of Standards, Washington, D.C., 1975. Order by

SD Catalog No. C13.10:396–1 ($1.25) from Superintendent of Documents, U.S. Government Printing Office, Washington, D. C. 20402.

Twenty-one Handbooks and Technical Compilations are reviewed in some detail, 9 information centers (including 5 foreign centers), 5 technical society data publications, and 7 trade association data publications are described giving scope and range of available information.

This source is cited because of the importance of mechanical properties of metals in design for corrosion service where elevated temperatures and pressures are experienced.

Diegle, R. B., and Boyd, W. K., *Critical Survey of Data Sources: Corrosion of Metals*, NBS Special Publication 396–3, U.S. Department of Commerce, National Bureau of Standards, Washington, D.C., 1976. Order by SD Catalog No. C13.10:396–3 from Superintendent of Documents, U.S. Government Printing Office, Washington, D.C. 20402. Current price (U.S.) $1.30.

Eighteen reference books on corrosion are reviewed as to cost, description, scope, input sources, size of data source, and remarks concerning adequacy of coverage. Fourteen information centers and 6 societies/trade associations and institutes are similarly noted and analyzed. In addition, 23 books are listed showing author(s), title, publisher, and date of publication.

Two recent compilations of interest are: (a) *Corrosion Inhibitors*, C. C. Nathan (ed.), National Association of Corrosion Engineers, Houston, Texas, 1973, a follow-up of *Corrosion Inhibitors*, J. I. Bregman, Macmillan, New York, 1963. (b) *Process Industries Corrosion*, Seminar Sponsored by NACE unit Committee T-5A, National Association of Corrosion Engineers, Houston, Texas, 1975.

Corrosion, Vols. 1–30 (1978). The regular journal of the National Association of Corrosion Engineers, P.O. Box 1499, Houston, Texas 77001.

In addition to the general sources listed above, the following books provide much information:

Evans, U. R., *Metallic Corrosion, Passivity and Protection*, Arnold, London, 1937.
Fontana, M. G. and Greene, N. D., *Corrosion Engineering*, McGraw-Hill, New York, 1967.
Hudson, J. C., *The Corrosion of Iron and Steel*, Chapman and Hall, London, 1940.
Pourbaix, M., *Thermodynamics of Dilute Aqueous Solutions with Applications to Electrochemistry and Corrosion*, Arnold, London, 1949.
Pourbaix, M., *Atlas of Electrochemical Equilibra in Aqueous Solutions*, Pergamon, Oxford, 1966. Also National Association of Corrosion Engineers, Houston, Texas, 1974.
Rabald, E., *Corrosion Guide*, Elsevier, New York, 1951.
Speller, F. N., *Corrosion Causes and Prevention*, 3rd ed., McGraw-Hill, New York, 1951.

IVY M. PARKER
WILLIAM A. CUNNINGHAM

Corrosion: Cathodic and Anodic Protection

Introduction

Cathodic and anodic protection are methods of corrosion control that directly use electrochemistry. Since corrosion is electrochemical in nature, it is logical that this be done.

Sir Humphrey Davy experimented with electrical current effects on corrosion and reported that copper could be protected from corrosion by coupling it to zinc or iron. This was in 1824—well before the electrochemical nature of corrosion was scientifically understood. This, however, did not prevent the use of cathodic protection for ship hulls and buoys shortly after the initial revelation of the usefulness of coupling these dissimilar metals together. Davey coupled iron slabs to the copper sheathing on the wooden ship hulls, and this successfully reduced corrosion. However, as a side effect the protected hulls became fouled with marine organisms. The copper salts generated by corrosion processes poison the fouling organisms. Therefore, the British Admiralty decided to abandon cathodic protection, but later, when steel hulls were used, zinc slabs were fitted to them [1].

Theoretically, cathodic protection can be applied to control corrosion on any metal in any electrically conductive medium. Practically, however, there are limitations due to current requirements. Cathodic protection has been widely applied to structures in the soil, structures and ship hulls in fresh and seawater, and storage vessels and process equipment handling a wide variety of corrosives.

Anodic protection is a much more recent development in corrosion control technology. In 1954 Edeleanu [2] discussed the possibility of using anodic protection on a plant-size scale. In 1960 a group at Continental Oil Co. disclosed successful field scale application of anodic protection [3–5]. These developments were based on the background developed by corrosion science.

Anodic protection is a much more limited control process in that only specific metal–environment combinations are suitable for anodic protection. Fortunately, the ubiquitous steel and stainless steel alloys can be anodically protected.

This article describes both cathodic and anodic protection. The application of those corrosion control methods will be described with concentration on process industry applications. By way of introduction, a brief discussion of the corrosion process, thermodynamics of corrosion, and kinetics, including polarization, will be given.

Background of Corrosion

Corrosion has been defined by many authors in slightly differing ways. All agree that it is a process by which a metal transforms from the elemental state to the combined state by reaction with the environment. Some include the qualifying term that it must be a destructive attack. There are some reactions with the environment that produce protective films or layers, and these should not be called corrosion. It seems that to describe corrosion fully and adequately, the qualifying term "destructive attack" must be included. Therefore, the definition stated by Uhlig [1] is a very good one: "Corrosion is the destructive attack of a metal by chemical or electrochemical reaction with its environment."

Thermodynamics

Corrosion is primarily electrochemical in nature, so at least a brief introduction to the fundamentals of electrochemistry is needed. Electrochemistry involves many topics which can be divided into two basic areas described by Bockris and Reddy [6] as "ionics" and "electrodics." Ionics describes the concern for ions in solution or melt, while electrodics is concerned with the region between an ionic and electronic conductor and the charge transfer which occurs across it. The primary act in electrochemistry is a transfer of charge across the interface between an electronic conductor (usually a metal) and the ionic conducting phase. The charge transfer involves the transfer of electrons into or out of the electronic conductor to or from ions in the ionic phase.

It is not possible to study this charge transfer by only one metal piece in a solution. It is necessary to have a second metal–solution interface in order to study the charge transfer process occurring on the first metal–solution interface. In addition, the two metals must be connected by a metallic path so the resulting system has an electronic path through the wire and an ionic path through the solution (see Fig. 1).

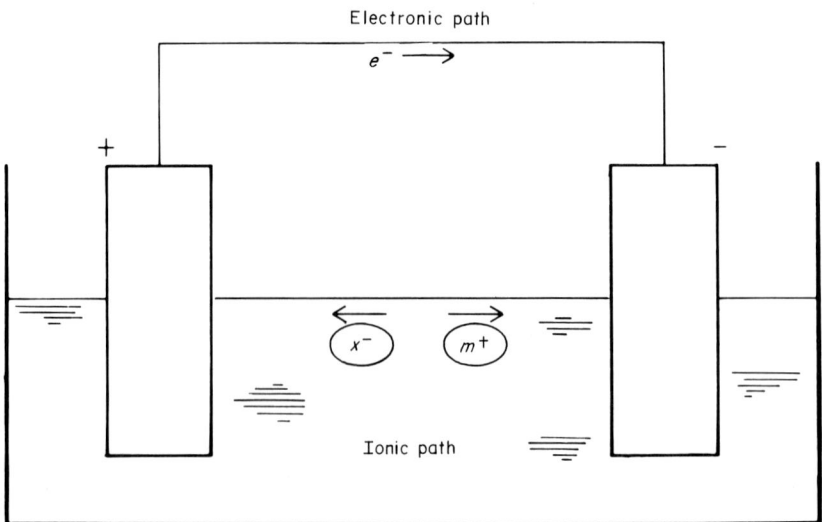

FIG. 1.   Electrochemical cell. Current is carried by electrons through the wire connecting the two electrodes. Current is carried by ions in the solution.

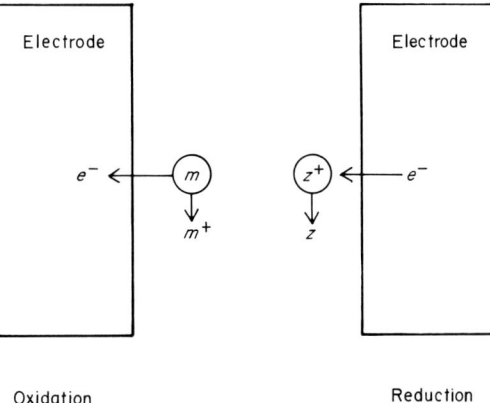

Oxidation   Reduction

FIG. 2. Basic electrochemical reactions. The two basic reactions of electrochemistry are
oxidation (transfer of an electron from an atom or ion to the electrode) and reduction
(transfer of an electron to an ion from the electrode).

When two different metals are placed in such a device, electrons and ions
will flow spontaneously. It is also possible to force electronic and ionic current
to flow through the circuits by connecting a dc source across the two electrodes.
These two cells have been described by Bockris and Reddy [6] as energy
producers and substance producers.

The energy producer is a system in which the current flows spontaneously
and when connected to a load will produce work. Battery and fuel cells are such
devices. The substance producer is the device resulting from connecting a dc
power source across the two electrodes. The charge transfer occurs in such a
way that ions or molecules are produced at the electrode surfaces. An example
of a useful device is the caustic–chlorine cell in which caustic and chlorine are
produced by electrolysis of a sodium chloride solution.

In both substance and energy producers, electron exchange reactions occur
at the electrode interfaces. An electron can be given *to* the electrode by an atom
or an ion in the reaction called oxidation. An electron can be transferred to an
atom or ion *from* the electrode in the reaction called reduction. The electrode at
which oxidation occurs is the anode and the electrode at which reduction occurs
is the cathode (Fig. 2).

When two electrodes are immersed in a solution and connected by an
electrical conductor, one electrode will be a site for oxidation reactions and the
other electrode will be a site for reduction reactions.

To summarize, an electrochemical cell is made up of two solution–metal
interfaces connected by a metallic path. Electrons flow through the metallic
path and ions flow through the solution. Oxidation occurs at one electrode and
reduction occurs at the other electrode. The rates of these reactions are equal
because there is no accumulation or storage of electrons (Fig. 3).

Both reduction and oxidation reactions can occur on a single metal surface
(Fig. 4). The metal must be immersed in an ionic conductor and the metal itself
acts as the electronic path. This type of cell is called the "energy waster-

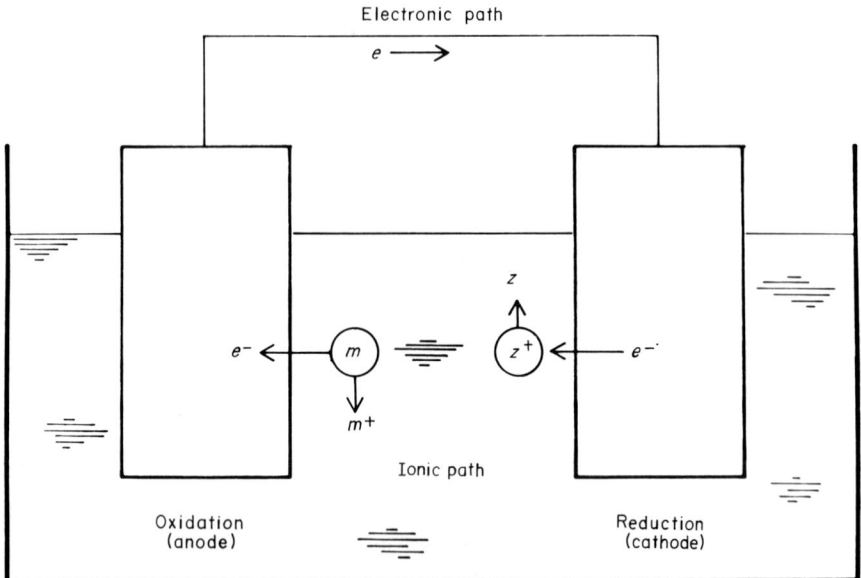

FIG. 3. Complete electrochemical cell. The current paths are shown with the two electrode reactions, oxidation and reduction. Oxidation occurs at the anode and reduction at the cathode. An electron is transferred at the anode for every electron transferred at the cathode.

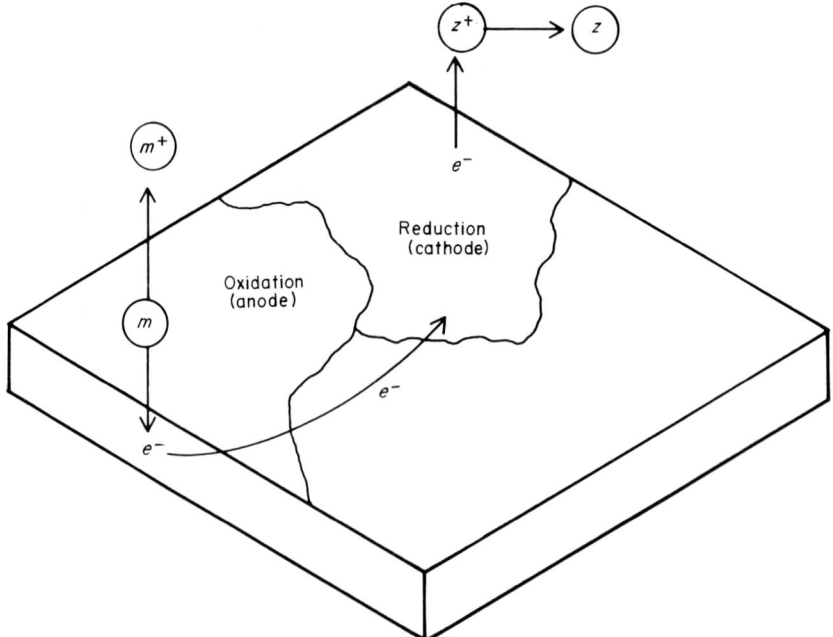

FIG. 4. Corrosion cell. Localized areas on a metal act as anodes while other areas act as cathodes. The metal is the electronic conductor and the corrosive solution is the ionic conductor.

substance destroyer" by Bockris and Reddy [6]. It is the corrosion cell. Therefore, an understanding of the electrochemical cell leads to understanding the electrochemical basis of corrosion.

The localized electrodes occur on the surface of the metal through a mechanism not fully understood. Investigators have postulated that surface defects or discontinuities lead to the localized electrodes. They do not necessarily remain at one place on the metal surface but may shift from location to location.

As mentioned above, when two different metals are connected while immersed in an ionic solution, one metal will oxidize (corrode) and a reduction reaction will occur at the other metal. The relative tendency for these reactions to occur has been experimentally determined for the elements under very special conditions. The emf series lists the electrode potential of each metal at $25°C$, 1 atm, and in equilibrium with its ions at unit activity. The potentials are those determined by measuring the electrode potential versus a standard hydrogen electrode. The hydrogen electrode is made of a platinum electrode over which hydrogen gas is bubbled at 1 atm while the platinum is immersed in a solution with unit activity of hydrogen ions.

These potentials are measured with a potentiometer or other high impedance meter such that no current is drawn from the electrodes. Under these conditions, the reaction

$$m \underset{\text{Oxidation}}{\overset{\text{Reduction}}{\rightleftarrows}} m^+ + e^-$$

is assumed to be at equilibrium, i.e., the rate of the oxidation reaction equals the rate of reduction. No net current is drawn from an electrode when it is at equilibrium.

There are two sign conventions used for the emf system. The system shown in Table 1 is known as the IUPAC system and is based on the sign of the electrode when connected to the hydrogen electrode. The zinc electrode would be a negative -0.763 V with respect to the hydrogen electrode when measured with a potentiometer. The American system presents the potentials with opposite signs to those shown in Table 1. The absolute values are identical. The signs in this system are based on the tendency for the reaction to occur as an oxidation reaction when connected to a hydrogen electrode. These different conventions can lead to confusion, but if the values are used carefully the calculated overall cell potentials will be independent of the half-cell sign convention.

Polarization

Corrosion cells and electrochemical devices do not operate at equilibrium because current is drawn from them. When current is drawn, the electrode potentials shift due to slow steps in the reaction mechanisms. This shift is called polarization.

The anode potential is shifted in the positive direction because the transfer of m^+ into solution lags the release of electrons. The cathode potential is shifted

TABLE 1 Emf Series (25°C, V vs normal hydrogen electrode)[a]

$Au = Au^{3+} + 3e^-$	$+1.498$
$O_2 + 4H^+ + 4e^- = 2H_2O$	$+1.229$
$Pt = Pt^{2+} + 2e^-$	$+1.2$
$Pd = Pd^{2+} + 2e^-$	$+0.987$
$Ag = Ag^+ + e^-$	$+0.799$
$2Hg = Hg_2^{2+} + 2e^-$	$+0.788$
$Fe^{3+} + e^- = Fe^{2+}$	$+0.771$
$O_2 + 2H_2O + 4e^- = 4OH$	$+0.401$
$Cu = Cu^{2+} + 2e^-$	$+0.337$
$Sn^{4+} + 2e^- = Sn^{2+}$	$+0.15$
$2H^+ + 2e^- = H_2$	0.000
$Pb = Pb^{2+} + 2e^-$	-0.126
$Sn = Sn^{2+} + 2e^-$	-0.136
$Ni = Ni^{2+} + 2e^-$	-0.250
$Co = Co^{2+} + 2e^-$	-0.277
$Cd = Cd^{2+} + 2e^-$	-0.403
$Fe = Fe^{2+} + 2e^-$	-0.440
$Cr = Cr^{3+} + 3e^-$	-0.744
$Zn = Zn^{2+} + 2e^-$	-0.763
$Al = Al^{3+} + 3e^-$	-1.662
$Mg = Mg^{2+} + 2e^-$	-2.363
$Na = Na^+ + e^-$	-2.714
$K = K^+ + e^-$	-2.925

[a]From A. J. de Bethune and N. A. S. Loud, *Standard Aqueous Electrode Potentials and Temperature Coefficients at 25°C*, Hampel, Skokie, Illinois, 1964.

in the negative direction because electrons are supplied at a greater rate than they are transferred to the ions. Polarization can be graphically described by plotting the electrode potential versus the logarithm of current density.

Figure 5 illustrates several factors which will only be mentioned here. However, a full discussion of these may be found in the books by Fontana and Greene [7] and Tomashov [8].

1. The currents labeled i_0 are exchange current densities. These are the currents equivalent to the rate of the reaction at equilibrium. The rate of the forward reaction equals the rate of the reverse reaction. Therefore, the point at ϕ_A or ϕ_C and i_0 locates the electrode condition as listed in the emf table.
2. The current i_C locates the point at which the rate of the anodic reaction equals the rate of the cathodic reaction. For a corroding metal piece, this point locates both the corrosion potential and the corrosion current.
3. The polarization is a function of the logarithm of the current density according to the Tafel equation:

$$\eta = \pm \beta \log (i/i_0)$$

where η = polarization = $(\phi_{measured} - \phi_{equilibrium})$
β = slope called Tafel slope
i = current at $\phi_{measured}$
i_0 = exchange current density

4. There is no accumulation of the electrons. Therefore the rate of reduction equals the rate of oxidation. This is also true for situations in which more than one oxidation or reduction reaction occur. In those cases the total oxidation rate equals the total rate of reduction.

For a corroding metal such as zinc in an acidic environment, the reaction which occurs on the localized cathodes is

$$2H^+ + 2e^- \rightarrow H_2$$

and the reaction on the localized anodes is as shown in Fig. 6:

$$Zn \rightarrow Zn^{2+} + 2e^-$$

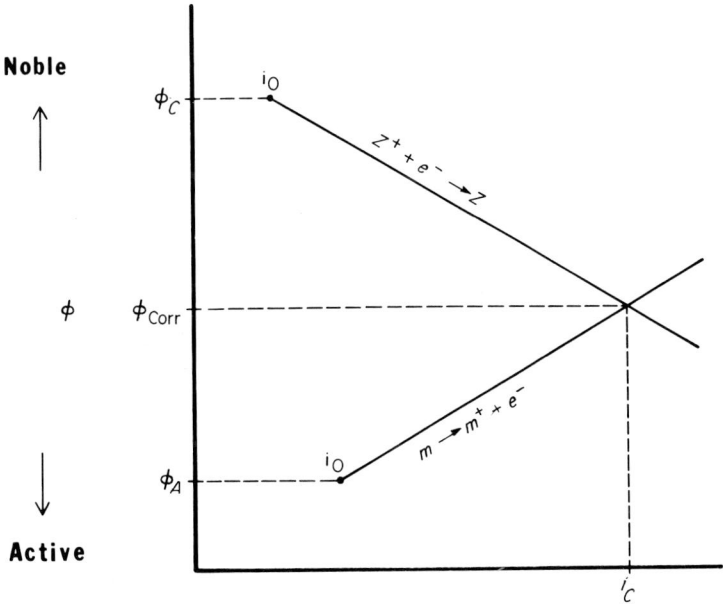

FIG. 5. Theoretical polarization diagram, potential versus logarithm of current density. The localized anodes are polarized to a more positive value and the localized cathodes are polarized to a more negative value. Exchange current densities i_0, corrosion current i_c, and corrosion potential ϕ_{Corr} are shown. The potentials ϕ_A and ϕ_C are the standard state potentials shown on the emf series.

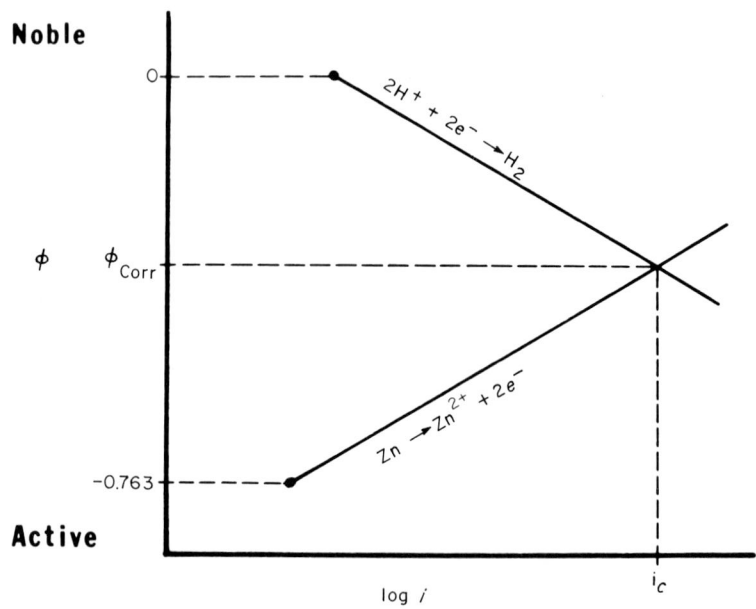

FIG. 6.   Polarization diagram for zinc. The potential values and reactions are shown for zinc in acidic solutions.

The potential of the hydrogen evolution reaction at equilibrium does not depend on the metal and is therefore zero. The equilibrium potential of the zinc is -0.763 V.

The corroding electrode will have a potential intermediate between those two potentials, and the corrosion current is illustrated by the polarization curves.

These curves are used as a graphical model of the corrosion mechanism. It is not possible to determine experimentally the curves as sketched. The measurements made, however, do allow these curves to be developed by extrapolation techniques.

Polarization measurements are made with the three-electrode system shown in Fig. 7.

The working electrode is the metal under study. The reference electrode is an electrochemical half-cell that has a fixed and well-known potential, e.g., the calomel cell. The auxiliary electrode is used to complete the current-carrying circuit. The potential of the working electrode is measured by measuring the potential difference between it and the reference electrode. Current is forced to flow between the auxiliary electrode and working electrode. Typical data taken with such an arrangement are illustrated in Fig. 8.

The nonlinear portions are typical of all systems and occur at potentials near the corrosion potential. The linear portions are described by the Tafel equations.

An electrode can be polarized by connecting it to another metal. Current

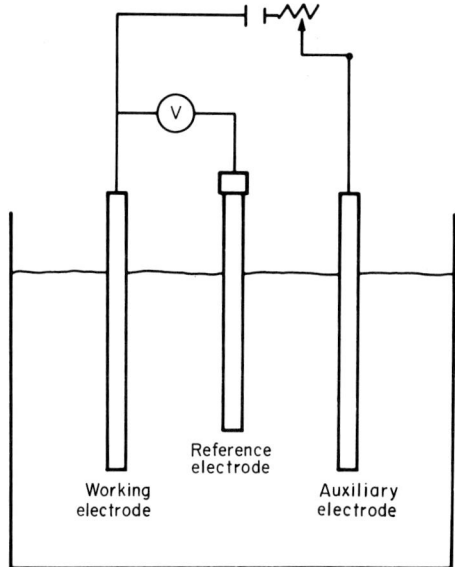

Polarization apparatus

FIG. 7. Polarization apparatus. The three-electrode electrochemical system is used for testing and corrosion control. The current flow between working electrode and auxiliary electrode will result in a shift in potential between working electrode and reference electrode.

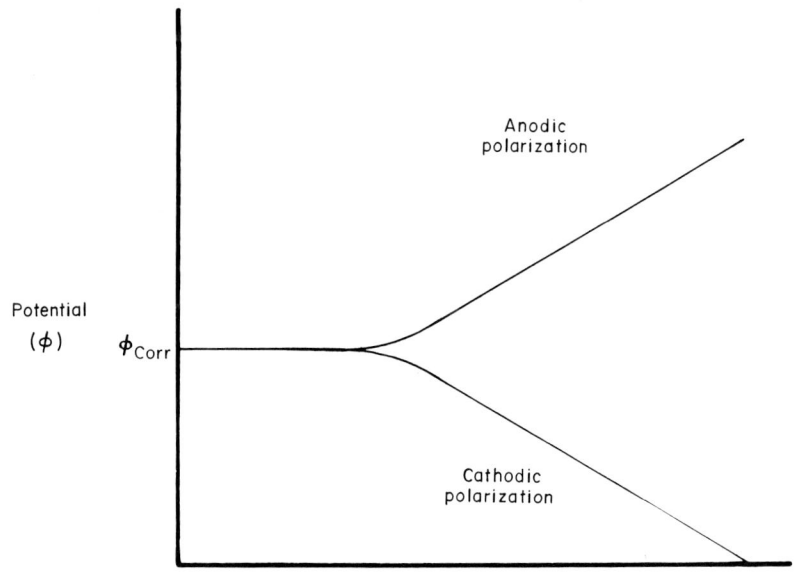

Log current density (log i)

FIG. 8. Polarization diagram. Anodic and cathodic polarization diagrams, as would be obtained using the apparatus in Fig. 7, are shown. Tafel behavior is seen in both anodic and cathodic curves.

will flow spontaneously due to the potential differences. A typical polarization can be modeled by examining the polarization diagrams.

The combination polarization diagrams in Fig. 9 may seem a bit complicated at first glance. They illustrate the behavior of the individual zinc and iron electrodes. In addition, the behavior of these two electrodes when they are coupled is also illustrated. Iron will corrode at $i_{C_{Fe}}$ and zinc will corrode at $i_{C_{Zn}}$ when they are individually immersed in the corrosive solution.

When the two are electrically connected, the potential of the iron electrode is shifted in the negative direction and the zinc potential is shifted in the positive direction until they are at the same potential. Notice the total oxidation rate includes the oxidation of the iron and zinc. The rate of iron oxidation ($i_{Corr_{Fe}}$) is lowered from the rate which occurred with the iron electrode alone. The zinc oxidation rate ($i_{Corr_{Zn}}$) is accelerated.

These background comments are included to orient the reader to the electrochemistry involved, which will hopefully make the description of cathodic and anodic protection theory easier to understand. More complete discussions of the electrochemistry of corrosion can be found in the books by Uhlig [1], Fontana and Greene [7], Tomashov [8], and Bockris and Reddy [6].

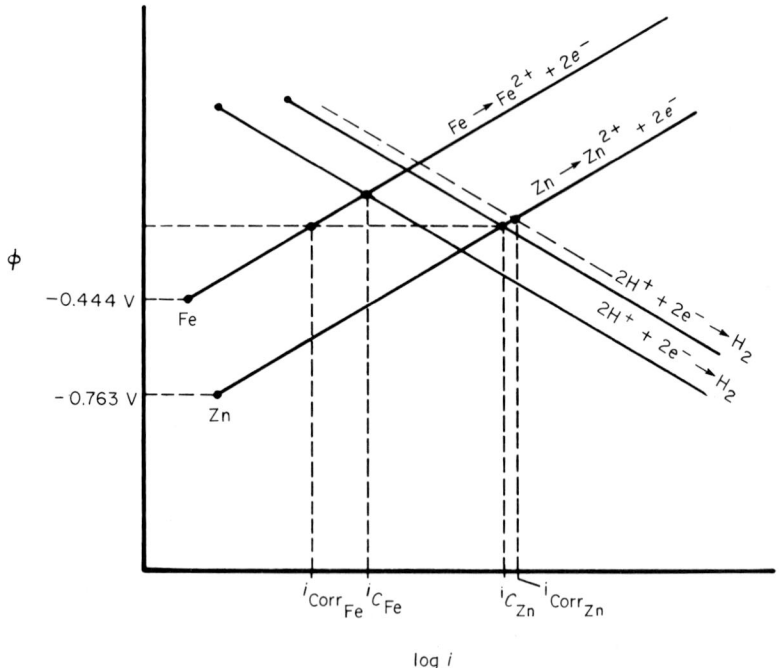

FIG. 9.   Theoretical polarization curves for zinc coupled to iron. The potential of the iron is shifted to a more negative value when it is coupled to zinc. The corrosion current of the iron is lowered from $i_{C_{Fe}}$ to $i_{Corr_{Fe}}$. This is the basis of cathodic protection by sacrificial anodes.

Cathodic Protection

As mentioned above, cathodic protection has been known to be effective in controlling corrosion since 1824. The widespread use of cathodic protection was coincident with the upsurge of the oil and gas industry. Pipelines required for the economic transportation of these products were subject to corrosion by soil. Cathodic protection was found to be an effective and economic method of controlling this corrosion and became a necessity on pipelines and on other buried structures. The application of cathodic protection to ship hulls has been a common practice for many years, especially by the navies of Great Britain and the United States. Descriptions of these types of applications may be found in books written by Applegate [9] and Morgan [10] and in the technical and professional trade literature. This discussion will neglect these more common applications of cathodic protection. Applications to process vessels, tanks, and equipment will be described since they should be of primary interest to those involved in the chemical process industries.

Theory and Background

The theory of cathodic protection has been described from several viewpoints. This discussion will concentrate on the polarization theories described in the introductory sections.

First, consider the situation in which current is impressed between two electrodes. One electrode is the metal to be cathodically protected (working electrodes) and the second is an auxiliary electrode. The potential of the working electrode is measured with respect to a reference electrode. This is the same system described in the section entitled "Polarization" and shown in Fig. 7. As noted in that section and shown in Fig. 8, if the current flow is impressed so that the working electrode is the cathode, the potential of that electrode becomes more negative. As shown in Fig. 10, the potential versus logarithm current follows the Tafel equation as the potential is shifted more cathodically.

Also shown superimposed on that diagram is the model polarization curve used to describe the electrochemical basis of corrosion. Notice that as the potential is shifted to become more active, the rates of the oxidation and reduction reactions may be determined from the appropriate curves. The oxidation or corrosion rate is seen to be reduced as the potential is shifted to more negative potentials. The corrosion reaction is completely halted when the potential reaches the equilibrium potential for the metal. This is the basis of cathodic protection when it is applied using impressed currents.

It is also possible to apply cathodic protection by coupling a relatively active metal to the metal of interest. The polarization diagram shown in Fig. 9 illustrates the basis for reduction in corrosion by this technique. As discussed above, the potentials of zinc and iron are shifted so they are equal. The polarization of each electrode is such that the zinc corrodes at a higher rate and iron corrodes at a lower rate. This same behavior is seen when any two metals are connected. The more active metal (on the emf series) corrodes and the more noble metal is protected.

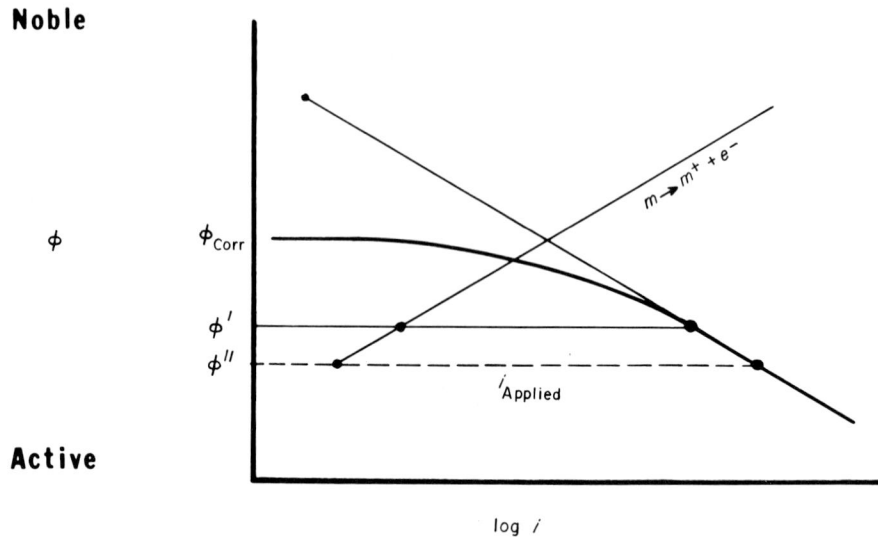

FIG. 10.   Cathodic polarization curve. The potential of the metal is shifted to ϕ' and ϕ'' by impressed currents using a system as shown in Fig. 7. The corrosion rate of the metal is the point the ϕ's intersect the theoretical anodic polarization curve. Corrosion is halted when the potential is at ϕ'' which is the standard state emf.

Application Requirements

Cathodic protection can be applied in two ways. Impressing current flow from an external dc power source such as a battery or a rectified ac power is one method. The second, use of sacrificial anodes, involves coupling a metal that is active in the emf series to the metal to be protected. This section describes the background information necessary to understand the application of these two methods of using cathodic protection.

Criteria for Protection

It is very important that the cathodic protection system is monitored in order to determine that it is working effectively. The ultimate measure of effectiveness is that the corrosion rate is lowered to an acceptable value or completely halted. Direct corrosion rate measurements using coupons or one of the many corrosion rate measuring devices can be used, but in most cases some period of time must elapse before the determination can be made. Measurement of the metal potential with respect to a reference electrode will indicate the electrochemical condition of the metal to be protected. The combination of this potential with the information available from the polarization diagrams can be used to set criteria for the effectiveness of the protection system. The potential measurements can be made quickly in most cases and provide information concerning the condition of the system immediately.

The literature contains several criteria based on these potential measurements. Some are based on the amount of shift in potential due to the current flow. Others are based on specific potential values. The criterion that is based most solidly on the electrochemical theory is the specific potential value. As discussed in the section entitled "Theory and Background," corrosion is completely halted when the potential of the metal is shifted to the value of the equilibrium potential for the metal reaction:

$$m \leftrightarrows m^{2+} + 2e^-$$

The value for iron is -0.44 as listed in the emf series shown in Table 1, based on the standard hydrogen electrode. It has been found experimentally that corrosion of steel in pH 9 conditions is halted if the potential is -0.53 V with respect to the hydrogen electrode [11]. In practice, the $Cu/CuSO_4$ reference electrode is used extensively for potential measurements of buried structures. This electrode has a potential of $+0.32$ V with respect to the standard hydrogen electrode. Therefore the steel potential compared to the $Cu/CuSO_4$ is -0.85 V when the metal is protected. This figure is used widely, and through practice has an almost magical aura to it. There can be conditions for which this value would not be suitable and other criteria must be used. As an example, lead is protected if the potential is -0.78 compared to the $Cu/CuSO_4$ electrode. Other metals and other solutions may require different potentials for successful halting of corrosion. In systems other than water–steel or soil–steel, the potential necessary for protection may need to be experimentally determined. The potentials determined in a laboratory study can be used directly in the field-scale application if the electrochemical conditions are faithfully reproduced in the lab. Potential versus corrosion rate measurements are the best criteria on which to base the proper choice of potential criteria. It is possible but not recommended that this type of information be obtained in the field application.

In many cases the potential value criterion for protection is only a minimum potential. Potentials more negative would also be indicative of a condition in which corrosion is halted. Steel in water would be protected at -1.200 V (compared to $Cu/CuSO_4$) just as well as at -0.85 V. However, it would require more current and could be more costly to maintain the potential at the more negative potential than at the minimum value.

There are systems in which the potential must be maintained within a given range to prevent damage by the cathodic protection currents. Aluminum is an amphoteric metal and will be corroded by both alkaline and acidic solutions. Hydrogen is evolved from the decomposition of water at the cathode (shown by the polarization curve) if the potentials are sufficiently elevated. The hydroxyl ion remaining after this decomposition is concentrated near the metal surface; this in turn increases the pH and tends to corrode the aluminum. Therefore the potential must be kept more positive than -1.00 V.

Cathodic protection is applied to coated structures to prevent corrosion of uncoated pinholes or holidays. If the potential is too negative, hydrogen evolution can be copious. This hydrogen gas can damage the coating by disruption of the metal coating bond, blister formation, and chemical attack. Therefore, on coated structures care must be taken to maintain the potentials close to the value at which corrosion is just halted.

TABLE 2   Current Requirements For Cathodic Protection

Metal	Environment	Conditions	Current Density (mA/ft^2)	Refs.
Steel	Seawater	Tidal	5–6	10
Steel	Seawater	Marine	10–20	10
Steel	Fresh water	Marine	1–3	9
Nickel	Halide salt solution	Stagnant	1–2	12
Steel	Pyroligneous acid	Stagnant	30	12
Steel	H_3PO_4	Stagnant	40–100	13
Steel	H_2SO_4	Stagnant	50,000	7

Current Requirements

The current requirements are dependent upon the metal, chemical environment, and temperature. Examples of some current density requirements are listed in Table 2.

Each new system must be tested to determine the current requirements to provide protection. The currents are also time dependent in that the requirements usually decrease with time. In some systems, especially seawater, a protective layer of precipitated salt forms as a result of the currents. In seawater a calcareous layer forms on the metal due to the increase in pH at the metal surface, as discussed above. Calcium and magnesium carbonates are precipitated at the elevated pH. These layers lower the current requirements simply by covering the exposed bare metal.

Current requirements also decrease with time due to the polarization of the metal surface. The metal surface is changed with time so that the current required to polarize the metal is reduced. Therefore, the current requirements determined after short time periods will be high and provide some degree of design safety factor.

System Selection

Cathodic protection can be applied by impressing direct current flow between an anode and the structure or by connection of an active metal to the structure. These are known as the impressed current and sacrificial anode methods, respectively. The initial selection to be made in design of a cathodic protection system is the choice between these two methods. Factors to be considered in this selection are cost, solution resistivity, control problems, current distribution, power availability, and installation considerations. These factors are discussed in this section.

In general, both capital costs and operating expenses will be lower for sacrificial anode systems than for impressed current systems. Maintenance will be the only operations cost for the sacrificial anode system. The impressed current system has some power consumption and may require more attention since it is a bit more complex. Therefore, except in unusual circumstances, total costs will be lower for the sacrificial anode method. However, there are other technical problems to be considered.

The resistivity of the electrolyte is a prime consideration in this selection. The voltage required for a given current flow is a direct function of the electrolyte resistivity. The voltage between the metals used as sacrificial anodes and the structure metals is fixed by their relative position in the emf series. The voltage between commonly used anode materials and steel are given in Table 3.

The values listed in Table 3 are the maximum, open circuit potentials between the anodes and steel. If the solution resistivity is high, these voltages may not be sufficient to achieve the necessary current flow.

It is possible to achieve any needed driving voltage with the impressed current systems by design of the dc power source.

Even distribution of current over the surface of the structure is necessary for complete protection. In some installations the structure geometry may be complex, and it may be easier to install sacrificial anodes throughout the structure than to use an impressed current system. This is especially true if portions of the structure are moving, e.g., traveling screens or arms. Some of the newer strip anodes available for both sacrificial and impressed current are particularly useful in achieving good current distribution.

It is obvious that availability of electrical power is one important factor in the choice between impressed current and sacrificial anode systems. In most chemical process plants and refineries, electrical power is readily available, so this factor is of no consequence. This does not preclude the use of sacrificial anodes if the system is complex or if power is not readily available.

Installation of the anodes and ancilliary equipment is another factor in the selection of the type of system to be used. It is possible to install the sacrificial anodes with no entrys into the vessel. The anodes may be clamped to the inside wall of the vessel with no hole cut through the wall, which may be a sizable advantage in some situations. An entry into the vessel is necessary if a reference electrode is used to monitor the vessel wall potential. It is necessary to have an entry for the impressed current anodes since wiring from the external power source must be connected to the anode.

It is necessary to control the current flow to the vessel wall in some applications. Therefore, the ease and effectiveness of this control is a factor in the selection process. The current in sacrificial anode systems can be controlled

TABLE 3 Potentials of Sacrificial Anodes to Steel

Metal	Potential Difference between Anode and Steel (V)
Zinc alloy	0.24
Magnesium alloy	0.75
Aluminum alloy	0.2–0.3

TABLE 4 Composition of Common Anode Alloys

Anode	Al	Cd	Cu	Hg	In	Fe	Mg	Mn	Ni	Pb	Si	Sn	Zn
Zinc [18]	0.1–0.5	0.25–0.15	0.005 max			0.05 max				0.006 max	0.125 max		Balance
Magnesium [18]													
H-1	5.3–6.7		0.05 max			0.003 max	Balance	0.15 max	0.003		0.3		2.5–3.5
High potential	0.01		0.02			0.03	Balance	0.5–1.3	0.001		0.05		
Aluminum [28]													
Zn-In	Balance	0.01			0.02	0.09							2–3
Zn-Sn	Balance											0.07–0.13	4–7
Zn-Hg	99.9			0.03–0.04		0.04							0.45–0.5

by adding external resistance between the anode and cathode. This type of control may be ineffective in systems where the solution composition and resistivity are changing. Several types of control may be incorporated in impressed current systems. The voltage between the anode and cathode is most easily controlled. The current level can also be controlled. The most complex and probably most effective method is to control the cathode (vessel wall) potential with respect to a reference electrode. These control methods are discussed in greater detail below.

Equipment

Cathodic protection systems usually consist of commercially manufactured components that have been collected for a given installation. It is possible to purchase these materials, engineer the system, and make the installation. There are also several companies that handle all aspects of this task and have the experience that is so valuable in this field.

    This section describes in a general way the type of equipment that is used for cathodic protection installations. Experience and information concerning a particular application are necessary in order to fully engineer a system.

Sacrificial Anodes

In principle, any metal more active than the corroding structure will protect it when electrically connected. However, in practice, alloys of magnesium, zinc, and aluminum are the only metals used in present-day applications. Tables 4 and 5 list some alloys currently used, with pertinent data which describe the operating characteristics.

    Of the three considered here, magnesium has the largest potential difference relative to steel. Pure magnesium does not perform as well as an anode because of local action corrosion cells. These act to corrode the magnesium so it delivers about one-half of the expected amount of current to the cathode. The

TABLE 5 Operational Characteristics of Sacrificial Anodes

Anode	Potential to Ag/AgCl (V)	Current Efficiencies (%)	Consumption (Ah/lb)
Zinc [18]	−1.05	95	354
Magnesium [18]			
H-1	−1.50	55	500
High potential	−1.68	55	500
Aluminum [28]			
Zn-In	−1.1	65	900
Zn-Sn	−1.1	50	1000
Zn-Hg	−1.10	95	1250

aluminum and zinc alloys listed do not have as high a driving potential to steel as the purer material, but some of them do have a higher current efficiency.

The current efficiency is calculated by assuming the reaction

$$Mg \rightarrow Mg^{2+} + 2e^-$$

This can be translated into ampere hours per pound by using the relationship of Faraday:

$$It = F\frac{W}{Eq.\ wt.}$$

where W = weight of metal
It = current × time
F = 96,500 C/equivalent
Eq. wt. = equivalent weight of metal

The theoretical capacity of magnesium is 1000 Ah/lb whereas in practice only about 500 Ah/lb is obtained. This may be caused by local action corrosion cells on the magnesium or it is possible that the following reaction also occurs:

$$Mg \rightarrow Mg^+ + e^-$$

The univalent ion then undergoes further oxidation to the divalent ion in such a way that the electron is not available for cathodic protection.

The purer magnesium alloy is less efficient than the material alloyed with aluminum and zinc.

Magnesium is used in systems that require relatively high currents or high resistivity. It would be a better choice for fresh water, for example, than for seawater. It is available in various shapes and sizes. One interesting shape is the ribbon or strip anode. The magnesium is cast around a steel wire for strength. This shape is useful for achieving good current distribution in installations with a complicated geometry.

Sir Humphrey Davy used zinc on the British Admiralty ships, which made it the first anodic material used. Pure zinc (99.993%) will perform well as an anode, but as a practical matter commercially pure zinc is used. The iron content must be very low. If it is above 15 ppm, a dense coating forms on the zinc surface, and this lowers the current flow. Aluminum and cadmium in small amounts reduce this effect. Therefore, the composition of the zinc used for cathodic protection contains some aluminum and cadmium since it is difficult to lower the iron content below the listed specification.

Zinc has a 95% efficiency and will deliver 350 Ah/lb. The driving potential is somewhat lower than that of magnesium and is therefore limited to low resistivity environments. It is best suited for seawater. It, too, is available in a large number of shapes and sizes.

Aluminum cannot be used in the pure form because it will passivate and halt current flow. Several alloys (shown in Tables 4 and 5) have been developed to overcome this deficiency. There are some variations of alloying element concentration depending upon the manufacturer [28]. Notice that although the Zn-Hg alloy has the highest current efficiency, the use of an alloy containing mercury must be carefully considered in these days of environmental concern. Aluminum alloys are used widely in both fresh and seawater applications. They are available in various sizes and shapes.

The size of the anodes used is based on mounting, space, and design life considerations. The space available in a vessel certainly can limit the shape and size of the anode. Mounting may also present problems due to internals, liquid level, or space. Design life is probably the most important factor in determining the size of the anode. The lifetime of the anode may be calculated from the current capacity of the particular material selected. The current drain must be estimated for this calculation. Calculation of the current obtained from an anode is dependent on electrolyte resistivity and anode geometry. These factors are briefly discussed below.

Impressed Current Anodes

The anodes used in the impressed current systems should not be affected by the currents, be mechanically tough, be capable of being formed into various shapes, and be low in cost. These ideal requirements cannot be met easily by any one material.

Table 6 lists several alloys used as impressed current anodes. The consumption of each electrode by the applied currents is also tabulated.

Scrap iron is the least expensive of the anode materials listed. However, it is corroded at a rapid rate by the impressed currents and must be replaced frequently. If it is used in storage or process vessels, iron contamination of the contents will also occur by this dissolution process. This material has been used for systems protecting underground structures and in marine environments, but no literature reference was found to its use in vessels.

Silicon-iron anodes were developed during the early 1950s and were

TABLE 6 Metals Used for Impressed Current Anodes

Metals	Consumption (lb/Ayr) [14]
Scrap iron-steel	20
Silicon-iron	0.5–2
Silicon-iron-chromium	0.25–1
Graphite	1.5–3
Lead-silver alloy	0.008
Lead-platinum-bielectrode	0.003
Platinized titanium	0.000019 (max)[a]
Platinized niobium	0.000019 (max)[a]
Platinized tantalum	0.000019 (max)[a]

[a]Estimated from Ref. 15.

TABLE 7 Analysis of Silicon-Iron Anodes

Element	Silicon-Iron	Silicon-Iron-Chromium
Silicon	14.4	14.40
Manganese	0.7	0.70
Carbon	0.95	1.0
Chromium	—	4.25
Iron	Balance	Balance

successful except in high chloride ion environments where they were pitted by the chlorine evolved at the anodes. Silicon-iron-chromium anodes were developed to overcome this problem. The analysis of each of these alloys is given in Table 7.

The anode life is excellent as can be determined from the consumption data in Table 6. Consumption is low due to the formation of a conductive oxide film on the surface of the anode. These electrodes are brittle and difficult to fabricate, and the electrical connections must be pressed into a hole cast into the anode. They must also be made carefully since any leakage in the insulation would allow the copper wire to act as an anode and be corroded very rapidly. The electrical connections are quickly broken under such circumstances and the system becomes inoperative.

Graphite anodes are extensively used in both underground and internal vessel cathodic protection systems. Those anodes are manufactured in a three-step process from petroleum coke. The coke is calcined, molded to shape using a pitch binder, and then fired at 2782°C to convert the amorphous carbon to the crystalline graphite [17]. The graphite is porous and is sealed with linseed oil in order to increase the anode life. The anodes are inert in that no reactions occur directly due to the action of the cathodic protection currents. Hydroxyl ions will be oxidized to oxygen gas and water at the graphite anode; chlorine gas will be evolved if the electrolyte contains chloride ions. Oxygen reacts with the graphite to form CO_2, and therefore the electrode does deteriorate with use. Chlorine gas does not react with the carbon but it does attack binders in the graphite.

Graphite anodes are installed in coke breeze backfill when used in underground installations. Several applications of the bare graphite in seawater and fresh water storage tanks have been described in the literature.

Pure lead cannot be used as an anode but several lead alloys have been tried. All the successful alloys form a conductive oxide layer on the surface of the lead. This layer prevents the deterioration of the lead. Two alloys have been described in the literature: one contains 1% silver and 6% antimony, while the second contains 2% silver [18]. Each alloy seems to perform satisfactorily in seawater.

Small microelectrodes of platinum imbedded in lead also provide a satisfactory lead peroxide coating so that the electrode can be used as an anode. One report describes these as being made from 1 to $\frac{1}{2}$ in. diameter lead with

0.030 in. diameter × 0.5 in. long platinum tacks imbedded at 6 to 12 in. intervals [14]. This report states that these anodes are used to protect the BART Tunnel in San Francisco Bay with low current densities to prevent damage to the peroxide film. Shrier, however, reports that current densities of up to 200 A/ft^2 were possible with this electrode [19]. Lead anodes are used in seawater but are quite heavy due to the high density of lead.

Several combinations of base metals with a thin coating of platinum are in use. These have been termed the precious metal anodes. They are expensive when compared to the other anodes but they are very durable and have long life. In most cases they also have low densities. Platinized titanium is a successful anode because an insulating oxide layer forms on the titanium. All current flows through the platinum portion of the electrode [14]. Current densities of 100 A/ft^2 of platinized area have been used. Originally about 100 μin. of platinum was used on the titanium, but recent experience has shown that 200 μin. or greater is necessary for long-term high current density operations [15]. The oxide film will be broken down if the electrode potential is above 8.5 to 12 V, and the electrode will be severly corroded. Applied voltages between anode and cathode can be much higher than this, even to 100 V, since the electrode potential may not necessarily be shifted to the breakdown potential. Platinized niobium and tantalum can also be used due to the same mechanism discussed for platinized titanium. The breakdown voltages for niobium is 40 to 50 V and for tantalum is up to 200 V. The cost of those electrodes is much greater than titanium; 18 times higher for niobium and 30 times for tantalum [18]. Titanium has an electrical resistivity comparable to stainless steel while niobium and tantalum are approximately one-third of that value. Therefore the use of titanium can lead to appreciable voltage drops in the electrode.

The choice of anode material is based on several factors: cost, compatability with environment, life, mechanical strength, difficulty in mounting, and availability. In each installation, several of the anodes discussed above may be used with no great difference in all these factors. Successful experience with a particular anode material in a given environment usually overrides or confirms these factors.

Table 8 lists the maximum current densities that can be used with several of the anodes discussed.

TABLE 8 Maximum Anode Current Densities

Material	Current	Density (A/ft^2)
Fe, Si-Cr [16]	0.25	Fresh Water
	4	Seawater
Graphite [12]	0.25	Bare
Lead-Ag-Sb alloy [18]	5–25	Seawater
Lead-Pt bielectrode [19]	200	
Platinized titanium [14]	100	

Reference Electrodes

It is imperative that the potential of the structure under protection be measured periodically or monitored continuously. It is necessary to make this measurement by comparing the structure to an electrode that has a well-known, stable potential. In most cases an electrochemical half-cell is used. In some cases another metal may be used. The electrochemical half-cell is best because the electrode is constructed so that a stable, well-known electrochemical environment is present.

Copper–Copper Sulfate Electrode. The most popular electrode used with cathodic protection installations is the saturated copper–copper sulfate electrochemical half-cell. Figure 11 illustrates a typical electrode used for underground structures. The electrode is based on the reaction

$$Cu \rightarrow Cu^{2+} + 2e^-$$

The potential of this electrode in saturated copper sulfate is 0.32 V compared to the standard hydrogen electrode. It is not suitable for immersion service in the form pictured in Fig. 11 because the copper sulfate solution weeps through the porous plug too rapidly.

Some commercial literature [20, 21] describes "permanent" copper–copper sulfate electrodes. These are designed so that the copper sulfate solution is in

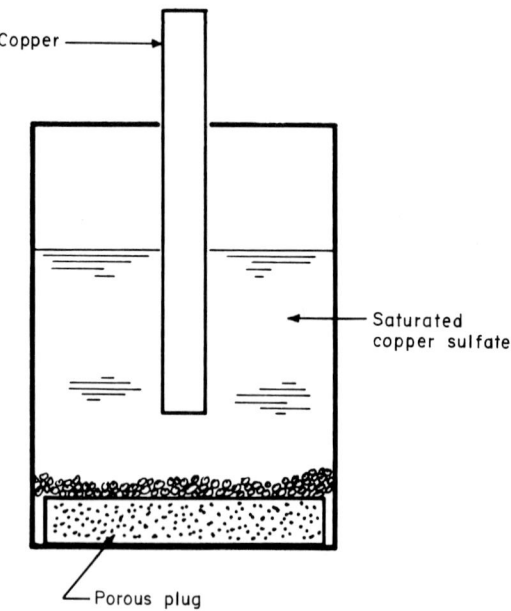

FIG. 11.   Copper sulfate reference electrode. This reference electrode is widely used for monitoring the potential of underground structures.

contact with the electrolyte in such a way that the solution will not "leak out" into the electrolyte at a rapid rate.

Silver–Silver Chloride.   The silver–silver chloride electrode is a solid electrode made by casting or electrochemically forming silver chloride on silver. This electrode can be considered to be permanent in that it can be installed and used in a system for many years without replacement. It is based on the electrochemical reaction

$$Ag + Cl^- \rightarrow AgCl + e^-$$

The standard electrode potential is $+0.222$ V with respect to the standard hydrogen electrode.

It has been most widely used in seawater applications.

Metals.   In some installations, metals such as zinc have been used as a reference electrode [10]. This is possible only if the zinc potential is stable in the environment. However, if the environment is not of constant composition, the zinc potential may vary and make the readings unreliable. Zinc has a potential of -1.1 V to copper–copper sulfate in the earth.

Rectifiers

Impressed current systems must have a source of direct current. Batteries have been used but require some means of recharging. Also, the current or voltage output of batteries cannot easily be controlled for varying electrolyte conditions. In most cathodic protection installations, the dc power is obtained by rectifying ac power. This section is a very brief introduction to this portion of the system.

The dictionary definition of rectifier refers to a device that converts ac to dc. In cathodic protection jargon, "rectifier" is used to describe the equipment which lowers input voltage to a usable level, converts ac to dc, and provides current protection and current control devices if they are used. This discussion uses the cathodic protection definition.

Each rectifier consists of a circuit breaker, step-down transformer, rectifying element, and, in some cases, a means of regulating the current output. The circuit breaker provides a means of protecting the rectifier from short circuits. The step-down transformer lowers the ac voltage to one that is suitable for the cathodic protection installation. It is usually equipped with taps on the secondary windings that allow the field operator to select the appropriate applied voltages. The rectifying element, also known as the stack, is constructed from selenium or silicon and converts the ac to dc. This is accomplished because these materials will allow electron flow in only one direction. These devices may be connected in several configurations to provide full wave (use of both positive and negative half-waves of the cycle) and half-wave rectification. The current control device is not used in all rectifiers. It is necessary for systems in which the structure potential or the current to the structure is controlled. A device called a saturable core reactor is usually used, but some units have silicon-controlled rectifiers. Voltmeters and ammeters are included in the rectifier cabinet so the devices may be easily checked by maintenance personnel.

All these components are usually enclosed in an air-cooled, weatherproof box. In cases where the installation is in a contaminated atmosphere or where safety rules dictate explosionproof equipment, the components are immersed in an oil bath. Several companies manufacture rectifiers, and it would be much less expensive to purchase one of these devices than to attempt to construct one. The various circuitry options available for rectifiers are discussed in the literature [22].

Control

Variables such as temperature, electrolyte composition and resistivity, and metal surface can affect the current requirements. Sacrificial anode systems cannot be automatically controlled to meet these changing conditions. The current output of these systems can be controlled manually by installing resistors in the circuit between the anodes and structure. Impressed current systems can be automatically controlled by use of the control devices mentioned in the section on rectifiers.

In many impressed current applications the voltage between the anode and the structure is the only protective system variable controlled. If the electrolyte resistivity changes, the applied current will also change. In many cases this is of no serious consequence, but in other applications it will lead to ineffective corrosion control.

Two criteria are used to control these systems: current and structure potential.

Constant current flow between the structure and anodes is used in some cases. This compensates for changing electrolyte resistivity. It is not entirely adequate if the current requirements change with time.

Constant structure potential is the best type of control from a theoretical viewpoint. The potential of the metal with respect to a reference electrode is controlled to a preset value by varying the current applied between the anodes and the structure.

Sudrabin lists the advantages of controlled potential cathodic protection [23]:

1. Full protection is continuously assumed
2. Excessive potentials are not applied across coatings to accelerate their degradation
3. Power demand is limited to the amount actually required to maintain the protective state

The equipment is more costly since additional electronic hardware is required. A more complete discussion of application, equipment, and results of using controlled potential may be obtained from the literature [23–25].

Applications

A few selected applications of cathodic protection will be of interest to those involved with chemical processing and petroleum refining. Two large areas of cathodic protection have been completely omitted: Underground pipelines, and structures and marine applications such as ship hulls and offshore structures. These omitted applications are certainly important technically and economically and have been fully discussed in the literature, but the scope of this article is limited to the chemical process industries.

The list of applications is not all-inclusive. It is included only to provide the reader with some idea of the types of corrosion problems that have been controlled by cathodic protection.

General Considerations

In designing any application of cathodic protection, several items must be considered. The important aspects are listed here:

1. Sacrificial anode or impressed current system. The selection of the type of system is made on the basis of cost, power availability, electolyte resistivity, and space considerations.
2. Current density. The total current required to protect the structure must be known. It is calculated by multiplying the required current densities by the exposed surface area. If a structure is only coated, the exposed surface area is used in the calculations. A good to excellent coating will cover 99 to 100% of the total surface area.
3. Resistivity of circuit. The total voltage required to force the amount of current through the circuit must be calculated. There are proven formulas that allow calculation of the circuit resistance for underground installations [26]. These have been modified to allow calculations for installations in vessel interiors [27]. This calculation will help to determine if it is possible to use sacrificial anodes or if impressed current anodes are needed. Also, the rectifier can be sized with this information combined with the current information.
4. Protective criteria. The potential to which the structure should be shifted should be determined prior to making the installation. The system can then be regulated or controlled so the structure will be maintained at this potential. In fresh and seawaters this potential is -0.85 V ($Cu-CuSO_4$). However, other chemical environments may require different protective potentials. These environments should be used in laboratory tests to determine the optimum potential. If coatings are present, it will be necessary to prevent the potentials from being made too negative to prevent coating damage. In water this potential is about -1.1 to -1.2 V with respect to the copper–copper sulfate electrode. Also, hydrogen entry into the metal may be accelerated if these more negative values are used.

Water Tanks

Corrosion of steel tanks containing fresh and salt waters is a problem which cathodic protection can solve. The techniques for protection are similar but differ because of electrolyte conductivity and protective current requirements.

Fresh Water.   Tanks containing fresh water, frequently drinking water, have been successfully protected with both impressed current or sacrificial anode cathodic protection systems. There are slight differences in the application of cathodic protection dependent upon temperature.

The current requirements for protection of steel in fresh waters are variable, depending upon temperature, oxygen content, and dissolved salt content. Several references give values that range from 0.1 to 10 mA/ft^2 [9, 10, 25, 31, 32] for tanks containing fresh waters. This current density range seems to be adequate for both cold and hot waters [10]. These values are based on exposed surface area. Thus for coated tanks the total current requirements are greatly lowered since 95 to 99% of the metal surface is covered if the coatings are well applied and in good condition.

Several anode materials can be used with the impressed current systems. The literature discusses applications which utilize aluminum, graphite, high silicon cast iron, and platinized niobium. The aluminum anodes are used for storage tanks that are exposed to freezing conditions in which the anodes are replaced annually [9, 31]. They are consumed at an appreciable rate; 6.5 to 14 lb/Ayr [9, 32]. These are used due to weight and cost considerations since the ice receding with the water level may break the anodes loose from the suspension wires. Permanent anodes (graphite, silicon iron, and platinized niobium) are used in cases where freezing does not occur. One manufacturer has developed a technique of mounting the platinized niobium from the bottom of the tanks on "floating rafts" to prevent ice damage. The distribution of the anodes in one fresh water tank is shown in Fig. 12. This illustrates the use of both platinized niobium and high silicon iron anodes.

Sacrificial anodes of magnesium and zinc have been used for fresh water tanks. However, zinc has been avoided for hot water tanks since the zinc–iron couple reverses polarity at about 140°F [10]. Magnesium anodes have been used successfully to protect domestic hot water tanks for many years [9].

Cathodic protection is used in many cases in conjunction with coatings on these water storage tanks. This greatly lowers the total current requirements, as discussed above. The coating life is also increased by the reduced corrosion of the uncoated spots on the tank wall. However, cathodic protection currents can harm the coatings if not controlled properly. Hydrogen evolution can occur behind the coatings and destroy the bond between the coating and metal. In order to prevent this damage, the metal potential should be kept less negative than about −1.0 V to copper–copper sulfate. Drisko states that potentials more negative than −1100 mV with vinyls and −1300 mV with epoxy coatings can result in damage [25]. Due to the possibilities of damaging the coatings and because the level varies widely in many storage tanks, controlled potential impressed current systems have been recommended by some engineers [25, 31].

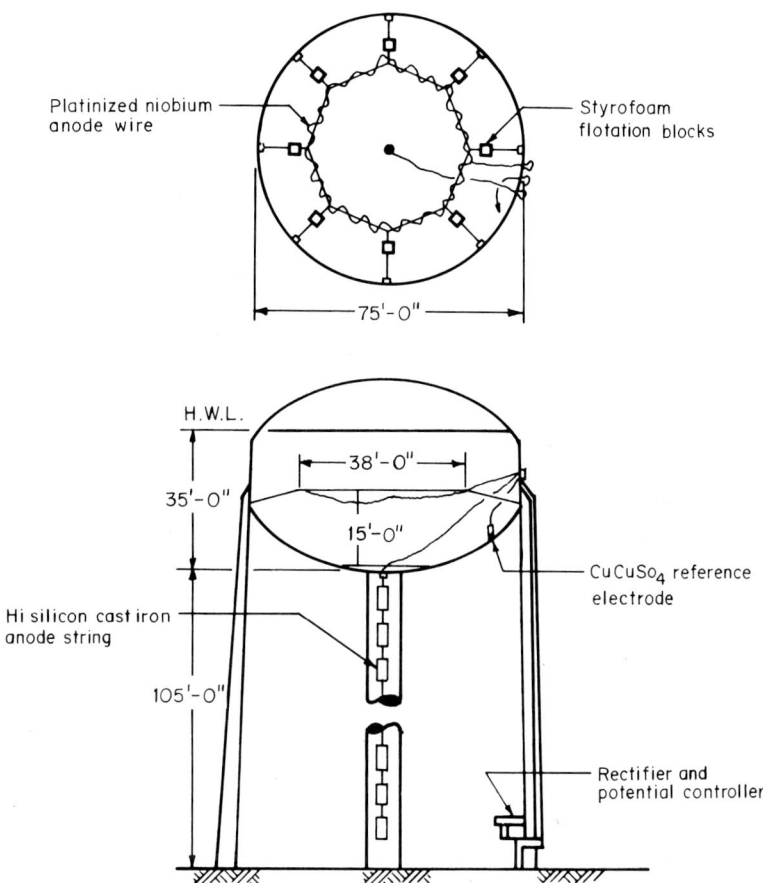

FIG. 12. Cathodically protected water storage vessel. A controlled potential system is used to cathodically protect this type of tank. Notice the anode positions used to distribute the current. (Courtesy Harco Corp., Medina, Ohio.)

These operate by varying the current levels to maintain the wall potential with respect to a reference electrode at an optimum potential.

Hot water tanks can provide special problems for the applications of cathodic protection. In some industrial size tanks, copper steam coils are used for heating. The copper–steel galvanic couple will result in corrosion of the steel wall unless protective measures are taken. The coil can be insulated from the shell, and cathodic protection can be applied to protect the shell. The copper coil can be coated with a calcareous coating as a result of the cathodic protection currents if the coil is not insulated from the shell. This has a deleterious effect on the heat transfer coefficient [30].

Drisko has described a system for a 420,000-gal fire-water storage tank which was internally coated with a coal-tar epoxy. High-silicon cast iron anodes were used with a rectifier that had a capacity of 24 V and 9 A. The rectifier was controlled so that the potential of the tank wall was maintained within 25 mV of the desired setting [25]. A silver–silver chloride reference electrode was used for the control electrode. The dc output was 2 V and 0.2 A over a 4-yr test period [25].

If the tanks are constructed of aluminum, precautions must be taken to prevent the potentials from being below -1.0 V with respect to copper–copper sulfate. The pH adjacent to the aluminum is shifted to high values due to the formation and evolution of hydrogen at these potentials. Since aluminum is amphoteric, it will be corroded by the high pH environment.

Salt Water. Carbon steel salt water storage tanks can be cathodically protected using either impressed current or sacrificial anode systems. However, Morgan [10] cautions that impressed current systems using permanent-type anodes may lead to damage in the vapor phase due to chlorine evolved at the anodes. A consumable anode such as iron or aluminum might prevent this type of damage since the reaction at the anode would be metal dissolution instead of chlorine evolution. Aluminum or zinc sacrificial anodes can be used in seawater since the conductivity is high.

A calcareous coating will build up on the walls of these tanks, causing increases in pH at the vessel wall. The thickness of this layer will increase with time and reduce the total current requirements since this tightly adhering coating covers a portion of the tank wall.

Heat Exchangers

The water side of heat exchangers can be cathodically protected. Investigators have reported using impressed currents and sacrificial anodes to protect the water boxes or headers of carbon steel and cast iron exchangers [10, 33–35] using both seawater and fresh water.

Lowe and Brand [35] describe impressed current protection of steel and cast iron heat exchangers using seawater as a coolant. They advocated use of a platinized titanium strip anode with automatic control of the exchanger potential with respect to a reference electrode.

The strip anode allowed the use of lower anode current densities to avoid chlorine evolution which can in itself cause corrosion of the equipment. In one exchanger located in a nuclear power station in England, they reported using 278 mA/ft^2 (based on tube plate area) to protect the water condenser water box.

They stated installation of protection on a corroded cast iron water box was complicated by the graphite residue remaining. Removal of the graphite by sand blasting followed by patching of pits and holes with epoxy proved to be satisfactory prior to the cathodic protection installation. In corroded steel boxes this cleaning and repair was not necessary. The corrosion scale is easily removed by the cathodic protection currents. The graphite residue is in itself conductive and therefore does not allow the ionic current to reach the metal surface.

Janssen [33] used magnesium anodes in the water box of heat exchangers which used seawater as the coolant. Those coolers had a carbon steel shell and admirality tubes. Severe corrosion occurred on the brass tubes and the steel parts prior to using the magnesium anode system. The steel parts were coated with a coal-tar epoxy in addition to the cathodic protection system. A current density of 20 mA/ft^2 (total cooling box surface area) was used. Satisfactory protection of the steel was obtained in all areas except those close to the brass tube sheets.

Heideman and Halford used strip anodes made of magnesium to control corrosion of the steel water boxes of condensers that used fresh water as a coolant. These proved to be more satisfactory in providing good current distribution than block-type anodes. The strip anode was $^3\!/_8 \times ^3\!/_4$ in. and contained a steel wire core. It was replaced annually and gave satisfactory protection [34].

Pipe Interiors

It is possible to protect the interior surfaces of pipes just as it is possible to protect tanks. The primary difficulty with this type of application is the distribution of the current along the length of the pipe wall. It would be possible to suspend a strip or wire anode along the length of the pipe to solve the current distribution, but this may not be practical. A more practical type of anode installation would be to provide entries at preselected distances along the length of the pipe. The distance between anodes must be selected so the pipe potential at a point midway between the anodes should be within the protected potential range. Also, if the pipe is coated, the potential of the pipe wall close to the anodes must not be more negative than the hydrogen evolution potential in order to prevent damage to the coating.

Morgan [10] considered this problem in detail for bare pipes carrying seawater. He derived an equation to calculate the maximum spread between anodes which fit his experimental data for bare pipes containing seawater at a low velocity. There was a logarithmic relationship between spacing and pipe diameter which ranged from 20 pipe diameters for 1-in. pipe to 4 pipe diameters for 50-in. pipe.

Groover and Peterson [36] studied the potential profiles inside pipes containing stagnant and flowing seawater. These pipes were 0.5 and 2 in. nominal diameter tubes and pipe made of 304SS, copper, and carbon steel. They found it was possible to protect 20 ft of pipe of all diameters and metals with one zinc anode if the seawater was stagnant. They found the oxygen content decreased in the water, and concluded that this was an important factor in achieving protection down the entire pipe length. When seawater was pumped through the pipes, protection was effective for only 6 ft or less from the anode, depending on alloy and pipe diameter.

It does seem possible to protect the interiors of pipes, but some practical problems certainly do exist. The anode spacing is a critical factor that must be evaluated for each system. Bare steel pipes containing seawater seem to be well understood, and this spacing may be predicted. Different electrolytes and pipes of different alloys and coatings add new factors that must be studied prior to making the spacing predictions.

Chemical Environments

There are large amounts of literature data and practical experience available for cathodic protection of vessels, pipes, and tanks containing fresh and seawaters. Equipment containing other electrolytes can also be protected cathodically but there are fewer applications of these.

Acids.   It is possible to protect steel cathodically from corrosion by inorganic acids such as sulfuric and phosphoric acids. However, the current densities required are high. One source lists the current to protect steel from "hot H_2SO_4" as 50 A/ft^2 [7]. Riggs [37] conducted laboratory tests with steel in reagent and wet process phosphoric acid. He found it was possible to achieve protection of steel over wide concentrations and temperature ranges. Table 9 summarizes his data.

These data indicate it is possible to protect steel in H_3PO_4 service, but the current densities are higher than encountered with waters. The impurities in the wet process acid seem to have an inhibiting action on the corrosion of steel. No reference has been found to a field installation of cathodic protection for a vessel containing phosphoric acid.

Schmidt and Brouwer [38] found that it was possible to protect steel from corrosion from pyroligneous acid, a mixture of aliphatic acids with other organic compounds from the destructive distillation of wood. It typically contains 8% acetic acid, 4% methanol, 7% of other organics, and 81% water. At pH 2.9 it was possible to protect steel at -0.78 V (SCE) at a current density of 30 mA/ft^2.

They also found it was possible to prevent intergranular corrosion of

TABLE 9 Cathodic Protection of Steel in Phosphoric Acid. Data from Riggs [37]

Concentration (%)	Acid	Temperature (°C)	Corrosion Rate (thousandths of inches per year)		C.D. mA/ft^2
			Unprotected	Protected	
85	Reagent	24	950	5	120
		36	3,800	28	37.2
		48	9,500	85	483
75	Reagent	24	1,800	5	65
		36	2,900	18.3	325
		48	6,800	74	929
55	Reagent	36	2,000	1.5	130
40	Reagent	36	1,900	3.0	176
20	Reagent	36	1,200	18.2	1,021
75	Wet process	24	260	0	40
		36	350	0	55
		46	430	0	72
		55	15,500	9.3	93

stainless steel by dilute sulfuric acid (1.5 to 2.5 pH) at 50 to 100°C. A current density of 12 to 15 mA/ft^2 maintained the potential at -0.200 to -0.250 V (SCE) and reduced the corrosion significantly. CAUTION: It can be dangerous to use cathodic protection for stainless steels since they depend on passivity to resist corrosion. Cathodic currents can destroy the passive film and greatly accelerate corrosion. Therefore, before using stainless steels with a cathodic protection system, the polarization characteristics in the corrosive environment must be well understood. This is discussed a bit more below.

Molten Salts.   Two papers discuss the use of cathodic protection in very aggressive molten salt environments. These laboratory studies indicate it may be possible to protect some metals cathodically in these systems.

Ito et al. [39] studied the protection of steel in a molten LiCl–KCl eutectic mixture at 450°C. They found it was possible to reduce the corrosion of bare metal and lower if not completely prevent corrosion of ceramic-coated steel in this mixture. They used a graphite anode and a $Cl_2{}^0$ reference electrode for their studies. The protected steel coupons and the test vessel were held at -2.5 V with respect to the chlorine electrode.

Matson et al. [40] studied the $NaF\text{-}LiF\text{-}ZrF_4$ molten salts that are used to dissolve uranium fuel elements prior to reprocessing under hydrogen fluoride. They found in the laboratory that it was possible to protect a 70Ni-18 Mo-7 Cr-5 Fe alloy when submerged in this molten material. However, interface corrosion was a problem not solved by cathodic protection. A side problem of a nickel sponge deposit formation on the anode was noted.

Paper Cylinder.   Hamner [41] reported on an interesting and unusual application of cathodic protection in the plant of H. Kiaer and Co., Fredrikstad, Sweden. The cast iron Yankee cylinder on a paper machine was protected by use of a stationary anode held close to the rotating cylinder. The cylinder was wet only on the surface covered by the paper web. The anode was placed on a dry portion of the cylinder. A unique anode assembly was developed that involved circulating an electrolyte between the platinized titanium anode and the cylinder surface. The protection achieved was dependent upon the residual effort of polarization after the cylinder rotated past the anode assembly to the paper web. Paper quality is dependent upon the smoothness of the cylinder, and it was improved after the protection system was installed.

Stainless Steel

Stainless steels are resistant to corrosion because the alloying elements nickel and chromium promote the formation of a passive film on the metal surface. The passive film can have defects in some environments, and these lead to pitting, stress-corrosion cracking, and intergranular corrosion. It is possible to control these types of corrosion by using cathodic protection, but care must be taken to prevent the destruction of the passive film. There is a range of potential over which the film is stable, and critical potentials within this range control

both pitting and stress-corrosion cracking. Uhlig [1] suggests cathodic protection for both stress-corrosion cracking and pitting as one method of control.
The potential of the metal is shifted toward the active direction but not so far as
to shift the potential out of the passive region.

Personal communications have indicated cathodic protection of stainless
steel is being used to prevent pitting in chloride-containing media.

Prevention of stress-corrosion cracking of 304SS by means of cathodic
protection was studied by Watanabe et al. [42]. They found it was possible to
prevent and halt stress corrosion of 304SS in boiling 42% $MgCl_2$. The potential
at which cathodic protection was successful depended on the stress loading of
the specimen. They found that the application of cathodic protection could be
achieved by coating the stainless steel with a hot-dipped aluminum layer
(aluminizing). They cited plant experiences of aluminizing 304SS used for heat
exchanger tubes and float valves in distillation trays. Cracking was reduced in
these applications.

Herrigel and Sargent [43] cite a similar result in reducing stress-corrosion
cracking of titanium in salt water environments (3.5% NaCl). They found
cracking could be prevented if the titanium potential was shifted to -1.2 to
-1.4 V (SCE). This shift was achieved by using metal-rich paints containing
aluminum or zinc.

Anodic Protection

Introduction

It is possible to use impressed currents to protect anodes, which may seem to be
a contradiction to the statement that the anode corrodes. Anodic protection is
possible due to the phenomena of passivity. Passivity has been defined in two
ways, but the definition of Fontana and Greene [7] will be used here. Simply, it
can be defined as a loss of chemical reactivity under certain environmental
conditions. Passivity is discussed in a bit more detail below. Anodic protection
is more limited than cathodic protection in its possible applications since the
environmental conditions for passivity are very specific and limited.

Anodic protection was first suggested by Edeleanu for plant scale applications. It was first successfully put into practice in 1958–1959 [3–5]. In the 20
years or so since then, anodic protection has been applied to a wide variety of
corrosion problems. This section discusses the theory, the engineering, and
reviews the applications of this method of corrosion control.

Theory

Anodic protection is an electrochemical method of corrosion control based on
an interesting anodic polarization behavior which may be studied by using an
experimental setup as illustrated in Fig. 7. The metal of interest, the working
electrode, is immersed in the corrosive solution with an auxiliary electrode and

a reference electrode. The reference electrode is some type of standard half cell, e.g., a calomel cell. The auxiliary electrode can be any type of metal but usually some noble material that will not contaminate the solution is chosen (platinum, for example).

Current flow between the working electrode and the auxiliary electrode is regulated in a manner to maintain the potential between the working and reference electrodes at given values. The potentiostat, an electronic apparatus that will do this automatically, is normally used for this type of testing.

The potential is shifted with respect to time either in a step-by-step or continuous sweep manner, and the current is recorded. Some metal–solution combinations will exhibit the anodic polarization behavior illustrated in Fig. 13. The potential of the electrode with no current flowing is known as the corrosion potential. As the potential is shifted more noble, the current potential curve follows Tafel behavior, i.e., $\eta = \beta \log (i/i_0)$. This is not unexpected. However, when the primary passive potential E_p is reached, a surprising reaction takes place which causes the current necessary for further shifting of the potential to decrease by several orders of magnitude. The reaction that occurs at that point has been and is still being studied extensively. There has been some debate concerning it, but all seem to agree that some sort of film is formed which lowers further current requirements. This film is known as the passive film, and the metal is in the passive state. The current required to shift the potential further toward noble remains low for a range which can vary from 100 mV to several volts depending upon the metal, corrosive solution, and temperature. The current then begins to increase due to oxygen evolution or other anodic reactions. In simplistic terms, the film breaks down. Three regions

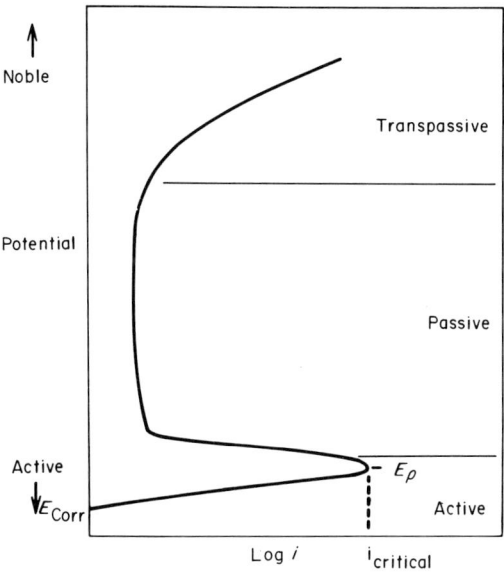

FIG. 13.   Anodic polarization curve. Some metal–solution systems have this type of anodic polarization behavior. This is termed active–passive behavior.

are labeled on the curve: active, passive, and transpassive. The metal corrodes in the active and transpassive regions but corrodes at a low rate in the passive region.

Metals such as Fe, Cr, Ni, Ti, and alloys of these display this polarization behavior in a variety of environments such as sulfuric acid, nitric acid, caustic, NH_4NO_3, urea, and oxalic acid. When a vessel, reactor, heat exchanger, etc. is polarized to a potential within the passive region, the corrosion of the equipment is decreased and it is under anodic protection.

Passivity

Passivity is a phenomenon that is valuable to the chemical process industry because it is the reason stainless steels are corrosion resistant. Several other metals and alloys also derive their corrosion resistance from this condition that is still not well understood. In fact, there is some disagreement even in defining passivity [1, 44]. The definition considered valid for this study states that passivity is the loss of chemical reactivity experienced by some metals in special environments. These metals and alloys, when passive, behave like the noble metals. This noble-like behavior is attributed to the presence of a thin film which is formed on the surface of the metal. There continues to be disagreement concerning the mechanism of the formation of this film. Two theories have been advanced and are vigorously defended by their proponents.

Uhlig [1] states that the film is formed by a chemisorbed layer of oxygen which displaces the water molecules from the metal surface. The oxygen layer may subsequently react with the base metal to form an oxide reaction product.

Bockris and Reddy [44], however, argue that the film is an oxide reaction product. The dissolution of the metal occurs first, then a salt or hydroxide with limited solubility is formed. This precursor subsequently undergoes some sort of change that allows it to be an electronic conductor, and it is then the passive film. This solid-state theory has, in the words of Bockris and Reddy, "... seemed to have gained the day."

The mechanism theory arguments have been in vogue for over 25 years and have not been completely resolved. However, fortunately for the user of passive alloys, passivating inhibitors, and anodic protection, this lack of agreement on mechanism does not prevent their successful use. Anodic protection therefore can be considered to be an applied use of electrochemistry to achieve passivity.

Passivity can be achieved by alloying and by chemical means. The natures of the passive films are very similar. Anodic protection can be used to form the passive film on metals in chemical systems that would normally be corrosive to the metal. At other times, anodic protection can be used to maintain the passivity of the metal so that process upsets or changes do not cause the metal to become active and corrode.

Application Background

Since anodic protection is a relatively new method of corrosion control, there are few guidelines and "rules of thumb" as have developed in cathodic protection. In addition, applications of anodic protection have been controlled in the United States and other countries by basic patents, most of them owned by Continental Oil Co. License to use these patents was sold to one company, Magna Corp.(now Rohrback Corp.), which has exclusive control of the sale of anodic protection installations throughout the United States and Canada.

Polarization Behavior

It is absolutely necessary to obtain an anodic polarization curve for the solution–metal system before using anodic protection. The curve must be of the type shown in Fig. 13 or it will be impossible to achieve protection. If the anodic polarization curve only exhibited Tafel behavior, the so-called protection system would severely corrode the vessel. In the early days of anodic protection each solution was sampled and used in the laboratory to obtain a polarization curve. However, it is now possible to predict the control potential for the most common applications, e.g., steel in concentrated sulfuric acid.

The curve provides the basic information concerning the potential at which the vessel should be protected. The values obtained in the laboratory can be used directly in the field. However, the electrode potentials may need to be changed to account for differences in reference electrode potentials if the electrode used in the plant is different from the electrode in the laboratory.

Values of the currents needed to obtain and maintain protection can be estimated from the polarization curve. These currents are time dependent and are usually larger than those determined in the plant. However, the data from the polarization curve are the information from which the engineering design must be made.

As an example, Banks and Sudbury [45] illustrated that the current density needed to obtain protection dropped from 0.29 to 0.02 A/ft^2 when the time was increased from 5 to 30 min. As a practical matter, the current density needed to obtain protection regulates the size of the power supply chosen for an installation unless special arrangements are made. It is possible to use a portable power supply, such as a dc welding machine, to establish protection, and then use the system power supply to provide the much lower currents needed to maintain protection. Times of up to 1 h have been used in field applications to obtain protection of equipment such as storage tanks. However, some applications, e.g., hot acid in stainless steels, must be repassivated in very short times if protection is lost, and therefore the system must be sized to deliver the proper current.

The current to maintain protection is also time dependent, decreasing with time. Therefore the currents to maintain passivity obtained with the polarization curves will usually be higher than the application values, and thus include an inherent safety factor.

Typical current densities to obtain and maintain protection (passivate) are given in Table 10. These vary widely, depending upon solution, temperature, and metal. The total current necessary to obtain protection may be decreased by using such techniques as establishing passivity with an initial low liquid level, thus decreasing the wetted surface area. The current to passivate the remainder

TABLE 10 Anodic Protection Current Requirements

Solution	Temperature ($^{\circ}$C)	Metal	Current (mA/ft^2) Passivate	Maintain	Refs.
Oleum	25	Steel	120	0.15	50
93–98% H_2SO_4	25	Steel	120	1.8	56
93–98% H_2SO_4	20	316SS	2		56
90–94% H_2SO_4	25	Steel		0.9	74
78% H_2SO_4	25	Steel	140	2.5	56
25% NH_4OH	25	Steel	1850	0.22	61
0.1 M oxalic acid	20	Steel		5.8	88
Fertilizer solutions: NH_3, NH_4NO_3, $(NH_4)_2CO_3$, $CO(NH_2)_2$	20	Steel		0.046	60
93% H_2SO_4	60	316L		0.4	82
67% H_2SO_4	25	317	164	0.09	47

of the tank as it is filled is usually not great since it takes a substantial time period to fill vessels.

Another factor that can be determined from the laboratory investigations is retention of the passive film without current flow. In some systems, fertilizer solutions in mild steel for example, once the metal potential is shifted into the passive region it will remain stable for days with no current flow. Other systems, e.g., hot sulfuric acid in stainless steel, will shift to a corrosive condition in a few seconds after current is removed. Mild steel in 93% sulfuric acid will maintain a constant electrode potential within the passive region for 30 to 45 min. This type of behavior allows the use of a less complicated and thus more economical on–off type of potential control. Fisher and Brady [46] found that an on–off controller applied current a decreasing percentage of the time until it was activated 10% of the time 12 h after 10% H_2SO_4 was added to a steel tank.

Electrodes

In each anodic protection system it is necessary to use at least one cathode and one reference electrode. Many materials have been used for these. The cathode should be a "permanent-type" electrode that is not dissolved by the solution or the currents impressed between the vessel wall and electrode. The reference electrode should have a potential that does not vary with time. The temperature characteristics should be well known. In most cases an electrochemical half-cell is used, but in others a metal can be used.

Cathode.   The cathodes used in most of the early installations of anodic protection were made of platinum-clad brass [3–5, 46–54]. The most widely used electrode was made of Scovill No. 20 brass clad with 0.025 in. of platinum

[47]. The platinum was selected to prevent any contamination of the system by more soluble electrodes. These electrodes are excellent electrochemically but are somewhat costly and the surface area is limited economically. The overall circuit resistance is proportional to the surface area so it is advantageous to use electrodes with a large surface area. In order to do this economically, a less costly metal must be used.

It is possible to use several metals with no problems of electrode corrosion since the cathodes will be cathodically protected. The metal to be used in a given environment should be polarized cathodically and the corrosion rate determined under these conditions. Table 11 lists some of the metals reported in the literature as cathode materials for anodic protection installations.

All these metals are used as rods, tubes, cables, or wires in the solutions described. An interesting departure from these has been suggested by Trusor and Kyruchkova [64]. They constructed an electrode using platinum-impregnated graphite similar to a fuel cell electrode. They term this electrode an "air" electrode.

The electrode size is chosen to conform geometrically to the vessel and also provide a large surface area. As mentioned above, the electrode surface area affects the circuit resistance and should be large to minimize total power consumption.

The electrode location is not a critical factor in simple geometries such as storage vessels. Stammen and Townsend [65] found it was possible to passivate vessels with the electrodes located asymmetrically in a cylindrical vessel. The passive area spreads completely around the circumference of the vessel with time.

In complex geometries such as heat exchangers, it is necessary to extend the

TABLE 11 Cathode Materials for Anodic Protection

Metal	Environment	Refs.
Platinum on brass	Various	3–5, 46–54
Steel	Kraft digester liquid	55
Illium G (nickel-based alloy) (Ni-Cr-Fe-Me-Co)	H_2SO_4, 78–105%	56
Silicon-cast iron	H_2SO_4, 89–100% 22% oleum	57
Copper	Hydroxylamine sulfate	58
Stainless steel	Liquid fertilizer	
	Nitrate, NH_3	59
	$(NH_4)_2CO_3 + NH_4NO_3$	60
	NH_4OH	61
Nickel plated	Chemical nickel plating solution	62
Hastelloy C	Nitrate fertilizer solutions	63
	Sulfuric acid	68
	Kraft digester liquid	55

electrode around the surface to be protected. For example, a wire electrode was used to protect the acid side of a spiral-type heat exchanger used for cooling sulfuric acid [66].

Multiple electrodes are used in large storage tanks to distribute the current better and to decrease circuit resistance.

Reference Electrodes.   Reference electrodes must be used with anodic protection systems since it is important that the vessel-wall potential be maintained at a preselected value. The reference electrode must have a constant electrochemical potential with respect to time. The temperature effect on the potential should be known and be of minimal value. In some installations the electrode potential should not be affected by changes in solution composition.

Standard electrochemical half-cells have been used in most applications of anodic protection. The theory of reference electrodes has been fully discussed by Ives and Janz [69].

The first field applications of anodic protection used the calomel cell as the reference electrode. This mercury–mercurous chloride electrode requires a potassium chloride solution bridge to make contact with the solution. The bridge was made of glass which presented many problems in a chemical plant since it was so fragile.

Silver–silver chloride has been used for many anodic protection applications. This is a particularly good electrode since it is solid and can be mounted inside the vessel with no salt bridge problems, but it is limited to those solutions in which it will not dissolve. It can be used in sulfuric acid up to 100%, but it cannot be used in oleum in which it is soluble.

Many other electrodes have been used in anodic protection installations; several of these are listed in Table 12 with the corrosive solutions in which they were used.

Kuzub et al. [67] have described an interesting electrode that has been widely used in the USSR. This is a mercury–mercurous sulfate paste mounted in poly(tetrafluorethylene) and uses no liquid salt bridge. This electrode is installed in the solution just as are solid electrodes.

TABLE 12   Reference Electrodes

Electrode	Medium	Refs.
Calomel	H_2SO_4, oleum, sulfuric acid	3–5, 47, 49
Ag-AgCl	H_2SO_4, fertilizer solutions, digesters	46, 47, 52, 59, 60
Mo/MoO_3	Green liquor (Na_2Co_3, etc)	54
Bismuth	NH_4OH	61
316SS	Nitrate, fertilizer solutions, oleum	56, 63
$Hg/HgSO_4$	H_2SO_4, hydroxylaminesulfate	57, 58, 67
Pt/PtO	H_2SO_4	53

TABLE 13 Anodic Protection of Sulfuric Acid Storage

Concentration (%)	Metal	Temperature	Refs.
Oleum	Steel	Ambient	47, 51
93	Steel	Ambient	54, 57, 74
98	Steel	Ambient	49, 54, 76, 87
Spent, various concentrations	Stainless steel	100–160°F	76, 77
100	Steel	30°C	46
77-oleum	Steel	Ambient	56
50	1kh18Nl0t	Ambient	73
90–94	Steel	Ambient	74
1.3 N H_2SO_4	18–8SS	70°C	86

An alloy or pure metal can be used as a reference electrode instead of an electrochemical half-cell. For example, stainless steel has been used as shown in Table 12. However, this metal must have a stable potential in the corrosive environment. These metal reference electrodes should be used only as a last resort since the potential is not nearly so reliable as that of the half-cell.

Applications

Anodic protection was first suggested as a corrosion control method in the chemical plant by Edeleanu in 1954 [2]. In 1960, several investigators with Continental Oil Co. reported the application of anodic protection in a petrochemical plant [3–5]. Since that time many vessels and pieces of equipment have been protected by this method. Tables 13–16 list many of the field applications discussed in the literature.

Sulfuric Acid Storage

Table 13 lists literature references to storage tank installations of sulfuric acid ranging from 50% to oleum.

TABLE 14 American Heat Reclaiming Spiral Heat Exchanger

Metal	316
Type	Spiral
Concentration and temperature	93–98% to 240°C
Corrosion rate	1 thousandth inch per year
Electrodes	Wire cathode reference, propriety

TABLE 15 Chemetics Heat Exchangers for H_2SO_4

Metal	316L
Type	Shell and tube
Flow pattern	H_2SO_4-shell side, water tube side
Concentrations and temperature	93% at 60°C (140°F), 99% at 110°C (23°F)
Corrosion rates	Unprotected, 200–400 thousandths inch per year Protected, 1 thousandth inch per year
Electrodes	Several cathodes, design propriety

It is possible to store 77% and higher concentrations of H_2SO_4 in steel tanks with little damage to the storage vessels. However, the iron content of the acid will increase with storage time at a rate of 10 ppm/d for 98% H_2SO_4 [49]. There are several chemical processes that require low iron content acid, and the acid manufacturer can obtain a higher price for the purer acid. Several investigators have found that anodic protection lowers the iron pickup to less than 1 ppm/d of storage for 93%–oleum in steel tanks [46, 48, 49, 57]. In addition, tank trucks used to transport the sulfuric acid are anodically protected to maintain the low iron content of the acid during delivery [53].

Storage and processing tanks for spent sulfuric acids have had their lives extended by being anodically protected [76, 77]. The acids were the by-products of chemical and petroleum processing.

More dilute solutions of sulfuric acid were contained in stainless steels but, due to temperature conditions, they were corrosive. Anodic protection lowered the corrosion of these metals [73, 86]. Supronov et al. [86] stated that the use of the inhibitor Katapin-A (reported to be $[C_nH_{2n+1} - C_6H_4 - N - C_5H_5]^+Cl^-$, where $n = 11$ to 13 [88]) greatly improved the degree of protection by anodic protection.

These systems all require a cathode or several cathodes in parallel, a reference electrode, potential sensing and control, and a dc power source. Figure 14 illustrates a typical storage tank installation.

TABLE 16 Heat Exchanger Applications

Concentration (%)	Metal	Temperature	Refs.
77 H_2SO_4	1kh18N9T	100–120°C	68
93, 98 H_2SO_4	316SS	240°C	81
93 H_2SO_4	316L	140°F	82
98 H_2SO_4	316L	230°F	82
3 H_2SO_4 with $NaSO_4$ and H_2S	Titanium	To 75°C	83, 84

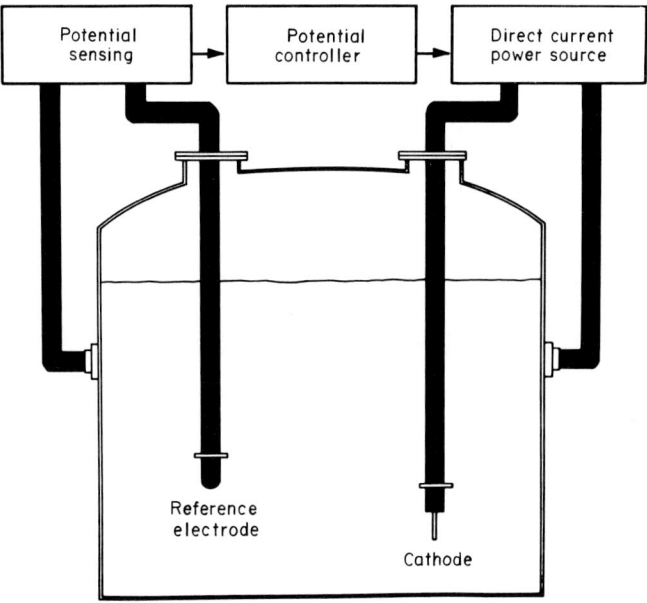

FIG. 14. Anodic protection system. The basic elements of anodic protection of a storage vessel are shown.

The cathodes and reference electrodes were discussed above. The potential control circuitry for the concentrated sulfuric acid tanks is predominately the on−off type. In 98% H_2SO_4−oleum the life of the passive film is lengthy, so several tanks can be protected by a single control and current supply system [48, 56]. A switch mechanism connects each tank to a control system that operates at a rate high enough to make a complete cycle before the potential shifts to an active value. Up to four vessels have been protected with one system [56].

The more corrosive systems cannot be protected with this type of control. A control that continually monitors the potential and impresses sufficient current between the cathode and vessel wall is required.

The 77% H_2SO_4−oleum storage tanks are the most common application of anodic protection in the United States. This seems to be the most successful system economically, and the company involved in promoting and selling anodic protection has concentrated its efforts on these applications [56].

Sulfuric Acid Heat Exchangers

Two companies have developed a package system for cooling the concentrated sulfuric acid produced in sulfuric acid manufacture [81, 82]. One company manufactures spiral heat exchangers with a wire cathode mounted internally during construction of the exchanger [81] (see Fig. 15). The cathode design is patented [66]. The corrosion rate is reduced to less than 1 mpy with anodic

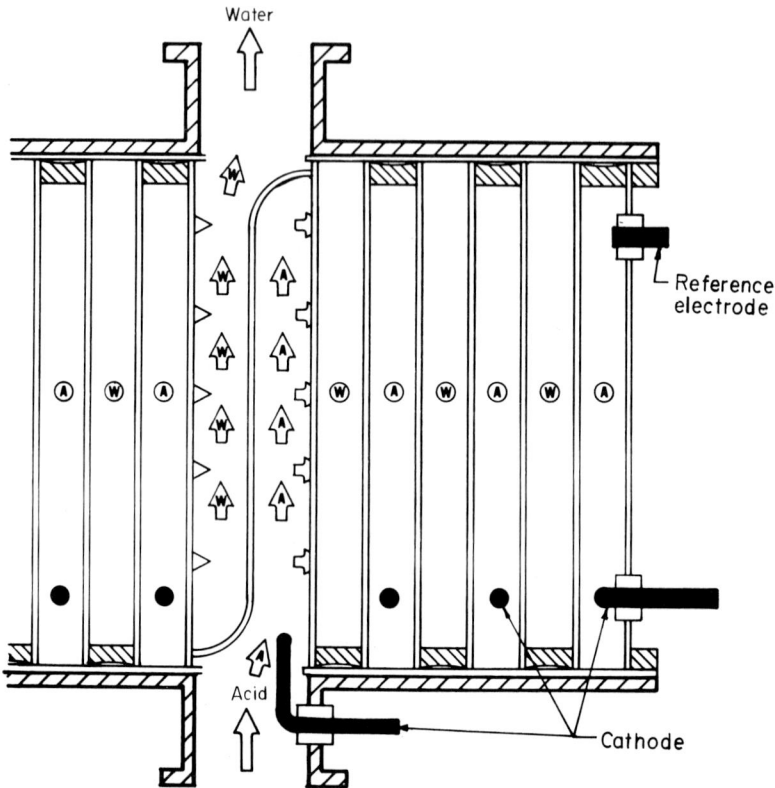

FIG. 15. Spiral-type heat exchanger, anodically protected. A wire cathode is wound in the exchanger as it is fabricated. This exchanger is used in cooling sulfuric acid. (Courtesy American Heat Reclaiming Corp., New York, New York.)

protection. Application of anodic protection to this type of heat exchanger was made in the early 1960s. The heat exchanger was installed in an acid plant on the shores of the Great Salt Lake and the cooling water contained a high concentration of chloride ions. The acid did not corrode the 304SS but the chloride ions caused failure of the exchanger by stress-corrosion cracking. The operation was a success but the patient died! Details of the spiral heat exchanger application are given in Table 14.

Another company has designed and now markets a shell and tube heat exchanger for this acid-cooling service. This exchanger includes cathode and reference electrodes mounted during construction of the exchanger [82]. Details of this application are given in Fig. 16 and in Table 15.

Titanium tubes in an exchanger handling dilute sulfuric acid from a rayon fiber spining bath have been anodically protected [83, 84]. The success of this application has led one company to construct an anodically protected evaporator for this service [84].

A summary of heat exchanger applications is given in Table 16.

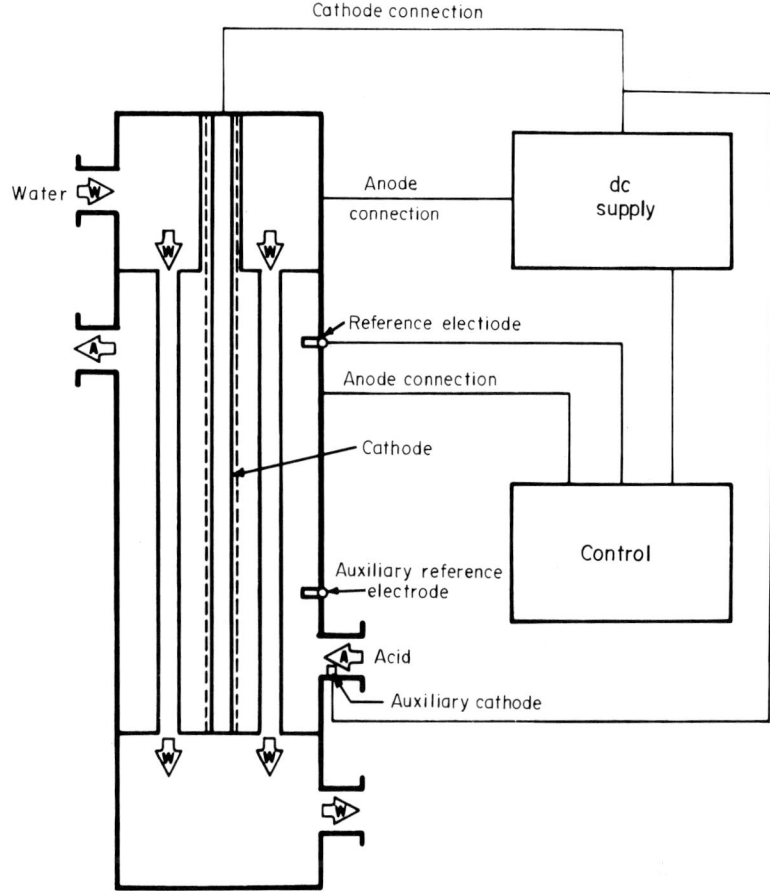

FIG. 16. Anodically protected shell and tube heat exchanger. This heat exchanger is sold with the anodic protection system as an integral part. (Courtesy Chemetics International, Toronto.)

Nitrate Fertilizer Solutions

Table 17 lists several references to protection of storage tanks containing fertilizer solutions. Steel can be protected anodically more economically than can stainless steel or aluminum. A fleet of tank cars used to transport these

TABLE 17 Fertilizer Solutions

Solution	Metal	Temperature	Refs.
NH_4NO_3, urea, $(NH_4)_2CO_3$, NH_3, H_2O	Steel	40°C	59, 60
NH_4NO_3, NH_3, urea, H_2O	Steel	80°F	63
NH_4NO_3, NH_3, H_2O	Steel	80	80
N-P-K fertilizer solutions (at 10% KCl)	0kh23N28M3D3T	Ambient	71

solutions was anodically protected [63]. This test was conducted with 136 steel cars during the 1966 fertilizer season and with 40 cars in 1967. These tests indicated corrosion was halted and that anodic protection of steel cars was economically preferred to aluminum cars.

At one time there were several storage vessels anodically protected in the United States. However, there are now no known installations in operation in this country [56]. However, several applications have been disclosed by investigators in the USSR [59, 60, 70, 80].

Miscellaneous Solutions

Table 18 lists several corrosive solutions that have been prevented from corroding metals by anodic protection. Most of these applications have been utilized due to the severity of the corrosion problem and because anodic protection proved to be the most economically attractive corrosion control method.

The installation described by Danielyan et al. of anodic protection for a vessel containing 25% NH_4OH is particularly interesting [61]. Table 10 indicates the current to passivate steel in this environment is 1.85 A/ft^2, which is practically impossible on a plant-scale storage vessel. This vessel has a 10,000 m^3 capacity (2.6×10^6 gal) and a wetted surface area of 2135 m^2 (23,000 ft^2). The vessel was first partially filled with a 4% ammonia solution which spontaneously passivated the walls. The anodic protection system was turned on, the vessel filled, and the ammonia concentration was slowly increased to the 25% level. The current to maintain was 0.22 mA/ft^2. This was an ingenious method of overcoming the large currents necessary to achieve protection.

TABLE 18 Miscellaneous Solutions

Solution	Metal	Temperature	Refs.
ClO_2-bleach filter wire	317SS	140°F	54
Recovery liquors from kraft plant			
Green storage tank	Steel	160°F	54
Kraft cooking liquor	Steel	350	55
Sulfuric acid reactor	304SS	160°F	5
Hydroxylamine sulfate storage	Steel	Ambient	75
	0kh21N5T	40°C	58
	1KH18N10T	Ambient	72
25% NH_4OH storage	Steel	5–25	61
Chemical nickel plating bath	kh18N10T	87–89°C	78, 79
	304SS		
NH_4HCO_3 storage			85

References

1. H. H. Uhlig, *Corrosion and Corrosion Control*, 2nd ed., Wiley, New York, 1971.
2. C. Edeleanu, *Metallurgia*, *50*, 113 (1954).
3. J. D. Sudbury, O. L. Riggs, and D. A. Shock, *Corrosion*, *16*, 47t–54t (1960).
4. D. A. Shock, O. L. Riggs, and J. D. Sudbury, *Corrosion*, *16*, 55t–58t (1960).
5. O. L. Riggs, M. Hutchison, and N. L. Conger, *Corrosion*, *16*, 58t–62t (1960).
6. J. O'M Bockris and A. K. N. Reddy, *Modern Electrochemistry*, Plenum, New York, 1970.
7. M. G. Fontana and N. D. Greene, *Corrosion Engineering*, McGraw-Hill, New York, 1967.
8. N. D. Tomashov, *Theory of Corrosion and Protection of Metals*, Macmillan, New York, 1966.
9. L. M. Applegate, *Cathodic Protection*, McGraw-Hill, New York, 1960.
10. J. H. Morgan, *Cathodic Protection*, Hill, London, 1959.
11. W. S. Schwerdtfuger and O. N. McDorman, *J. Res. Natl. Bur. Stand.*, *47*,(2), (August 1951).
12. H. W. Schmidt and A. A. Brouwer, *Mater. Prot.*, *1*(2), 26 (1962).
13. O. L. Riggs, *Mater. Prot. Perform.*, *9*(10), 21 (1970).
14. E. W. Dreyman, *Mater. Prot. Perform.*, *11*(9), 17 (1972).
15. M. A. Warne and P. C. S. Howard, *Mater. Perform.*, *15*(3), 39 (1976).
16. W. T. Bryan, *Mater. Prot. Perform.*, *9*(9), 25 (1970).
17. G. D. Brady, *Mater. Prot. Perform.*, *10*(10), 20 (1971).
18. G. L. Doremus and J. G. Davis, *Mater. Prot.*, *6*(1), 30 (1967).
19. L. L. Shreir, *Corrosion*, *17* 118t (1961).
20. Wallace and Tiernan Division, Penn Walt Corp., 25 Main Street, Belleville, New Jersey 07109.
21. Harco Corp., 1055 W. Smith Road, Medina, Ohio 44256.
22. Staff Feature, *Mater. Prot.*, *4*(4), 75 (1965).
23. L. P. Sudrabin, *Mater. Prot.*, *2*(2), 64 (1963).
24. R. Ferry, *Mater. Prot.*, *7*(8), 26 (1968).
25. R. W. Drisko, *Mater. Prot.*, *12*(11), 24 (1973).
26. D. A. Tefankjian, *Mater. Prot. Perform.*, *11*(11), 50 (1972).
27. G. E. Moller, J. T. Patrick, and J. W. Caldwell, *Mater. Prot.*, *1*(2), 46 (1962).
28. T. S. Lennox, R. E. Groover, and M. H. Peterson, *Mater. Perform.*, *10*(9), 39 (1971).
29. J. B. Bushman, *Harco "TASC" III Automatic Cathodic Protection for Water Storage Tanks and Treatment Equipment*, Paper HC-242, Harco Corp., Medina, Ohio.
30. E. W. Dreyman, *Mater. Prot.*, *1*(2), 58 (1962).
31. *Cathodic Protection of Steel Water Storage Tanks*, T-3G-5, Proposed NACE std. Draft No. 1, June 1976.
32. Harco Corp., Medina, Ohio.
33. W. S. Janssen, *Mater. Prot.*, *1*(1), 42 (1962).
34. W. A. Heideman and H. Halford, *Mater. Prot.*, *1*(2), 15 (1962).
35. R. A. Lowe and J. W. L. F. Brand, *Mater. Prot.*, *9*(11), 45 (1970).
36. R. E. Groover and M. H. Peterson, *Mater. Perform.*, *13*(11), 24 (1974).
37. O. L. Riggs, *Mater. Prot.*, *9*(10), 21 (1974).
38. H. W. Schmidt and A. A. Brouwer, *Mater. Prot.*, *1*(2), 26 (1962).
39. Y. Ito, T. Ohmori, Z. Takehara, and S. Yoshizawa, in *5th International Congress on Metal Corrosion*, NACE, Houston, 1974, p. 1029.

40. L. K. Matson, E. F. Stephan, P. D. Moller, W. K. Boyd, and R. P. Milford, *Corrosion*, *22*, 194 (1966).

41. W. E. Hamner, *Mater. Prot.*, *8*(2), 48 (1969).

42. M. Watansbe, Y. Mukai, T. Kawamura, M. Watanable, and H. Kosai, in 5th *International Congress on Metal Corrosion*, NACE, Houston, 1974, p. 370.

43. H. R. Herrigel and J. C. Sargent, *Mater. Prot.*, *7*(10), 26 (1968).

44. J. O'M. Bockris and A. K. N. Reddy, *Modern Electrochemistry*, Vol. 2, Plenum, New York, 1970.

45. W. P. Banks and J. D. Sudbury, *Corrosion*, *19*, 300t (1963).

46. A. O. Fisher and J. F. Brady, *Corrosion*, *19*, 37t (1963).

47. C. E. Locke, M. Hutchison, and N. L. Conger, *Chem. Eng. Prog.*, *56*(11), 50 (1960).

48. Staff, *Mater. Prot.*, *2*(9), 69 (1963).

49. J. D. Sudbury, C. E. Locke, and D. Coldiron, *Chem. Process.*, p. 23 (February 11, 1963).

50. J. D. Sudbury and C. E. Locke, *Oil Gas J.*, *61*(43), 111, (1963).

51. J. D. Sudbury and C. E. Locke, *Chem. Eng.*, *70*(23), 268 (1963).

52. J. C. Redden, *Chem. Process.*, *27*(14), 82 (October 1964).

53. C. E. Locke, *Mater. Prot.*, *4*(3), 59 (1965).

54. C. E. Locke, *Tappi*, *49*(1) 61A (1966).

55. T. R. B. Watson, *Mater. Prot.*, *3*(6) 54 (1964).

56. Magna Corp., Private Communication, 1976.

57. Ya. M. Kelotyrkin, V. A. Makarov, V. M. Novakovskii, L. L. Faingold, Kh. L. Tseitlin, V. M. Kornev, and R. F. Plakhova, *Zashch. Met.*, *7*(6), 722 (1971).

58. V. S. Kuzub, V. A. Makarov, A. I. Tsinman, V. K. Sokolov, and Ya. M. Kolotyrkin, *Zashch. Met.*, *4*(4), 362 (1968).

59. L. G. Kuzub, V. I. Gnezdilova, and V. S. Kuzub, *Zashch. Met.*, *4*(5), 564 (1968).

60. V. S. Kuzub, V. G. Moisa, L. G. Kuzub, V. I. Gnezdilova, N. K. Koz'menko, and L. A. Danielyan, *Zashch. Met.*, *7*(3), 361 (1971).

61. L. A. Danielyan, A. I. Tsinman, V. S. Kuzub, V. G. Moisa, and N. N. Statsenko, *Zashch. Met.*, *9*(4), 492 (1973).

62. A. V. Ryabchenkov, V. I. Velemitsina, V. A. Makorov, and L. N. Alekseeva, *Zashch. Met.*, *7*(6), 718 (1971).

63. W. P. Banks and M. Hutchison, *Mater. Prot.*, *8*(2), 31 (1969).

64. G. N. Trusor and E'. Ya. Kryuchkova, *Zashch. Met.*, *10*(4), 440 (1974).

65. A. M. Stammen and C. R. Townsend, *Corrosion*, *23*, 343 (1967).

66. C. E. Locke and G. D. Harral, U.S. Patent 3,409,530 (November 5, 1968).

67. V. S. Kuzub, A. I. Tsinman, V. K. Sokolov, and V. A. Makarov, *Zashch. Met.*, *5*(1), 56 (1969).

68. Ya. M. Kolotyrkin, V. A. Makarov, V. S. Kuzub, A. I. Tsinman, and L. G. Kuzub, *Zashch. Met.*, *1*(5), 598 (1965).

69. D. J. G. Ives and G. J. Janz, *Reference Electrodes—Theory and Practice*, Academic, New York, 1961.

70. R. L. Every and W. P. Banks, *Corrosion*, *23*, 151 (1967).

71. V. S. Kuzub, C. S. Novitskii, L. B. Golorneva, and V. P. Rebrunov, *Khim. Promst.* (*Moscow*), *1974*(8), 609; *Chem. Abstr.*, *82*, 65939a (1975).

72. V. S. Kuzub, *Tr. Ukr. Resp. Konf. Elektrokhim.*, *1st 1973*, *2*(8), 8–93 (1973); *Chem. Abstr.*, *81* 85146x (1974).

73. V. S. Kuzub, N. N. Statsenko, L. G. Kuzub, and V. G. Moisa, *Khim. Tekhnol.* (*Kiev*), *1974*(1), 63; *Chem. Abstr.*, *81*, 57504t (1974).

74. V. S. Kuzub, N. N. Statsenko, L. G. Kuzub, V. G. Moisa, E. A. Kryzhnil, I. A. Palanuer, P. F. Gurtounik, and V. K. Malevich, *Koks Khim.*, *1973*(10), 48; *Chem. Abstr.*, *80*, 9733a (1974).

75. V. S. Kuzub, V. G. Moisa, N. N. Statsenko, N. K. Koz'menko, and E. A. Kryzhnii, *Khim. Prom.*, *49*(4), 267 (1973); *Chem. Abstr.*, *79*, 26408v (1972).

76. J. C. Redden, *Mater. Prot.*, *5*(2), 51 (1966).

77. C. E. Locke, W. P. Banks, and E. C. French, *Mater. Prot.*, *3*(6), 50 (1964).

78. A. V. Ryabchenkov, V. I. Velumitsina, V. A. Makarov, and L. N. Alekseuva, *Zashch. Met.*, *7*(6), 718 (1971).

79. C. E. Locke, U.S. Patent 3,375,178 (March 26, 1963).

80. J. D. Sudbury, W. P. Banks, and C. E. Locke, in *2nd International Congress on Metal Corrosion*, NACE, Houston, 1966, p. 267.

81. American Heat Reclaiming Corp., New York, Private Communication, 1976.

82. Chemetics International Ltd., Montreal, Private Communication, 1976.

83. P. E. Morgan and L. S. Evans, *Mater. Prot.*, *4*(1), 60 (1965).

84. L. S. Evans, P. C. S. Hayfield, and M. C. McMorris, in *Proceedings of the 4th International Congress on Metal Corrosion*, NACE, Houston, 1972, p. 625.

85. Chi-Cheng Chen, *Hua Hsueh Tung Pao*, *1974*(2), 117; *Chem. Abstr.*, *83*, 30415v (1974).

86. N. A. Supronov, Kh. Freid, and N. S. Baburina, *Tr. Ivanov. Khim-Tekhnol. Inst. 1970*(12), 180; *Chem. Abstr.*, *79*, 142218g (1973).

87. P. Neufeld and R. C. Williamson, *Corros. Sci.*, *5*(9), 605 (1965).

88. R. B. Perry and F. F. Lyle, *Mater. Perform.*, *15*(8), 38 (1976).

89. L. D. Perrigo, *Mater. Prot.*, *5*(3), 73 (1966).

CARL E. LOCKE

Corundum and Emery Supply–Demand Relationships

The important supply and demand 1975 data for these components are obtained from *Minerals in the U.S. Economy* (*1977 Issue*), Bureau of Mines, U.S. Department of Interior. See Figs. 1 and 2 and Table 1.

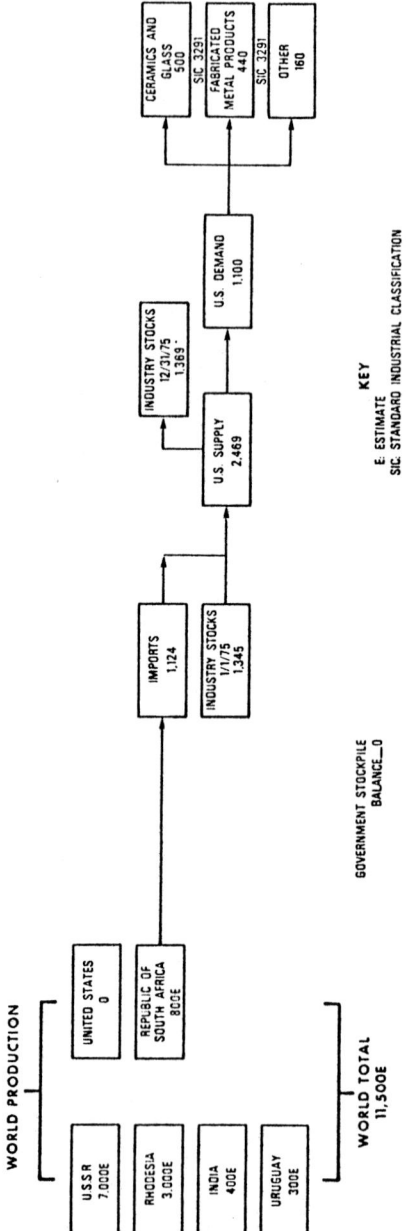

FIG. 1. Corundum supply–demand relationships, 1975. Short tons of corundum.

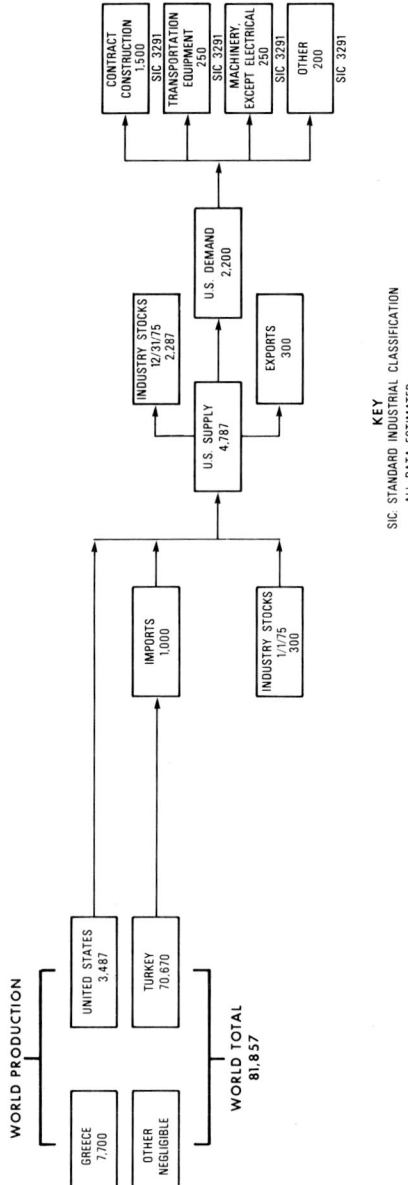

FIG. 2. Emery supply–demand relationships, 1975. Short tons of emery.

TABLE 1 Corundum Supply–Demand Relationships, 1966–1975 (short tons)[a]

	1966	1967	1968	1969	1970	1971	1972	1973	1974	1975
World production:										
United States	—	—	—	—	—	—	—	—	—	—
Rest of world	11,000	11,000	11,000	10,000	10,927	10,861	11,516	11,775	11,775	11,500
Total	11,000	11,000	11,000	10,000	10,927	10,861	11,516	11,775	11,775	11,500
Components of U.S. supply:										
Shipments of government stockpile excesses						1,964				
Imports	3,000	2,000	5,575	—	—	48	—	520	2,275	1,124
Industry stocks, January 1	2,000	2,400	2,000	4,975	2,975	1,788	1,800	300	70	1,345
Total U.S. supply	5,000	4,400	7,575	4,975	2,975	3,800	1,800	820	2,345	2,469
Distribution of U.S. supply:										
Industry stocks, December 31	2,400	2,000	4,975	2,975	975	1,800	300	70	1,345	1,369
Industrial demand	2,600	2,400	2,600	2,000	2,000	2,000	1,500	750	1,000	1,100
U.S. demand pattern:										
Ceramics and glass:										
Optical instruments and lenses	1,500	1,300	1,500	900	900	900	675	338	450	495
Fabricated metal products	800	800	800	800	800	800	600	300	400	440
Other	300	300	300	300	300	300	225	112	150	165
Total U.S. primary demand	2,600	2,400	2,600	2,000	2,000	2,000	1,500	750	1,000	1,100

[a]Import figures are as reported. All other figures are estimates.

Cosmetics

Cosmetics are the end products of cosmetic chemistry, a not too well-defined science that blends the skills of specialists in the chemical, physical, biological, and medical fields. Present-day practice has evolved from the relatively simple cold cream and glycerin-and-rose water formulations to scientific products created by research and development laboratories well staffed with able and creative scientists [1]. Some products reach into the drug field, creating labeling and regulatory problems. New materials with interesting properties become available as the industry expands. Information on the raw materials used in the manufacture of cosmetics can be obtained from several references [2, 3] and from the literature available from suppliers to the industry. The Society of Cosmetic Chemists, a professional society, has its own journal, published monthly, and regular meetings and seminars keep its members up-to-date on research developments in the industry. The Cosmetic, Toiletry and Fragrance Association keeps its member companies informed on current legislation that affects the industry, issues standard specifications for cosmetic materials [4], and has regular meetings at which papers of interest to the industry are presented. A cosmetic research institute is under consideration by all elements of the industry.

The contribution of the sciences to cosmetics has been inestimable. Through the introduction of scientific methods—accurate measurements, uniform procedures, specifications for raw materials, and the testing of finished products—the making of cosmetics has been taken out of the class of the hit-or-miss, rule-of-thumb arts, and established as one of the many branches of chemical technology.

Cosmetology is a branch of applied science dealing with the external embellishment of the person through the use of cosmetic products and treatments. As an organized study, cosmetology dates from 1895 when the first school to teach the use of cosmetics in treatments for the skin and hair was established in Chicago. The educational movement spread rapidly, first in private schools but, after 1922, also in many vocational and industrial high schools [5].

Legal Status.   The first laws pertaining to cosmetics and cosmetology were those passed in various states for the education and professional practices of operators in beauty shops. Between 1919 and 1948, 47 states and what were then the Territories of Alaska and Hawaii passed such laws. Cosmetic products were under no control, except that of local sanitary codes (notably in the City of New York), until 1938, when the Federal Food and Drug Act of 1906 was revised to include them. This law now controls composition, directions for use, labeling, adulteration, and misbranding of cosmetics and borderline products. The Wheeler-Lea Act, through the Federal Trade Commission, protects both competitive manufacturers and consumers through the control of advertising claims. The Robinson-Patman Act, originally planned to control chain stores, affects beauty shops as outlets for cosmetic products. As the three federal laws apply only to articles in interstate commerce, they have served as models for laws enacted by many individual states to control the manufacturing and trade practices within their own borders. Information on the new revisions of the Federal Food and Drug Act can be obtained from the Food and Drug Administration, Washington, D.C. [6].

The Anatomy of Skin and Its Protective Function

The primary function of skin (Fig. 1) is to protect its encased organism. Thousands of intricate biologic processes continually occurring in this unique organ are designed to perform this function. It is uniquely adapted to prevent injury from a multiplicity of noxious environmental influences. It is a magnificent barrier to water loss yet, when needed, it can provide excess water to the surface to maintain an even internal temperature. It is an extremely efficient regenerative organ. It manufactures, in the form of keratin, one of the most flexible yet durable, resistant structures known in nature. It has created, in the form of melanin, one of nature's most effective screens to ultraviolet light. Its viable cells in the epidermis have developed physical attachments to each other which only the severest forces can disrupt. It has adapted intricate physical and chemical means to withstand successfully the continuous on-slaught of innumerable microorganisms. Many diseases of the skin are directly related to disturbances of its protective forces.

Terminology

The skin is divided into dermis and epidermis. The following layers are recognized in the epidermis.

The outermost layer of skin is a tough horny material called the stratum corneum. The process by which this layer is formed from viable epidermal cells beneath is known as the cornification process. The proximal part of the stratum corneum is apparently a specialized structure with different properties from the horny material constituting the rest of the stratum corneum above. There is no official term for this area, but because of its remarkable role in preventing absorption of agents through the skin, it is now commonly called the barrier area.

The cellular layer immediately beneath the stratum corneum is prominent because it contains unique basophilic granules and is called the granular layer.

Below the granular layer lies the cellular epidermis. This consists of the stratum Malpighii and a layer of palisaded cells known as the dermal–epidermal junction.

The connective tissue which, along with its blood vessels and nerves, is located between subcutaneous fat and epidermis is known as the dermis.

The specialized epithelial derivatives in the dermis, such as sweat glands and

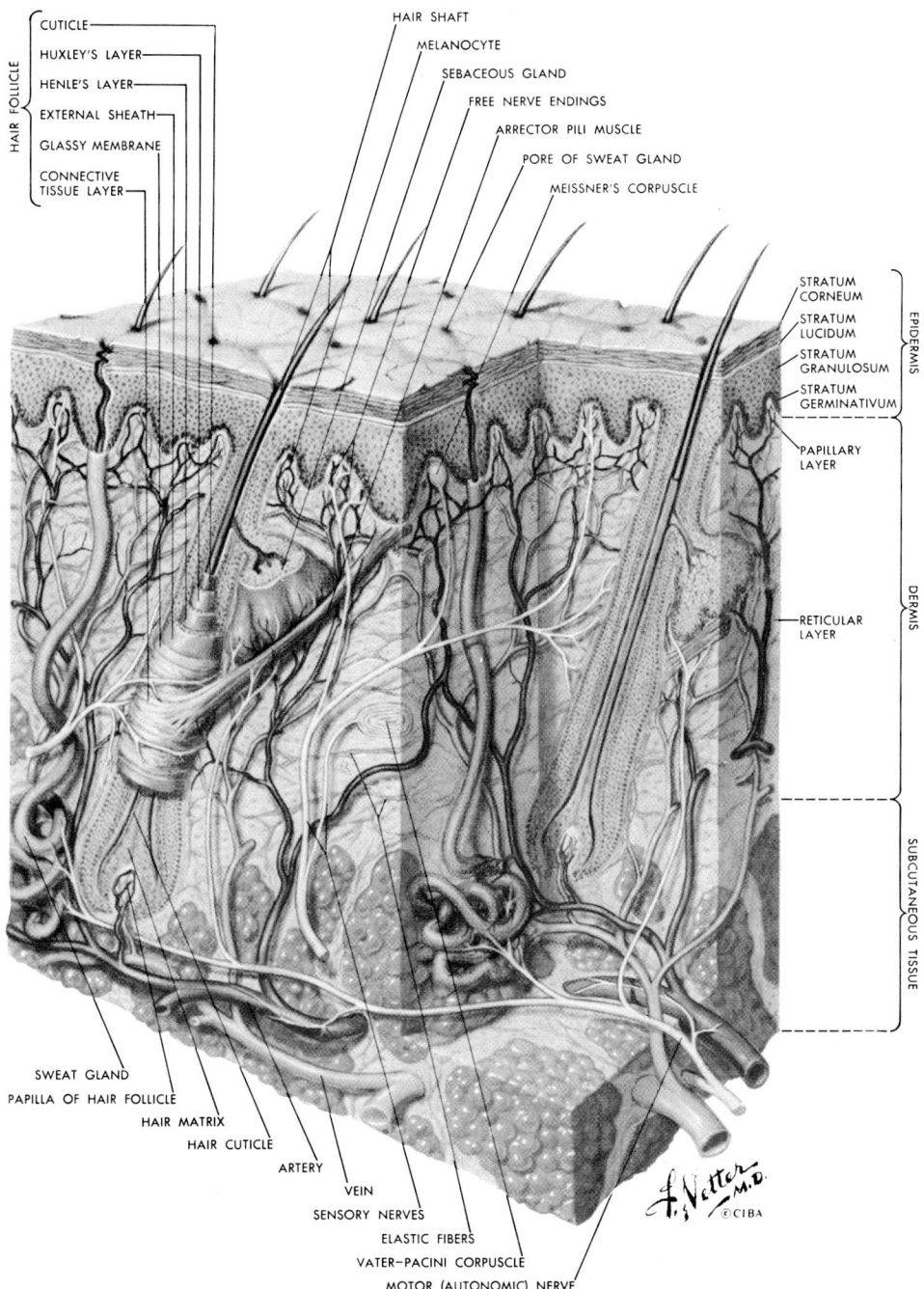

HAIR FOLLICLE
CUTICLE
HUXLEY'S LAYER
HENLE'S LAYER
EXTERNAL SHEATH
GLASSY MEMBRANE
CONNECTIVE TISSUE LAYER

HAIR SHAFT
MELANOCYTE
SEBACEOUS GLAND
FREE NERVE ENDINGS
ARRECTOR PILI MUSCLE
PORE OF SWEAT GLAND
MEISSNER'S CORPUSCLE

STRATUM CORNEUM
STRATUM LUCIDUM
STRATUM GRANULOSUM
STRATUM GERMINATIVUM
EPIDERMIS

PAPILLARY LAYER

RETICULAR LAYER
DERMIS

SUBCUTANEOUS TISSUE

SWEAT GLAND
PAPILLA OF HAIR FOLLICLE
HAIR MATRIX
HAIR CUTICLE
ARTERY
VEIN
SENSORY NERVES
ELASTIC FIBERS
VATER–PACINI CORPUSCLE
MOTOR (AUTONOMIC) NERVE

F. Netter M.D.
©CIBA

FIG. 1. Skin. (Copyright © 1967 CIBA Pharmaceutical Company, Division of CIBA-GEIGY Corporation. Reproduced, with permission, from *Clinical Symposia*, illustrated by Frank H. Netter, M.D. All rights reserved.)

ducts and the hair follicle with its attached sebaceous gland, are frequently referred to as epidermal appendages.

Surface Film

The first line of defense of the skin against its environment is a thin film of emulsified material spread rather evenly over its entire surface. The components of this complex film are contributed by sebaceous glands, sweat glands, and the products of cornification.

Cosmetic Materials [7]

Materials are grouped under headings that reflect their main function and purpose. Many appear under several headings because the majority of cosmetic raw materials can be claimed to be multifunctional.

This is not intended to be a complete collection of materials used in cosmetic applications.

Antigelling Agents.   These are materials which will prevent an increase in viscosity (usually on aging) of liquid systems. These materials function variously, due to their ability to modify hydrogen-bonding characteristics, inhibit crystallization, effect interfacial tension, etc.

Examples :

Phosphate ether-esters
Polyoxyethylated (POE) lanolin

Anti-Irritants.   These are materials which have a palliative effect on irritation caused by other externally applied materials.

Example :

Polyvinylpyrrolidone (PVP)

Dispersing Agents.   These are materials capable of breaking up agglomerates of insoluble or dispersed particles, rendering mixtures more uniform, and often reducing viscosity and improving flow properties. The vehicle in which the agglomerates are suspended, and the materials of which the agglomerates consist, determine to a large extent the type of material which will act as a dispersing agent.

Examples:

Various fatty acids esters
Various fatty alcohols (e.g., oleyl alcohol)

Emulsifying Waxes. Emulsifying waxes are proprietary self-emulsifying, self-bodying materials which produce thick, stable oil-in-water emulsions. They can be nonionic, anionic, or cationic in nature.

Example :

Polawax (Croda) ethoxylated stearyl alcohol

Emollients and Moisturizers. In general, these components fall into two categories; those which function by a purely physical mechanism and those which act by a physiological mechanism. In either case the material must be capable of preventing the skin from losing its flexibility. The flexibility of normal healthy skin is generally reduced by dehydration, which manifests itself by chapping or roughness of the skin. These materials are generally capable of preventing dehydration or bringing about rehydration of the skin (i.e., imparting and/or maintaining a water balance on the skin surface).

Examples :

Acetylated lanolin alcohols
Lecithin
Isopropyl esters
Lanolin esters
Fatty alcohols

Emulsifiers: Oil-in-Water (O/W). Oil-in-water emulsifying agents produce emulsions in which the continuous phase is hydrophilic in character (water, etc.), hence such emulsions are generally dispersible in water and will conduct electricity. The surfactants which are capable of producing such emulsions usually have a hydrophilic–lypophilic balance (HLB) of above 6.0 or preferably 7, the hydrophilic portion of their molecules being predominant. (Between HLB 5 and 7, many surfactants will function as either W/O or O/W emulsifiers, depending on their manipulation.)

Examples :

Polyethylene glycol (PEG) 300 distearate	Nonionic	7.3 HLB
Sorbitan monolaurate	Nonionic	8.6
PEG 400 distearate	Nonionic	9.3
Triethanolamine stearate	Anionic	12.0
PEG 6000 monolaurate	Nonionic	19.2

Emulsifiers: Water-in-Oil (W/O). Water-in-oil emulsifiers produce emulsions in which the continuous phase is lipophilic in character (oil, wax, fat, etc.), hence such emulsions are not generally dispersible in water and do not conduct electricity. The surfactants which are capable of producing such emulsions usually have an HLB below 6.0 and preferably below 5. The lipophilic portion of their molecules is predominant. (Between HLB 5 to 7, many surfactants will function as either W/O or O/W emulsifiers, depending on their manipulation.)

Examples:

Lanolin alcohols	Nonionic	~1.0 HLB
Ethylene glycol monostearate self-emulsifying (S/E)	Nonionic	2.0
Propylene glycol monostearate S/E	Anionic	3.2
Sorbitan mono-oleate	Nonionic	4.3
PEG 200 dilaurate	Nonionic	6.0

Emulsifiers for Electrolytes. Preparations containing high concentrations of electrolytes, i.e., antiperspirants, depilatories, and neutralizing creams, cannot normally be emulsified with the conventional anionic emulsifiers. Special emulsifiers, generally nonionic or cationic in nature, may therefore be employed. Nonionics containing ether linkages are probably the most suitable types of surfactants for very strong electrolytes because they are not prone to hydrolysis, as are some esters.

Nonionic surfactants (ethylene oxide condensates) possess inverse solubility characteristics, i.e., the surfactant becomes less soluble with an increase in temperature. These compounds are also salted out of solution by strong electrolytes.

Examples:

POE lanolin alcohols (ethers)	Nonionic
POE fatty alcohols (ethers)	Nonionic

Foaming Agents. These are any surface-active materials which when agitated in solution, usually aqueous, produce a high degree of foaming.

Examples:

Sarcosinates (e.g., sodium lauroyl sarcosinate)	Anionic
Fatty alcohol ether sulfates (e.g., sodium lauryl ether sulfates)	Anionic

Foam Stabilizers. These compounds are primarily used to stabilize the foam produced by foaming surfactants.

Example:

Alkylolamides (lauric diethanolamide)	Nonionic

Gelling Agents. Many modern cosmetics are offered in the form of gels based on metallic soaps, colloidal dispersions, or micelle systems.

Examples:

Alkylolamides (e.g., lauric diethanolamide)
Phosphate ether/ester
Carboxy vinyl polymers

Pearling Agents. Pearling agents are those compounds included in cosmetic preparations to give the end product a pearly sheen. Certain fatty chemicals will crystallize out of detergent/surfactant solutions to give this effect. Inorganic materials in the form of plate-like particles which reflect light are also employed in a variety of both hydrophilic and lipophilic systems.

Examples:

Diethylene glycol monostearate
Stearic acid
Bismuth oxychloride

Plasticizers.   Plasticizers are used to render a brittle film or mass more plastic. They are found mainly in such cosmetic preparations as nail varnish and hair lacquer. Other cosmetics in which plasticizers are usable are those of high wax content which require plasticizing to prevent cracking and to enable the preparation to be spread more easily on the skin, i.e., lipsticks, mascara, etc.

Examples:

Alkoxylated lanolin
Fatty alcohols (oleyl alcohol)
Acetyl triethyl citrate

Sequestrants.   These materials are capable of complexing with metal ions which might otherwise catalyze or react with components in a formula to cause modification or degradation of the end product.

Examples:

Ethylenediaminetetraacetic acid
Tetrasodium salt of ethylenediaminetetraacetic acid

Solubilizers.   These surfactants enable normally water-insoluble ingredients to be solubilized in water to give bright transparent solutions. Only surfactants having HLB's of approximately 8 and above are applicable.

This phenomenon is brought about by the formation of micelles. A micelle is an aggregation of single molecules or monomers of certain types which, when dissolved in water, form particles of colloidal dimensions.

Examples:

POE sorbitan esters
POE lanolin alcohols
POE fatty alcohols

Stabilizers.   In oil-in-water emulsion systems certain emulsifiers, generally predominantly lipophilic in nature, are included to act as stabilizers, minimizing creaming and sedimentation, i.e., they act as auxiliary emulsifiers. Materials which increase the thermodynamic stability and viscosity of the external phase of an emulsion also exert a stabilizing influence. They act as a physical barrier, preventing particles of the internal phase from coalescing. In O/W systems, fatty waxlike materials and soaps of polyvalent metals may be employed.

Examples:

Ethylene glycol monostearate
Sorbitan esters
Lanolin esters
Lanolin alcohols

Superfatting Agents.   The function of a superfatting agent is to prevent defatting of the skin or hair brought about by solvent or detergent action.

Examples:

Ethoxylated acetylated lanolin alcohols
Special modified lecithin
Phosphate ether/esters

Natural Products.   Natural products are raw materials which are extracted from naturally occurring sources.

Examples:

Triglycerides (vegetable oils and fats)
Lanolin
Natural waxes (e.g., beeswax, carnauba wax)

Emulsions [8, 9]

Definitions

An emulsion is a two-phase system consisting of two incompletely miscible liquids, the one being dispersed as finite globules in the other. The dispersed, discontinuous, or internal phase is the liquid that is broken up into globules. The surrounding liquid is known as the continuous or external phase. A suspension is a two-phase system closely related to an emulsion, in which the dispersed phase is a solid. A foam is a two-phase system, similar to an emulsion, where the dispersed phase is a gas. An aerosol is the inverse of a foam, air being the continuous phase and liquid being the dispersed phase. An emulsifying agent is a material usually added to one of the phases to promote ease of formation and stability of the dispersion.

Industry is most concerned with the emulsification of oil and water. Special designations have been devised for this system to indicate which is the dispersed and which the continuous phase. Oil-in-water (O/W) emulsions have the oil as the dispersed phase in water as the continuous phase. In water-in-oil (W/O) emulsions, water is dispersed in oil, which is the external (continuous) phase.

Properties of Emulsions

The properties that are most apparent, and thus are usually most important, are ease of dilution, viscosity, color, and stability. For a given type of emulsification equipment, these properties depend upon: (1) the properties of the continuous phase, (2) the ratio of the external to the internal phase, (3) the particle size of the emulsion, (4) the relationship of the continuous phase to the particles (including ionic charges), and (5) the properties of the discontinuous phase. In any given emulsion the properties will depend upon which liquid constitutes the external phase, or whether the emulsion is O/W or W/O. The resulting emulsion type is controlled by: (1) the emulsifier, type, and amount; (2) the ratio of ingredients; and (3) the order of addition of ingredients while mixing.

The dispersibility (solubility) of an emulsion is determined by the continuous phase. Thus, if the continuous phase is water-soluble, the emulsion may be diluted with water; conversely, if the continuous phase is oil-soluble, the emulsion may be diluted with oil. The ease with which an emulsion may be diluted may be increased by decreasing the viscosity of the emulsion.

The viscosity of an emulsion when the continuous phase is in excess is essentially the viscosity of the continuous phase. As the proportion of internal phase increases, the viscosity of the emulsion increases to a point where the emulsion is no longer fluid. When the volume of the internal phase exceeds the volume of the external phase, the emulsion particles become crowded and the apparent viscosity is partially "structural viscosity."

The stability of an emulsion depends upon the particle size; the difference in density of the two phases; the viscosity of the continuous phase and of the completed emulsion; the charges on the particles; the nature, effectiveness, and amount of the emulsifier used; and conditions of storage, including high and low temperatures, agitation and vibration, and dilution or evaporation during storage or use. The stability of an emulsion is affected by almost all factors involved in its formulation and preparation. In formulas containing sizable amounts of emulsifier, stability is predominantly a function of the type and concentration of emulsifier. As a matter of strict definition, an emulsion is stable as long as the particles of the internal phase do not coalesce.

Emulsifiers

Emulsifiers may be divided according to their behavior into ionic or nonionic. The ionic type of emulsifier is composed of an organic lipophilic group and a hydrophilic group. The ionic types may be further divided into anionic and cationic, depending upon the nature of the ion-active group. The lipophilic portion of the molecule is usually considered to be the surface-active portion.

Nonionic emulsifiers are completely covalent and show no apparent tendency to ionize. They may, therefore, be combined with other nonionic surface-active agents and with either anionic or cationic agents as well. The nonionic emulsifiers are likewise more immune to the action of electrolytes than the anionic surface-active agents.

The solubility of an emulsifier is of the greatest importance in the preparation of emulsifiable concentrates.

Emulsifiers, being surface-active agents, lower surface and interfacial tensions and increase the tendency of their solution to spread.

Emulsifiers: Oil-in-Water (O/W).   Oil-in-water emulsifying agents produce emulsions in which the continuous phase is hydrophilic in character (water, etc.), hence such emulsions are generally dispersible in water and will conduct electricity. The surfactants which are capable of producing such emulsions usually have an HLB of above 6.0 or preferably 7.0, the hydrophilic portion of their molecules being predominant. (Between HLB 5 and 7, many surfactants will function as either W/O or O/W emulsifiers, depending on their manipulation.)

Examples:

PEG 300 distearate	Nonionic	7.3 HLB
Sorbitan monolaurate	Nonionic	8.6
PEG 400 distearate	Nonionic	9.3
Triethanolamine stearate	Anionic	12.0
PEG 6000 monolaurate	Nonionic	19.2

Emulsifiers: Water-in-Oil (W/O).   Water-in-oil emulsifiers produce emulsions in which the continuous phase is lipophilic in character (oil, wax, fat, etc.). Hence, such emulsions are not generally dispersible in water and do not conduct electricity. The surfactants which are capable of producing such emulsions usually have an HLB below 6.0 and preferably below 5. The lipophilic portion of their molecules is predominant. (Between HLB 5 and 7, many surfactants will function as either W/O or O/W emulsifiers, depending on their manipulation.)

Examples:

Lanolin alcohols	Nonionic	~1.0 HLB
Ethylene glycol monostearate S/E	Nonionic	2.0
Propylene glycol monostearate S/E	Anionic	3.2
Sorbitan monooleate	Nonionic	4.3
PEG 200 dilaurate	Nonionic	6.0

Creams [10]

Materials used in creams may be prepared in O/W or in W/O emulsions. The aesthetic effect and degree of emolliency depend to a great extent on the emulsion type as well as the emulsion composition. Oil-in-water emulsions produce a cooling effect on application to the skin due to water evaporation. Water-in-oil emulsions will not produce this cooling effect since water evaporation is retarded by the occlusive film of the oil in the continuous phase.

The classical example of a cream was the USP "Unguentum Aquae Rosae" which was prepared as follows:

3.0	Beeswax
11.8	Spermaceti
40.2	Sweet almond oil
45.0	Rose water

In 1890 the formula was changed to:

12.1	Beeswax
12.6	Spermaceti
55.4	Sweet almond oil
0.5	Borax
19.4	Rose water

This was the basic formula for the familiar cold cream that is now made with mineral oil as a replacement for the vegetable oil. Its occlusive action aided in rehydration of the corneum when allowed to remain on the skin for an appreciable length of time. The mineral oil tends to remove skin surface lipids, due to solvent action, when applied for a short period of time, and partial replacement with a vegetable oil is needed. These emulsions are W/O since they are low in water content and the emulsifier is sodium cerotate formed by reaction of borax and free cerotic acid in the beeswax. If the water content is raised to approximately 45% or more, the composition changes to a O/W emulsion.

Nonionic emulsifiers, such as glyceryl monostearate, propylene glycol, and polyethylene glycol esters of fatty acids, sorbitol, and ethoxylated sorbitol esters of fatty acids, are used to prepare creams that have stability at acid pH as well as alkaline pH.

Anionic emulsifiers, such as the amine soaps prepared from fatty acid reactions with various amines (e.g., triethanolamine), are popular in preparing creams that are usually slightly alkaline. Most of these creams are of the O/W type. Water-in-oil creams can be prepared with anionic soaps formed in situ such as calcium and magnesium soaps of fatty acids.

Cationic emulsifiers are used in the preparation of emulsion systems that may increase deposition of the emulsion on negatively charged surfaces such as skin and hair. A popular cream prepared with cationic emulsifiers has the following composition:

0.10	Antioxidant
3.00	Cetyl alcohol
3.00	Stearyl alcohol
3.00	Dewaxed lanolin
3.00	Mineral oil
0.15	N (Colamino formyl methyl) pyridinium chloride*
1.20	N (Stearyl colamino formyl methyl) pyridinium chloride[†]
0.15	Preservative: Methyl and propyl paraben (5:1)
4.00	Isopropyl myristate
6.00	Propylene glycol
76.05	Distilled or deionized water
0.35	Perfume

100.00

*Emcol E-607 (Witco Chemical).
[†]Emcol E-607S (Witco Chemical).

Vanishing Cream

Vanishing cream can be considered to be an emulsion of a free fatty acid (usually stearic acid) in a nonalkaline medium. The basic ingredients are:

Stearic acid	15–20%
Glycerin	8–12%
Alkali (KOH)	0.5–1.5%
Water	65–75%
Preservative	q.s.
Perfume	q.s.

Of the stearic acid used, about 15 to 25% is saponified and the remaining stearic acid remains as free stearic acid.

Manufacturing Procedure

The oils, waxes, emulsifiers, and other oil-soluble components are heated to 75°C in a steam-jacketed kettle. The water-soluble components (alkalis, alkanolamines, polyhydric alcohols, and preservatives) are dissolved in the aqueous phase and heated to 75°C in another steam-jacketed kettle. To allow for evaporation of water during the heating and emulsification, about 3 to 5% excess water (based on formula weight) is added. The exact amount may be determined after the first production batch.

The procedure for preparing O/W and W/O emulsions is to add the inner phase (at 75°C) very slowly to the outer phase of the emulsion (at 75°C), stirring constantly and homogenizing to assure efficient emulsification. An alternate method of preparing finely dispersed O/W emulsions is to add the aqueous phase to the oils. Initially the relatively low concentration of water forms a W/O emulsion according to the phase-volume relationship. The slow addition and emulsification of the water increases the viscosity of the system while the oil phase expands to a maximum. At this point the continuous oil phase breaks up into minute droplets as emulsion inversion occurs, characterized by a sudden decrease in viscosity. This emulsification technique proceeds smoothly at the critical inversion point in a well-balanced, low oil–wax system, but frequently it will cause coagulation in a high oil–wax emulsion. The conventional procedure of adding the inner phase to the outer is preferable for creams and lotions. When a satisfactory product is finally prepared in this manner, the emulsion inversion procedure may be used to determine which of the two methods yields the more stable O/W system.

The rates of addition and mechanical agitation of the dispersed phase are critically important in determining the efficiency of emulsification. This may vary from a completely dispersed inner phase in a well-emulsified system to a mixed emulsion in a poorly emulsified system, the latter being due to an excessive rate of addition of inner phase and to inadequate stirring. This in turn affects the consistency, viscosity, and stability of creams and lotions.

Total stirring times and cooling rates are important factors in lotion viscosity, cream consistency, and emulsion stability. If experimental formulas are made in vessels which are not equipped with a heating and cooling jacket, longer stirring times are necessary for air-cooling. Full-scale production is usually in jacketed equipment which introduces a variable in the physical factors contributing to emulsion preparation. If cooling is started too soon after emulsification is completed, crystallization of the higher-melting waxes may occur.

The temperature of the cream or lotion at which the perfume oils are added is another factor contributing to emulsion instability. The addition of perfume to a W/O emulsion proceeds smoothly due to its solubility in the external phase. In O/W systems the oils must break through the continuous aqueous phase to be emulsified.

If the cream is to be hot-poured, it is stirred to 5°C above the congealing point, color solutions are added (if required), and the cream is maintained at that temperature (with occasional stirring) during the filling procedure. If cold-filling is preferred, the cream is stirred to 35°C, color solutions are added (if required), and then filling proceeds at room temperature.

Lotions

The oils and waxes of lotions are identical to those of an emollient cream but are lower in concentration. An O/W emollient lotion usually contains more water than the corresponding cream, whereas the W/O type may have the same water content, with oily components replacing part of the waxlike materials. These lotions are preferred for use during the day in order to apply a lighter or less oily emollient film to the skin. However, they may be formulated to contain the same concentration of oil phase ordinarily used in creams. The sales appeal of some emollient lotions derives partly from their convenience in use and partly from the greater variety of package design possible for liquid emulsions.

The formulation of these products, in all emulsifier and emulsion types, is similar in scope to the development of emollient creams. The emulsion must be stable at elevated temperatures (45 to 50°C) for whatever period is deemed necessary and at room temperatures for a minimum of 1 year. An additional check on emulsion stability is the freeze–thaw test. The lotion is subjected to a temperature of −5°C for 24 h and then allowed to return to room temperature, at which time it should be stable and pourable.

An example of an all-purpose hand, face, and body lotion prepared with an anionic emulsifier is the following:

4.00	Stearic acid, triple pressed
1.50	Lanolin anhydrous
1.50	Mineral oil
1.00	Cetyl alcohol
0.80	Triethanolamine

0.25   Preservative: Methyl and propyl paraben (5:1)
90.60   Distilled or deionized water
0.35   Perfume
_____
100.00

An example of an all-purpose hand, face, and body lotion prepared with cationic emulsifiers is the following:

 1.00   Cetyl alcohol
 0.50   Stearyl alcohol
 1.00   Lanolin anhydrous
 4.00   Mineral oil
 0.10 N (Colamino formyl methyl) pyridinium chloride
 0.80 N (Stearyl colamino formyl methyl) pyridinium chloride
 0.25   Preservative: Methyl and propyl paraben (5:1)
 6.00   Propylene glycol
74.80   Distilled or deionized water
 0.125   Sodium chloride
 0.125   Sodium benzoate
11.00   Distilled or deionized water
 0.30   Perfume
_____
100.00

Deodorants and Antiperspirants

Deodorant and antiperspirant products are marketed as aerosols, creams, gels, lotions, powders, soaps, and sticks [10a, 11].

Active Ingredients

Aluminum chloride, because of its low pH, may cause fabric damage and skin irritation. This has led to the development of various basic aluminum compounds. The most widely used compound is aluminum chlorhydroxide [11]. Other compounds that have been or are now used are basic aluminum bromide, iodide, and nitrate; basic aluminum hydroxychloride-zirconyl hydroxy oxychloride with and without glycine. The use of zirconium salts in aerosol antiperspirants was banned in 1977.

Antiperspirant Roll-On
40.60–33.60   Methocel, 65HG 400 cP, 3% solution[*,†]
36.00–30.00   Chlorhydrol, 50% solution[‡]

4.65	Propylene glycol
16.25–29.25	Alcohol SDA #40
2.50	Solubilized perfume: Solubilizer and perfume oil (4:1)

100.00

*Methocel Solution

0.1	Methyl paraben
3.0	Methocel 65HG 400 cP
96.9	Distilled or deionized water

100.0

†Methyl cellulose.
‡Aluminum chlorhydroxide complex (Reheis Chemical Co.)

Deodorant-Antiperspirant Sticks

For deodorant sticks, sodium stearate is the primary gelling agent—about 7 to 9% by weight. The grade employed depends on the fatty acid used in making the stearate.

Derivatives such as Croda's acetylated sucrose distearates also can be used but are more commonly employed for antiperspirant sticks to gel cetyl alcohol-based formulations (typical use level: 28%).

There are three basic types of stick formulations for antiperspirants: Hydroxyethyl stearamide, which produces clear-melting, homogeneous sticks; stearamides and cetyl alcohol dry-powder dispersions of ACH; and stearamides coupled with propoxylated alcohol. Stearamide wax content is 26 to 29%; the active concentration is 20 to 25%.

Volatile silicones, the siloxanes, are employed in the newer antiperspirant sticks at levels as high as 40 to 50%, and the tetrameric and pentameric cyclic compounds are preferred because of their volatility. When the stick is applied to the skin, the silicones evaporate, leaving the active compounds on the skin in a nonsticky film. The compounds also act as lubricants and emollients.

The silicones function as a processing aid, reducing cracking and crumbling of the stick.

The silicones are also useful for pump antiperspirants, acting as a lubricant for valves and a suspending agent for the active material.

A typical deodorant stick formula is:

By Weight

A.	0.50	Irgasan DP 300 (Ciba Geigy)
	0.15	Trichlorocarbanilide (Monsanto)
	17.00	Propylene glycol
	75.35	Alcohol SDA #40
B.	6.00	Sodium stearate, purified USP
	1.00	Perfume

100.00

A typical antiperspirant stick formula is:

By Weight

A.	3.0	Distilled or deionized water
	9.0	Propylene glycol
	20.0	Rehydrol powder (Reheis Chemical Co.)
B.	15.0	Witconal APM* (Witco Chemical)
C.	27.5	Witcamide 70H[†] (Witco Chemical)
	0.5	Ceraphyl 50[‡] (Van Dyk)
D.	24.0	Silicone F-218[§] (Dow Corning)
	1.0	Perfume
	_____	
	100.0	

*Propoxylated myristyl alcohol.
[†]Stearic acid monoethanolamide (special Witco grade).
[‡]Myristyl lactate
[§]Polydimethylsiloxane (low viscosity, volatile).

A dry compressed antiperspirant stick formula is:

25.00%	Ultrafine micro dry powdered aluminum chlorohydrate (Reheis Chemical Co.)
9.00	Talc
65.75	Avicel Ph-105*
0.25	Magnesium stearate
_____	
100.00%	

*Microcrystalline cellulose.

Emulsion Forms: Cream and Lotion Antiperspirants [74, 75]

Most emulsions are nonionic systems, as the nonionics have low irritation levels, make more stable and uniform systems, and produce a more "elegant" system.

An example of a cream antiperspirant formula is:

7.0	Veegum K* (Vanderbilt)
61.5	Distilled or deionized water
1.0	Germall 115[†] (Sutton Laboratories Preservative)
4.0	Glyceryl monostearate–self-emulsifying (Arlacel 165-ICI)
26.0	Chlorhydrol,[‡] 50% (Reheis Chemical Co.)
0.5	Perfume
_____	
100.0	

*Magnesium aluminum silicate.
[†]Imidazolidinyl urea.
[‡]Aluminum chlorhydroxide complex.

Sunscreens [12]

The sun emits energy in a grossly continuous band throughout the electromagnetic spectrum. Passing through the upper atmosphere, the shorter wavelengths are absorbed by atmospheric components so that at sea level the radiation extends from a cutoff near 290 nm through the near ultraviolet to the conventional end of the ultraviolet range, which is near 400 nm. The intensity of the radiation varies nonlinearly throughout this range.

The production of erythema and the subsequent production of melanin pigment are both maximum with 296.7 nm radiation. As the wavelength of irradiation increases, both responses fall rapidly, so that 10 μW/cm^2 of 307 nm radiation, 100 μW/cm^2 at 314 nm, 1,000 μW at 330 nm, and 10,000 μW at 340 nm are each equivalent to the effect of 1 μW/cm^2 of 296.7 nm radiation in the production of an erythema.

A unit of erythemal flux, the E-viton, is equivalent to the erythema induced by 10 μW/cm^2 of 296.7 nm radiation. The response of the skin to an E-viton (or viton) is constant: irradiation by 10 vitons for 1 h produces the same erythemal response as does 5 vitons for 2 h. On this basis also, the effects due to irradiation by different wavelengths are additive. It takes about 20 min exposure to midsummer sunlight to produce a minimum perceptible erythema (MPE) on "normal" Caucasian skin.

MPE varies with each person. Ultraviolet energy is expressed in energy units called E-vitons, with 1 E-viton equivalent to 10 μW of energy at 2967 Å. To produce MPE on average untanned skin, 40 E-viton min/cm^2 are required:

1 E-viton/cm^2 acting for 40 min produces an MPE on 1 cm^2	
10	4
40	1

Sunscreen Products [13, 14]

The cutaneous changes commonly attributed to aging are, for the most part, a reflection of the damaging effects of prolonged exposure to sunlight. A simple visual comparison of chronically exposed facial skin and habitually covered areas like the buttocks will quickly convince the skeptic of the validity of this statement. Changes such as wrinkling, coarseness, dryness, mottling, flaccidity, and blemishes never develop in protected skin.

Another important result of chronic sun exposure is the development of cutaneous premalignant and malignant lesions. At least three different types of skin cancer (basel cell cancer, squamous cell cancer, melanoma) are believed to occur with higher frequency in sun-exposed skin. The experimental induction of epithelial tumors in animals with artificial UV light sources leaves little doubt that such radiation is carcinogenic.

The skin has remarkable adaptive capabilities. It can protect itself against further acute sunburn by rapidly increasing the amount of melanin production (tanning) and by increasing the thickness of the protective barrier, the stratum corneum. However, individuals who sunburn easily and tan poorly cannot rely on these natural responses.

Sunscreens are of two types: physical and chemical. Physical screening agents, such as titanium dioxide and zinc oxide, are opaque materials that block and scatter light, and thus act as mechanical barriers. Their action is nonselective. Chemical screening agents, which act by absorbing UV light, offer selective protection against certain UV wave bands, depending on their absorption spectrum. Anthranilates, cinnamates, benzyl and homomenthyl salicylate, and P-aminobenzoic acid (PABA) and its ester derivatives belong in this category and have maximal absorption in the sunburn region (UVB). An ideal preparation would resist wash-off from swimming or sweating and would not come off during exercise or rubbing. Maintenance of significant protection following swimming or sweating would suggest that the chemical is substantive or can diffuse into the horny skin layer. PABA and its esters are moderately substantive, and appreciable protection remains after intensive sweating and brief periods of immersion in water (5 to 10 min).

It is highly important, especially during the first few days of exposure, to reapply PABA-type sunscreens after swimming. For significant residual protection, PABA has to be applied *at least 2 h prior* to immersion. This probably reflects the time required for penetration into the stratum corneum. With repeated daily applications, there is a gradual build-up of the drug in the horny layer, resulting in significantly increased protection. Sunscreens with much greater resistance to swimming can be formulated with acrylate polymers which leave flexible films on the surface. A chemical sunscreen incorporated in the formulation will tend to be bound in the film and remain in place despite sweating and bathing.

Absorption

The literature on most UV absorbers reports the absorption characteristics of a specific level of absorber solution at various wavelengths of UV using a spectrophotometer for the determination. This assumes a direct relationship between precise instrument determined properties of a sunscreen and the less than precise skin properties of human beings, whose skin color varies from Caucasian white to yellow, tan, and shades of brown. Differences in melanin content of skin affects absorption characteristics of UV light.

Sunscreens

Chemical Sunscreens

2 Ethoxyethyl *p*-methoxycinnamate (Giv-tan F, Givaudan)
Menthyl anthranilate
Homo-menthyl salicylate
Glyceryl *p*-amino-benzoate (Escalol 106, Van Dyk)
Isobutyl *p*-amino-benzoate (Cycloform)

Isoamyl-*p*-*N*,*N*-dimethylaminobenzoate (Escalol 506, Van Dyk)
2-Hydroxy-4-methoxybenzophenone 5-sulfonic acid (Uvinul MS-40, GAF)
2,2′-dihydroxy-4-methoxybenzophenone (Uvinul D-49, GAF)
2-Hydroxy-4-methoxybenzophenone (Uvinul M-40, GAF)
p-Aminobenzoic acid
Mono- and dihydroxypropyl isomer of ethyl *p*-aminobenzoate (Amerscreen P, Amerchol)
2-Ethylhexyl *p*-dimethylaminobenzoate (Escalol 507, Van Dyk)

Natural Oils as Sunscreens

The UV absorbence is measured at 290 to 340 nm. The performance in relation to concentration (in order of performance) is

1. Mink oil
2. Avocado oil
3. Sweet almond oil
4. Sesame oil
5. Persic oil
6. Safflower oil
7. Peanut oil
8. Jojoba oil
9. Coconut oil
10. Olive oil

Physical Sunscreens

These are opaque materials such as titanium dioxide and zinc oxide.

Perfume

Because of extensive application and skin exposure, care is needed in perfume selection regarding irritation and sensitization.

Exposure to sun and heat means careful testing for stability.

The concentration should be sufficient to cover the base product odor, with a slight excess for aesthetic purposes.

Examples of sunscreen formulations are the following:

Clear Sunscreen Lotion

20.00	Alcohol denatured #40
5.00	*p*-Aminobenzoic acid
74.50	Polyalkylene glycol*
0.50	Perfume
100.00	

*Ucon LB-625 (Union Carbide).

Opaque Sunscreen Lotion

10.00	Sesame oil, preserved*
10.00	Mineral oil[†]
0.50	Cetyl alcohol
3.00	Stearic acid, triple pressed
5.00	Sunscreen: *p*-Aminobenzoic acid
0.50	Triethanolamine, 98%
2.90	Propylene glycol
0.25	Preservative: Methyl and propyl paraben (5:1)
67.60	Distilled or deionized water
0.25	Perfume

100.00

*Contains antioxidant.
[†]Saybolt viscosity at 100°F is 65/75.

A number of sunscreen agents are incorporated in sunscreen formulas in the form of emulsions, dispersions, and clear solutions for protection against the erythema-producing range of UV light. Information on these is contained in a number of patents [72, 73].

Aerosols

The word aerosol covers nearly all products packaged under pressure and dispensed from a container as a result of pressure. In colloid chemistry an aerosol is a suspension of small particles in air or gas with the radius of the particles less than 50 μm [15].

In a bulletin released in 1966, the Aerosol Scientific Committee of the Chemical Specialties Manufacturers Association defined an aerosol product as a liquid, solid, gas, or a mixture that is discharged by a propellant force of liquified and/or nonliquified compressed gas, usually from a disposable-type container and through a valve [21, 22].

The three main components of aerosol products are the propellants, solvents, and the active ingredients. The liquified gas propellants have primarily been the chlorofluorocarbons (Freons by DuPont are an example) which were banned in 1978 for use in all personal and household aerosol products. Exceptions were pharmaceutical aerosol products such as asthma spray aerosols. Compressed gases and liquified hydrocarbons, such as butane and propane, became the primary propellants in revised formulations [19, 20].

Liquefied Gas Propellants [16, 17]

With the exception of fluorocarbon propellants, hydrocarbon propellants are by far the largest liquefied gas propellant used today. Hydrocarbon propellants, as the name implies, are composed of two most basic parts of organic matter: hydrogen and carbon. The three most common hydrocarbon propellants used are propane, isobutane, and *n*-butane. These propellants are highly purified liquefied petroleum gases that have been derived from a specially selected natural gas stream and further processed to produce a special grade for the aerosol industry.

The specifications call for a material which is at least 95% pure, virtually free of all moisture, with no unsaturated compounds and no offensive odor.

Most of the aerosol packagers who use hydrocarbon propellants use a blend of two of the above materials to produce the pressure required. Some of the most common blends are A-46, A-70, A-85, and A-92. The "A" designator indicates Aerosol Grade, and the numerals indicate the pressure in pounds per square inch gauge at 70°F.

The aerosol package is essentially the same for all products regardless of the liquefied gas propellant that is used. The propellant exists in both a vapor and liquid form, with the liquid boiling and turning to vapor as the can is evacuated.

Since hydrocarbon propellants are lighter than water, they tend to float on top of the water in a water-based product as opposed to the fluorocarbon propellants which tend to collect at the bottom of the can.

Typical Hydrocarbon-Filled Aerosols

Household Products. These include germicidal disinfectants, air fresheners, insecticides, insect repellants, glass cleaners, dust mop treatments, furniture polishes, oven cleaners, tile cleaners, starches, fabric finishes, butane lighters, upholstery shampoos, suede cleaners, and ant and roach sprays.

Personal Care. These include shave creams, hair sprays, deodorants, antiperspirants, powder sprays, colognes, perfumes, and hand, face, and body foams.

Air Deodorant Gels. Air deodorant gels are solid gel products containing 80 to 90% water, 1 to 10% of a suitable fragrance, and 1 to 4% of a gelling agent, solubilizers, and solvents. In most of these gels the perfume is dispersed in the aqueous media and the gelling agent is present in a percentage that will make a firm gel substantially devoid of syneresis. The gel is usually packaged in an airtight container, and the gel releases its air deodorant principles when exposed to the air.

A typical formula for a deodorant gel is:

4.0	Water-soluble alginate
0.3	Water-soluble preservative
81.7	Water
4.0	Perfume
10.0	Aqueous color solution

The patents for this type of product make use of thickeners and gellants such as gelatin, locust bean gum, and cellulose derivatives. In addition, small percentages of alcohol and glycols are used [71].

Color [23–26]

The certified dyes and toners used in makeup preparations are synthetic, organic products.

Straight colors are referred to in the trade as "primary colors," "primary dyes," and "primaries." Strictly speaking, straight colors include both primaries and "lakes" of such primary colors. A "lake" is an insoluble color made by extending a primary color on a substratum, such as aluminum hydroxide, barium sulfate, a mixture of aluminum hydroxide and barium sulfate, aluminum benzoate, and zinc oxide.

The principal primary colors are red, blue, and yellow. Practically every hue can be produced by blending red, yellow and blue in proper proportions (Fig. 2). Hue, in color, is a quality whereby one color differs from another. Brightness is a measure of the reflectance value of a dye. Strength is a measure of the tinting or coloring value of a dye.

The "lakes" vary in pure dye content (strength) from 2 to 80%. The primaries contain over 80% pure dye. Primary colors are frequently used in diluted forms. Lakes are one form of diluted colors; mixtures are another.

Mixtures are made by mixing two or more primary colors, and are sometimes referred to as "primary mixtures." Mixtures of two primary colors are referred to as secondary colors, and mixtures of three primary colors are referred to as tertiary colors.

The initials FD&C, D&C, and External D&C are part of the name of straight colors, and tell at a glance for what use the color has been certified. The name of a mixture usually tells nothing of its permitted use, but if there is a restriction in the use of the color, a statement to that effect is required on the label.

For safety, colors which have been certified for cosmetic use must contain not more than 0.002% of lead, not more than 0.0002% of arsenic, and not more than 0.003% of heavy metals other than lead and arsenic.

The majority of cosmetic colors are certified coal-tar colors. However, a few natural coloring agents are used in limited amounts to color foods, drugs, and cosmetics. The more important ones are alkanet, annatto, carotene, chlorophyl, cochineal, saffron, and henna. These require very little comment since their use in cosmetics is limited.

Inorganic Colors

Except for the white pigments such as titanium dioxide, zinc oxide, and talcum, the inorganic colors are used in a number of cosmetic products. They usually have excellent lightfastness and complete insolubility in solvents and aqueous solutions.

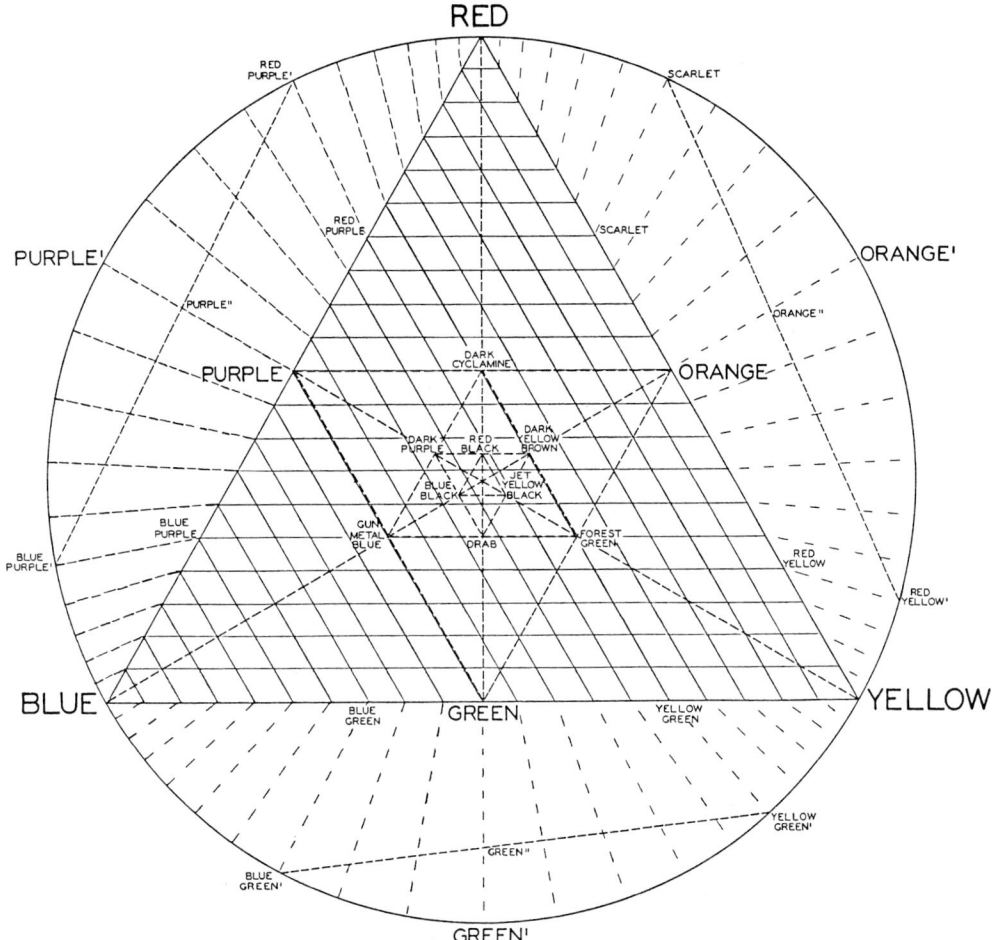

FIG. 2. This chart serves as a color matching aid or stencil by orienting color relationships. It illustrates the development of the major secondary and tertiary hues by blending the three primaries—red, yellow, blue. Points on the outer circle are of equal brightness as they are equidistant from the neutral black center, the point where all visible radiant energy is absorbed. By mentally locating a color on this chart, the colorist can decide whether to add reddish, yellowish or bluish corrections to approach some desired hue.

The naturally occurring colored minerals depend for their color upon the presence of oxide of iron. They are known by such names as ochre, umber, and sienna. Because they are naturally occurring products, they possess certain inherent disadvantages. They show greater variation in color and tinting power than one gets with manufactured pigments. The nature and amount of impurities in them may vary and may, from time to time, exclude their use in a cosmetic product.

The most important inorganic colors or pigments are the iron oxides, chrome oxide greens, ultramarine blues and pinks, and the carbon blacks. Iron oxide is sold in the form of yellow hydrated iron oxide (or ochre), brown iron oxide, red iron oxide, and black iron oxide.

The yellow iron oxide is made by an alkaline precipitation from a ferrous salt, followed by an oxidation. The product is approximately 85% Fe_2O_3, and is supplied in shades from light lemon yellow to a deep orange.

The brown iron oxides are made by treating a mixture of red, yellow, and black iron salt with alkali and partially oxidizing the precipitate. The brown oxide is a mixture of Fe_2O_3 and Fe_3O_4. The red iron oxides are usually made by heating the precipitated yellow iron oxides. A light to dark red can be obtained, depending on heating conditions, and the product is 96 to 98% Fe_2O_3.

The black iron oxide results from a precipitation of Fe_3O_4 under carefully controlled conditions. The chromium oxide green is pure Cr_2O_3. The anhydrous pigment has a dull green shade, and the hydrated pigment shows a bluish tint. This type of pigment is used in eye shadows and eye makeup.

Carbon blacks are obtained from vegetable or natural gas origin. These blacks find wide cosmetic application, particularly in eye shadow and eye makeup.

The ultramarine blues and pinks are made from a chemical composition containing sulfur, sodium carbonate, charcoal, pitch or rosin, and several other ingredients, carefully heated for a lengthy period of time. The finely pulverized pigment gives shades of ultramarine blue from the pink to the green side, depending on its silica content.

It is well known that many dyes are pH indicators. Such dyes will change in hue with pH variations. Those which do not change much in hue will be affected in other important properties, and it is therefore important to check on pH conditions in a cosmetic product. The inorganic colors are not as much affected by pH conditions as are the coal tar dyes.

All water or organo-soluble dyes, and most insoluble dyes and pigments, are destroyed or chemically changed by the photochemical action of UV, visual light, IR, and other short wave radiations. The degree of sensitivity varies with the dye and with the conditions.

The dyes change in tinctorial strength and hue, becoming progressively weaker and usually duller, as the destruction of the dyestuff progresses. This is considered normal fading action. The ability of the certified dyes to resist this chemical change, due to photochemical action, has been charted for a number of the dyes and is available in table form.

The hue of some dyes will vary with the percentage of moisture present. The term "bleeding" is used to refer to the strengthening of the hue of a color in the presence of moisture, particularly in moist powders that are stored for a long period of time. When moisture is absent, photolytic effects are much retarded. Absence of air, oxygen, hydrogen, and other oxidizing or reducing reagents also retards fading. The presence of titanium dioxide greatly accelerates the fading action.

The factors influencing the judging of color effects are:

1. Color perception
2. Light sources
3. Location of color laboratories

4. Light intensities
5. Eye strain, fatigue, or color memory
6. Mass-tone, top-tone, under-tone, and over-tone
7. Transparency, translucency, and opacity
8. Fluorescence and iridescence
9. pH changes

Face Powders [27]

Face powder, both the loose and the popular cake or pressed powder, is a blend of white pigments tinted and perfumed, to be applied to the face and body. The function of a face powder is to impart a smooth, velvetlike finish to the skin by masking any shine due to the secretions of the sebaceous and sweat glands. To obtain this effect, the powder must not be too transparent to mask the shine, neither must it be too opaque to give a masklike appearance. In addition, it must possess reasonable lasting properties in order to avoid the necessity of frequent repowdering. In other words, it must be adherent to the skin and reasonably resistant to the mixed secretions of the skin. Finally, through intimate contact of its perfume-laden particles over a warm and relatively large area, it should disseminate a pleasing odor.

No single substance possesses all the desired properties of covering power, slip, absorbency, adhesiveness, and bloom. Therefore, a modern face powder is a blend of several constituents, each one chosen for some specific quality. The materials most commonly used can be grouped in the following manner:

1. Covering power: To cover skin defects such as enlarged pores and skin shine. Examples: titanium dioxide, zinc oxide, kaolin, and magnesium oxide.
2. Slip: To assist in spreading and to give the characteristic smooth feeling. Example: talc, zinc stearate, magnesium stearate, and starch.
3. Absorbency: To absorb sebaceous or oily secretions and perspiration, and thus reduce shine. Examples: precipitated chalk, magnesium carbonate, starch, and kaolin.
4. Adherence: To improve clinging to the face. Example: metallic soaps such as magnesium and zinc stearates and, in addition, the incorporation of small quantities of oil or fatty materials with the powder base.
5. Bloom: To give a smooth, velvetlike appearance to the skin. A matte effect. Examples: chalk and starch.

A small percentage of purified titanium dioxide should be used in face powders *if* it is one of the ingredients of a formulation. Titanium dioxide, due to its extremely fine particle size, provides unusually great hiding or covering power, and the actual percentage used varies, usually from 5 to 8%.

Kaolins are hydrous-aluminum silicates. Kaolins occur in many parts of the world, but the whitest are the English kaolins. These are geographically primary kaolins, meaning that they were formed in the identical location where they are now mined. Less contamination from foreign matter occurs in the primary kaolins than if they are of the secondary or transported class. Kaolin is strip-mined, and to produce suitable grades for cosmetics it is normally washed in water. The kaolin, as it occurs in the mine, contains varying amounts of water. The kaolin is mixed with water and made into a thin slip or slurry. The kaolin slurry is maintained at such a consistency that the heavy foreign impurities, such as free quartz, will drop to the bottom of the floatation tanks. Mica and other materials having a lighter specific gravity are removed with the top water. The clean slurry is then either pumped into a filter press or treated on drum filters to remove the major portion of the water. The kaolin is next dried and pulverized to what is classified as "air-floated" fineness.

Air classification is more accurate than screening and permits more controlled segregation of the finer particles. The "air-floated" materials generally can be considered to be 200 to 325 mesh, depending upon the air stream adjustment. Kaolin in the classification of "air-floated" is used for cosmetics. Kaolin is used to give adherence, but large amounts are not used in face powders because it will develop a claylike odor when it is wet.

Magnesium carbonate is made from magnesium hydroxide by differential precipitation. The density or bulk of the magnesium carbonate is controlled by the reprecipitation and the drying process. Control of this process will give considerable variation to the final product, resulting in a light or a heavy material. Magnesium carbonate is frequently used as a perfume carrier with certain reservations. Phenols, phenolic materials, and acid materials present in perfume compounds will be affected by the presence of magnesium carbonate.

Magnesium oxide is also prepared from magnesium hydroxide by calcining or heating. The rate of drying, rate of heat treating, plus temperature makes a difference in the specific gravity of the magnesium oxide. Magnesium oxide is sold as heavy and light magnesium oxide, depending on the method of manufacture. Magnesium oxide is also prepared from magnesium carbonate at times in order to control impurities normally present in the magnesium salts.

Most *calcium carbonates* used by the cosmetic industry are the chemically precipitated kinds. These vary considerably in their bulk and are selected for density and flowability. The use of calcium carbonate varies considerably, and it is frequently included with other white pigments. The precipitated form can be prepared by either of two processes. The first is by the reaction of lime water and carbon dioxide, and the second is the reaction between sodium carbonate and lime.

In both cases a pure calcium carbonate with amorphous particles is produced. In general, the precipitated calcium carbonate is cleaner and brighter than the natural calcium carbonate. While the precipitated type has higher chemical purity, it also has greater alkaline reactivity. If the precipitated calcium carbonate is not washed carefully, it is also subject to greater variations of pH. It is possible to obtain special grades of exceptional fineness, accurately balanced to prevent harshness, possessing good absorption and grease-resisting properties, in varying densities according to the purpose required. The grades of natural-type calcium carbonate of most interest for cosmetics are those which are water-ground and in which the particle size is controlled. Natural-type calcium carbonates are produced from mining amorphous chalk such as is found in the "White Cliffs of Dover" in England or from deposits found in France.

English chalk is processed by both wet and dry grinding. In the case of the wet grinding, the selection of grades is made by water classification and sedimentation. All the French chalks are dry-ground. The majority of chalks lack slip and must be modified by the inclusion of other constituents.

Starch is a carbohydrate elaborated and synthesized in the living plant. It is polymerized dextrose. Starch is seldom used in face powders manufactured in the United States, but is very popular in Europe.

Zinc oxide is prepared by calcining or heating zinc metal under carefully controlled conditions. The resultant product is a white, odorless, amorphous powder. The zinc oxide of the desired particle size for optimum covering power will retain less oil than other materials specially prepared for this purpose, but its oil absorption capacity is in the range of other face powder constituents. As face powder absorbs moisture and oil from the skin, its covering power decreases.

Zinc oxide has a fairly high refractive index under the conditions described above, and loses less opacity proportionately than materials of low refractive index. Zinc oxide is slightly drying and slightly antiseptic. It has also been determined that zinc oxide has sun-screening properties.

Formulation

The basic property of face powder is to conceal shine and endow the skin with a new and better looking color or tint. The finer the particle size of face powder, the greater the diffusion of light, and the better the bloom, with correspondingly less shine. The reason for this is that the light is reflected in many different angles, somewhat like from the frosted surface of glass which seems whiter and less shiny than regular glass because of greater light diffusion.

Within certain limits, the smaller the powder particles, the higher the covering power and the better the tinting. A common method of determining fineness of particles is the bite test, which consists of grating the ingredients between the teeth. The gritty threshold on grating chalk powder between the teeth has been placed at approximately 12 μm. The materials used in a face powder should be passed through at least a 300-mesh sieve or, more properly, through a micronizing or micropulverizing machine.

The following is a procedure commonly used in manufacturing loose face powder:

1. Screen all basic materials into a Day Spiral (ribbon) Mixer, and blend these materials for a period of time.
2. Add the inorganic colors and lakes or toners which have previously been diluted with either some of the mixed base material or with a portion of the talcum and screened.

3. Add the perfume, which has previously been diluted and thoroughly mixed with a portion of the mixed base material or the talcum and screened.
4. Blend well in the spiral mixer and transfer to a 300-mesh silk or stainless steel screen. Pass the mixed material through the screen several times and check the batch for color. A more efficient and rapid method of screening the batch consists of passing the material through a Fitzpatrick Comminuter or a micronizer, using a suitable screen.
5. If the color check indicates that adjustments are necessary, add the estimated or required amount of color dilutions and, in addition, add the necessary amount of perfume to compensate for the additions.
6. When the color has reached the stage where it corresponds closely with the standard, the batch should be again thoroughly mixed in the spiral mixer, and the entire batch passed through the 300-mesh screen or the Fitzpatrick Comminuter or a micronizer.
7. The screened material can now be stored in a cool, dry place, or can be sent to the packaging room.

The manufacturing procedure for compact powder closely parallels the procedure used for manufacturing the loose powder, except that the moistening agent must be added slowly and mixed well in the base materials. If water is used in the moistening of the base materials, it has generally been found that the water content of the mixed powder should be reduced by drying to approximately $2\frac{1}{2}$ to 3% of water content.

Very few manufacturers use water in the moistening of the powder because it lengthens the process considerably due to the time involved in drying the powder, and it also may cause some "bleeding" or extending of some of the lakes used in coloring the material or finished batch. The finished powder, having been passed by the control laboratory for shade, can be stored in the same way that the loose powder is stored, but it has been found that agitation is necessary before the powder is placed in the pans for pressing. This is due to the inclusion of air, which results from the mixing and screening process, and may cause breakage of the compact powder unless it is vibrated before it is passed through the press.

Vibration does not necessarily remove all the air, and the process known as "prepress" is a part of the manufacturing procedure in making a compact powder. Prepress consists essentially of applying moderate pressure to the material placed in the pans and then releasing the pressure and passing the cakes on to the hydraulic press, at which point sufficient pressure is applied to make a hard, compact powder cake. Breakage of the compact cake is a very important factor for rejection. One firm tests the cake of compressed powder by dropping it from a height of 1 ft onto a thin metal plate, cushioned by a few sheets of paper. The cake of compressed powder has to withstand more than 10 drops from this height to pass the control test.

A loose face powder is a dry mixture of pigments, inert materials, and perfume. The most common reason why loose face powder loses its saleability is due to the loss of perfume. The stability of the perfume in a particular face powder can be readily checked by using the method of minimum quantity to discover whether perfume and powder are not reacting with each other.

For this test one uses not more than 0.1% of perfume. A good perfume and suitable powder should maintain a slight but agreeable perfumed note for at least 2 days while kept in a vessel open to the air. The usual failures are the disappearance of all of the perfume or the appearance of a phenolic or medicated odor.

The stability of the pigment used is of great concern to every manufacturer of face powder. Not all shades can be produced with earth colors exclusively, and it is necessary to use some organic lakes and toners. These should be used very sparingly because they are not light stable for any length of time. A simple test to check the light stability is to place a puff on the smoothened surface of loose face powder and then put the sample into bright sunlight for about 30 min. Any appreciable bleaching action of the unprotected area will indicate possible trouble.

Zinc stearate may develop an odor on aging, and it is necessary to use a good grade of zinc stearate for face powder. Kaolin may develop a claylike odor when moistened, and magnesium carbonate may have a deleterious effect on the perfume. All materials used in the preparation of face powders must be carefully tested to prevent the inclusion of impurities. The same precautionary measures taken in preparing loose face powder must be observed in the manufacture of compact or pressed face powder. In addition, care must be taken in the choice of ingredients since the compact face powders usually contain moistening ingredients.

Color Correction

If more color is needed, the color strength with respect to the base becomes too strong even though the hue or color match is still not the same as the standard. It is therefore necessary to add more of the base formula, allowing for the diluent used in the colors, to compensate for the color addition and to keep the ratio of color to base at a predetermined figure. Perfume additions are also necessary, and all corrections should be sifted and well blended with the entire batch.

Color Match

Color matching of loose face powder is generally performed by placing a small amount of the standard powder on a white pad of paper and next to it a small amount of the batch being prepared. The powder in both mounds is very carefully pressed and drawn down the surface of the pad until a smooth, even layer is formed of the two mounds side by side. Both can then be examined in the proper light and checked for color strength. Corrections are generally determined visually, and there is no exact formula for the amount of correction needed. Pressed powder is generally compared to a standard by rubbing a small amount of the surface of the pressed cake on a pledget of clean white cotton, and both the standard and the sample are then compared side by side in the proper light. A small amount of the powder can also be rubbed on the nonhairy part of the arm and the color match checked in this way. Here again, corrections are somewhat empirical and are determined visually.

The designations "light," "medium," and "heavy" as applied to a face powder denote covering power or opacifying qualities, not weight. In the absence of exceptional skin defects a dry skin will require a light powder, but since such a skin is lacking in grease, either the powder must possess increased powers of adherence or a powder base must be used. A greasy skin, on the other hand, will require a powder of high opacity and grease resistance.

Adherence is supplied by the skin itself, and in this case a powder containing sufficient grease-resistant constituents such as kaolin, together with a drying agent such as zinc oxide, may be used. The latter powder would be termed a heavy powder. The average woman prefers a light to medium-weight powder, since both go well with dry or normal skin.

A precautionary note is advisable on the possibility of off-odors from the glues used in making loose powder boxes, powder puffs, and in gluing the metal pan to the compact surface. Some concerns make a standard practice of perfuming the powder puff before it is placed in the container.

Dusting Powders, Sachets, Talcs

Dusting powders, sachets, and talcs require a fragrance oil with great stability toward oxidation and polymerization. The immense surface area of exposure offered by powders and the alkalinity of their vehicles favor both these chemical reactions which can be highly destructive of an odor. The essential oils develop terpene-type rancidity odors. Aldehydes are destroyed either by oxidation or polymerization. Eugenol and vanillin develop discoloration in contact with the alkaline surfaces and air. These are only a few of the difficulties encountered among the powder products.

Toilet Powders

Most of the information contained in the discussion on face powders will apply to such toilet powders as talcum powder, after-shave talcum, body powder, dusting powder, and baby powder. The formulations for the after-shave talcum and dusting powder are usually very simple and, in some cases, a good grade of talcum is all that is necessary.

The colors used in talcum powders vary, but a very light pink or flesh color is the most popular. Perfume is important in the dusting powder and, to a lesser extent, in the after-shave talcum.

Medicated Powders

The antiseptic powders, foot powders, and deodorant powders can be grouped together and considered as powder formulations containing germicidal compounds or antiseptics, including such materials as the quaternary ammonium salts and the phenols, oxyquinoline sulfate, oxyquinoline benzoate, hydroxyquinoline salts and many other germicidal and antiseptic compounds.

Powder Formulas

Loose Face Powder

Kaolin	3.0%	
Talc (Italian)		
Talc (Sierra)	64.0	
Magnesium stearate	1.5	Base
Magnesium carbonate	0.5	
Zinc oxide	15.0	
Corn starch	10.0	
D&C Red #2 (lake), 20% in talc		
D&C Red #3 (lake), 10% in talc		
D&C Orange #4 (lake), 10% in talc	5.0	Color dilutions
Yellow iron oxide, 20% in talc		
Brown iron oxide, 20% in talc		
Perfume	1.0	

Cake or Pressed Powder

Kaolin	10.0	
Zinc stearate	5.0	
Zinc oxide	10.0	Base
Magnesium stearate	5.0	
Talc (French)	61.4	
Mineral oil light	2.0	
Cetyl alcohol	1.0	Binding agent
Lanolin	0.3	
D&C Orange #4 (lake), 10% in talc	2.7	
D&C Red #2 (lake), 20% in talc	0.8	Color dilutions
Brown iron oxide, 20% in talc	1.0	
Perfume	0.8	

Lipstick

A lipstick is a solid fatty base product containing dissolved and suspended color materials. A good lipstick must possess a certain maximum and minimum of thixotropy; that is, it must soften enough to yield a smooth, even application with a minimum of pressure. It should not be necessary to apply a lipstick more than every 4 to 6 h. The applied film, no matter how thin, should to some extent be impervious to the mild abrasion encountered during eating, drinking, or smoking.

The lipstick should be of such composition as to color only that portion of the lip to which it is applied, and not bleed, streak, or feather into the surrounding tissue of the mouth, thus giving the effect of smearing. It should be of a consistency that can be quickly and easily applied, and the perfume should be carefully compounded so that it does not have an unpleasant taste.

The percentage of perfume used in a lipstick varies between 1 and 2%, the better quality lipsticks containing 2% of perfume.

Color in lipsticks is produced either by the use of insoluble lakes or oil-soluble dyes of the Eosin group, or both.

It is of interest that D&C Red No. 21, when dispersed in a lipstick and applied to the lips, will be converted by the alkali in the blood to D&C Red No. 22, which is a different shade.

The bromo acids are used to produce indelibility in the applied film. From 1 to 5% of these dyes are used in lipsticks, but ordinarily, 2 or 3% is adequate. The various other pigments are used in concentrations ranging from 5 to 15%, with 10% the average.

The bromo acids, as they are called, are bromo derivatives of fluorescein. There are two of these compounds that are used more generally—the dibromo- and the tetrabromofluorescein. The di compound produces a yellow-red color while the tetra compound will give a more purple stain. The two are usually used in combination with each other, as neither will produce desired results when used separately.

In formulating a lipstick, a highly refined grade of castor oil is one of the most common ingredients. It is used primarily because it imparts an inherent viscosity to the molded stick and, secondly, because it has a small solvent action on the bromo acids.

Castor oil will dissolve less than 0.3% of the bromo acids, and its solvent action is due to the presence of free hydroxyl groups in the glycerides of ricinoleic acid which make up a large percentage of castor oil.

In the so-called "high stain" sticks, a solvent such as polyethylene glycol 400 is used. The polyethylene glycol is not soluble in castor oil. A mutual solvent or coupling agent is therefore necessary, and propylene glycol monoricinoleate is used for this purpose. Isopropyl myristate, isopropyl palmitate, butyl stearate, and some of the higher esters of the mono- and dihydric alcohols are commonly used in lipsticks, both for their solvent action on the bromo acids and to reduce the thickness of the castor oil film [82].

Lanolin is used for its emollient properties and for a degree of tackiness and drag. Various derivatives of lanolin are also employed. Acetylated lanolin is more soluble in castor oil than lanolin itself. Lanolin oils, which are solvent extractions of lanolin, eliminate much of the tack of lanolin. Lanolin aids in holding the lipstick mass in a uniform homogenous mixture. Lanolin, used in the proper ratio, will help to prevent sweating of the solvent oils and confer a certain amount of protection against abrupt temperature changes.

Various waxes are used to impart different characteristics, such as hardness, thixotropy, melting point, and ease of application. These waxes are used in amounts regulated by the nature of the other components making up the finished stick.

Carnauba wax is responsible for stiffening the stick, and if used in the proper proportions will impart to the stick exactly the right degree of thixotropy. Its melting point is very high, much higher than any of the other natural waxes. Candelilla wax is used for the same reason, but to a lesser extent. Too much Carnauba will cause a granular texture in the finished lipstick, and it is best to keep its concentration as low as possible, usually less than 5%.

Ceresin and ozokerite are both good stiffening agents, with ceresin exhibiting the most stiffening action. Since ceresin is merely a mixture of ozokerite and paraffin, the same effect can be obtained in using these two waxes in combination. Ceresin can have a variety of melting points, depending upon the ratio of ozokerite to paraffin, thus affecting the melting point of the lipstick. Ozokerite is used more often to raise the melting point. A combination of ozokerite and carnauba is most efficient.

Beeswax can be used to raise the melting point. It is a good binder, helping to produce a homogeneous mass. In large amounts it will cause granulation and a dulling effect.

Manufacture of Lipsticks

In manufacturing lipsticks, the colors are mixed with part of the oil, usually in a Day Pony Mixer, and then passed through a 3-roll mill. The color and oil are ground a sufficient number of times to achieve a color distribution of 7 or better on a grind gauge.

The base material is heated in a steam jacketed kettle to 90°C, and the ground colors are added to the base material in the kettle. The batch is stirred with a high-speed mixer (Eppenbach) for several hours, and a sample is then drawn for color comparison with a standard. Any adjustment in shade is made by using the ground colors. Base material is added in proportion to the amount of color used in the correction to keep the consistency approximately equal to that in the original formulation.

When the color match is satisfactory, the lipstick mass is drawn off from the kettle and passed through a 200-mesh stainless steel screen. It is collected in a molding pan and slabbed for future molding. Prior to molding, the mass is melted with slow agitation in a melting kettle to remove entrapped air.

Temperature control is important throughout the processing, with care being taken to avoid excessive temperature in the kettle. The warmed lipstick mass is poured into warm molds, and the molds are passed through a cooling chamber. After passing through the cooling chamber, the lipsticks are removed from the molds, inserted in the lipstick-holders and flamed by passing them rapidly through a Bunsen gas flame, and are then ready for packaging and distribution.

The tendency to sweat in a lipstick is directly proportional to the lack of uniformity of pigment distribution. The texture of a lipstick depends on what type of pigment aggregation is dominant. The texture is coarse if flocs of loosely bound pigment particles are present, and fine if there are dense agglomerates of pigment particles.

Gas that is adsorbed on the surface of pigment particles diffuses slowly into the surrounding lipstick mass. A marked improvement is noticed in a pigment dispersion that is placed under vacuum. Dispersion of pigment involves a complicated combination of surface phenomenon—adsorption, wetting, surface tension, interfacial tension, and formation of phase boundaries. Electrical charges must also be considered. Mechanical force alone cannot produce a satisfactory dispersion [28, 30].

Control Methods

Control in manufacturing lipsticks consists essentially of a thorough check of raw materials and a color check against established standards [29].

A check is made on the heat resistance of the finished product when kept for 24 h in a constant temperature box at a temperature of 55°C to determine any tendency to droop or distort.

A hardness test may be done using a penetrometer. Usually five readings are taken. The average is calculated and compared with a standard.

A tendency for the stick to sweat or bleed can be checked by putting the filled lipstick containers in a desiccator containing water. The desiccator is kept in a constant temperature box at a temperature of 45°C.

Shade and intensity of stain are best judged on the skin by comparison with the standard. Skin tone and undertone are also checked on the skin. Mass tone is judged by comparison with a standard.

A typical lipstick formula follows:

4.5	Carnauba wax
11.6	Ozokerite 160
1.7	Cetyl alcohol
5.0	Lanolin anhydrous
26.3	Wecobee S (cocoa butter substitute), hydrogenated vegetable oil (PVO)
49.6	Castor oil
0.1	Preservative
0.2	Antioxidant
1.0	Perfume
100.0	

To this base formula add the necessary pigment pastes for a finished lipstick.

Mascara and Eye Makeup [31–33]

For the formulation of products for use in the area of the eyes, there is a broad limitation imposed by law. The limitation of the choice of colorants to natural dyes and inorganic and carbon pigments is made necessary by the close proximity of these products to the eye during application and use. The commercially available pigments used are the varying shades of the iron oxides, the different carbon blacks, and the ultramarine blues.

The area of the eyes is defined as the area enclosed within the circumference of the supra-orbital ridge and the infra-orbital ridge (eye socket), including the eyebrow, the skin below the eyebrow, the eyelids and the eyelashes, the conjunctival sac of the eyes, the eyeball, and the soft aereolar tissue that lies within the perimeter of the infra-orbital ridge.

The accentuation of the eyes is best achieved by emphasizing the eyelashes, which is the function of the mascara. The addition of the color to the lashes makes them appear more highly colored and longer, since in many instances the ends of the lashes tend to be rather blondish as compared to the rest of the hair length.

The film applied should show good resistance to moisture, since the eyes may tear and cause smudging. It should be resistant to gentle abrasion, as would happen if the hand should be brushed across the eyes. The best known form of mascara is the cake mascara which is used in conjunction with a wet brush. This mascara depends on the use of an oil-soluble soap which, on gentle brushing with water, forms an emulsion containing the colorants and the rest of the mascara vehicle. After application of the emulsion to the lashes, the film is allowed to dry. In this way the lashes are colored, and the colors are protected by the rest of the dehydrated vehicle.

In this system the greater the amount of soap used in the mascara formulation the easier it will be to form an emulsion with water. The rest of the mascara formula must be balanced in such a way that the hydrophobic or water-resistant part of the formula protects the film, which has been applied to the eyelashes, from washing off when exposed to water during the wearing period. Waxes impart part of the water resistance.

The ideal formula contains a low soap concentration and a high wax and fatty materials content to give the mascara film a full waterproof property. Enough soap is formed so that when the brush is wet and rubbed vigorously on the cake, an emulsion is formed. This emulsion is transferred to the eyelashes. The more vigorous mixing needed to make the emulsion, the more the film will resist accidental contact with water when on the eyelashes.

Triethanolamine soap is the most commonly used hydrophilic substance. These soaps are readily soluble in waxes or other esters. Carnauba wax is probably the major wax constituent of the formula. Its choice is dictated by its high melting point and its inclination toward producing a highly water-repellent film.

Waxes of lower melting point are included so that actual production of the mascara may be made easier. Easy flowability of the system is insured by having an adequate amount of the wax phase. It is inadvisable to use a pigment concentration in excess of 10% unless the pigments are easily wetted by the oil phase.

The mascara is usually made by incorporating the pigments into the hot wax and soap blend, heating and mixing for several hours until thorough incorporation has been achieved. A color check is then made of this blended material. If passable, it is poured into pans and allowed to cool. The cold mass is transferred to a 3-roll mill and ground to form a smooth homogeneous mass. This mass is remelted and poured onto warm molds. The molds are passed through a cooling chamber and the cake mascara is removed and packaged [31–33].

A prototype formula for cake mascara is:

18.75	Stearic acid, triple press
8.35	Triethanolamine
34.20	Carnauba wax
20.85	Beeswax
q.s.	Preservative

This base formula is mixed with pigment pastes.

Cream Mascaras

In cream mascara the water has already been incorporated to provide a mascara ready for use without previous trituration. There are some advantages in this type of preparation since a much higher concentration of pigment can be incorporated here than is possible in a cake. In addition, the optimum amount of water may be added rather than leaving this to the discretion of the consumer, which would be the case in using the cake.

The pigments used in cream mascara are ground into the base material, and the fine grind is then mixed with the necessary water and humectants to form the cream.

A formula for a roll-on mascara follows:

By Weight

46.05	Soltrol #130 (Phillips Petroleum, hydrocarbon solvent)
7.50	Beeswax, white
8.25	Carnuba wax #1 refined
0.10	Propyl-*p*-hydrobenzoate
4.00	Arlacel C (ICI–U.S., glyceryl monostearate S/E)
0.40	Tween 60 (ICI–U.S., POE (20) sorbitan stearate)
0.40	Borax-5 H20
0.05	Methyl-*p*-hydrobenzoate
33.25	Distilled or deionized water

100.00

Color is added as pigments as required.

Eye Shadows

The colors used in eye shadows are somewhat limited because of proximity of use to the eye. The carbon blacks, ultramarine blue, and the various yellows, browns, and reds of the iron oxide pigments are common colors in eye shadows. There is usually a higher concentration of pigments used here as compared to mascaras, running as high as 25%.

The base into which the colors are ground usually consists of beeswax, ozokerite, mineral oil, lanolin, and petrolatum. The method of manufacture consists of melting the waxes and oils together, then adding the colors and titanium dioxide, the latter usually used to impart the desired opacity, and grinding in a 3-roll mill.

Eye shadows that have a metallic luster are made by incorporating very finely ground aluminum, bronze, silver, or gold powders. The completed formula is heated and stirred slowly to remove occluded air and is then poured into the container.

A prototype formula for an eye shadow is:

24.0	Ozokerite, 80° C
1.7	Paraffin
14.0	Petrolatum, 55/75 Saybolt viscosity at 210° F
40.0	Mineral oil, 65/75 Saybolt viscosity at 100° F
0.3	Propyl parahydroxy benzoate
20.0	Chromic oxide hydrated (40% in petrolatum)
———	
100.0	

Nail Products

Nail Lacquer [34–36]

The formula ingredients are usually a resin, plasticizer, solvents, and pigment.

Resins can be nitrocellulose or dinitrocellulose, also known as pyroxylin. Nitrocellulose films have a tendency to shrink, and surface adhesion may be only moderate. To overcome this, other resins are added to impart adhesion and gloss, and plasticizers are added to reduce shrinkage and make the film flexible.

The solvents used will influence the ease of application of the lacquer, its rate of drying and hardening, and the final characteristics of the film. The preferred solvents for nail lacquers are mixtures of low and medium boiling point solvents. Solvents are the alcohols, aromatic hydrocarbons, and aliphatic hydrocarbons.

Plasticizers may be added singly or in combinations of two or more, depending on the formulation. High molecular weight esters are used as well as castor oil. The esters can be dibutyl and dioctyl phthalate, triethyl citrate, and acetyl tributyl citrate.

The pigments used in nail lacquer formulations are carbon black, iron oxides, chromium oxides, ultramarines, metallic powders (gold, bronze, aluminum, copper), aluminum and calcium lakes of FD&C and D&C blue, red, yellow, and orange, and titanium dioxide. Transparent systems require the use of solvent-soluble colors such as the D&C red, green, yellow, and violet.

Pearl pigment can be the natural pearl essence or guanine (2-amino-6-hydroxy purine). The pearly luster is due to (1) the simultaneous reflection of light not only from the upper layers of the transparent crystals but from crystal planes at different depths, and (2) the alignment of the crystals in the same direction. The crystals remain in alignment when the film hardens. The synthetic pearl is usually a bismuth salt.

A formula for nail lacquer follows:

	$\%w/w$
Nitrocellulose "$\frac{1}{2}$ s"	10.0
n-Butyl acetate	12.8
Isopropanol	5.0
n-Amyl acetate	11.1
Ethyl acetate	11.1
Camphor (plasticizer)	2.5
Dibutyl phthalate	2.5
Aryl sulfonamide–formaldehyde resin	10.0
Toluene	30.5
Ethanol	1.5
Bentonite	3.0

Lacquer Removers

Lacquer removers may be made from any of a number of solvents available. In the majority of cases, any of the solvents used in the formulation of the actual enamels can be used as removers. The fastest acting seems to be acetone, but because of its high volatility rate, ethyl acetate has come into favor as well. Many of the lacquer removers contain an additive designed to leave a film on the nail, since the solvent itself has so high a degreasing tendency. Various oils and emollient compounds are added to the lacquer removers, such as castor oil, lanolins, and lanolin derivatives.

A formula for a nail polish remover follows:

Acetone	30.0%
Ethyl acetate	40.0
Sesame oil	1.0
Dibutyl phthalate	10.0
Carbitol	19.0
Perfume (q.s.)	

Cuticle Removers and Softeners

Cuticle removers and cuticle softeners usually consist of a dilute solution of alkali in water with some glycerin or other humectant added to keep the water from evaporating too easily. Potassium hydroxide, the alkali usually used, results in a somewhat harsh preparation, but it is effective. Trisodium phosphate, triethanolamine, and some of the quaternary ammonium salts have been used as replacements for potassium hydroxide. These are not as effective as the alkali, but do a fair job of softening the cuticle.

Permanent Wave Preparations

The subject matter and history as well as the chemistry of permanent waving is massive, involved, and very technical. Our attempt to reduce this topic to a very few short paragraphs is doing an injustice to the great efforts of the pioneers in this important work.

In the theory of permanent waving it is well known that hair is stretched in the permanent waving process and fixed into a new permanent position with the aid of waving solutions, curling rods, and heat.

To produce a wave, hair must be stretched so as to produce a strain on one surface or side of the hair shaft. This is accomplished by winding the hair so as to produce a difference in the degree of strain on the outer side of the hair shaft as compared to the inner side, without which no wave can be produced. Thus a round hair shaft is more or less flattened on one side, and if this effect be fixed or made permanent, a lasting wave will be produced. This is why natural curly hair is slightly flattened on one side.

The second reason for the development of a permanent wave requires a short discussion of hair chemistry whereby the hair, after being wound on a suitable rod, is subjected to the action of moisture and an alkali. A chemical reaction termed "hydrolysis" occurs, and certain close linkages in the molecular structure or chain are broken down and new unstable links are formed which tend to maintain the hair in its new position. Finally, a point is reached where the set becomes permanent and the process can no longer be reversed. This is the so-called "permanent wave." On setting the hair, it will always strive to return to its new shape or form.

Some of the earlier methods of waving hair were carried out by an operator. Two of the most famous were known as the "Nessler" method and the "Croquinole" method. These processes also involved curling the hair on special rods and then treating the hair with alkali solutions to weaken the hair before curling, with or without the aid of heat.

The Nessler method used a volatile alkali such as ammonium hydroxide, whereas the Croquinole method used fixed alkalis. These methods also had their particular or specific method of winding the hair on the rods.

The two general procedures for permanent wave products are:

1. The hot method in which the hair is treated with an alkaline sulfite solution and then wound around a rod having a small diameter. Heat is then applied to produce a permanent wave. The heat applied can be accomplished or obtained via the use of electrical heaters or chemicals which develop an exothermic heat of reaction.

2. The cold process in which the hair is similarly wrapped around a rod of suitable diameter, either prior to or after treatment with an alkaline reducing agent such as ammonium thioglycolate. In this method the hair is transformed from its straight or normal form into the desired wave form without the application of external heat. When this has been achieved, as determined by the inspection of a test curl, the waves are fixed or set by the application of a suitable neutralizing or oxidizing agent. Some of the products on the market today omit the use of the oxidizing agent, permitting the air to serve as the oxidizer and do the final neutralizing of the waving solution. In hair undergoing a cold wave treatment in which the tress is saturated with ammonium thioglycolate, the following simplified reaction takes place:

$$R-S-S-R + 2HSCH_2COONH_4 \underset{\text{Oxidation}}{\overset{\text{Reduction}}{\rightleftharpoons}}$$

Cystine
link
(hair),
keratin disulfide   Ammonium
thioglycolate

$$R-SH \qquad\qquad + R-S-S-CH_2COONH_4$$

Sulfhydryl
cysteine
link   Mixed disulfide cystine
terminal group

$$\updownarrow$$

$$R-S-S-CH_2COONH_4 \qquad R-SH$$

Aside from the two methods of permanent waving described above, there are several matters of importance and interest which must be given consideration. First, the shampooing of the hair prior to waving it. The action of the detergent is mainly used to clean the hair prior to the application of the waving solution. It is also important because the detergent often combines with and modifies some of the chemical groupings of the keratin of the hair, thereby influencing the subsequent action of the waving chemical during the actual waving process.

A second consideration is the size of the diameter of the rod upon which the hair is wound. If one will imagine a rod of small diameter as compared to a rod of large diameter, it is quite evident that the curl produced with the rod of smaller diameter will produce a smaller wave having a greater number of undulations per linear inch of hair.

A third consideration involves the fineness or curliness of the hair to be waved. The diameter of the hair shaft will make a difference in the degree of curl produced following the waving process. The finer the hair, the more surface will be exposed to the waving solution, thereby producing a more thorough and permanent wave. The converse of this is also true.

The procedure for giving a permanent wave is as follows: The hair is first given a shampoo in the usual manner and is then wetted with the ammonium thioglycolate solution before it is wrapped around the rod or curlers made of material which will not react with the thioglycolate solution. The hair must be thoroughly moistened with the waving solution. In practical terms, the degree of alteration of and the reduction of hair in the cold wave process depends upon the amount of thioglycolate solution which diffuses into the hair and upon the rate at which this is accomplished.

At the end of the required period, which may vary from 10 min to 1 h, and before the hair is unwound from the curlers, it is treated with the arresting or neutralizing agent which contains an oxidizing compound, the function of which is to stop the action of the thioglycolic solution. This action-stopping solution may consist of an acidified solution of hydrogen peroxide or another oxidizing agent, and it should be left on the hair for a period of time sufficient for it to penetrate the hair shaft and check the reducing action of the thioglycolate.

Other neutralizing solutions consist of sodium or potassium bromate or sodium perborate monohydrate. A chemical neutralizer is not used in some commercial hair waving preparations. In these preparations the hair is permitted to dry in the air, with the air acting as the oxidizing or neutralizing agent.

The cold wave solution commonly sold to the public is a 6.25 to 6.5% solution of ammonium thioglycolate made alkaline with ammonia to a pH of 9.25 to 9.5, and which is free of metallic impurities which, if present, would inactivate the solution. For this reason the manufacture of these cold wave thioglycolate solutions should be carried on in suitable vessels made from glass or porcelain, hard rubber, Saran, or stoneware.

Minute traces of iron or copper will accelerate the decomposition of the thioglycolate solution. In fact, traces of iron, which produce a red-violet coloration in the solution, serve as a basis of the analytical test to determine the assay of the thioglycolate solution.

Some typical formulas for wave lotions follow:

	I	II
Thioglycolic acid, 75% strength, g	6.0	9.0
Ammonium hydroxide, 28% strength, cc	6.0	9.0
Wetting agent (Brij 35 or		
triethanolamine lauryl sulfate), g	0.1	0.1
Distilled water, cc	100.0	100.0
Perfume, g	0.25	0.33

Additional amounts of a clouding or opacifying agent may be used as desired. However, a most careful selection and shelf test is essential to provide a suitable material for this purpose.

Some typical formulas for neutralizing lotions follow:

Hydrogen peroxide, 6% solution, cc	60	Use this solution as is
Tartaric acid, g	36	and pour through the hair
Distilled water, cc	120	repeatedly

Sodium perborate monohydrate, %	92	Use this powder mixture at
Citric or tartaric acid, %	7	the rate of $\frac{1}{2}$ oz to 1
Sodium lauryl sulfate, %	1	qt of water, and the solution
		poured and rinsed
		through the hair repeatedly

Hair Straighteners

Thioglycolate Straighteners

Though the thioglycolate straighteners are essentially the same as products utilized for hair waving, they differ in that the straighteners are appreciably more viscous in order to assist in maintaining the hair in a straight position while being softened. After application, the hair must be repeatedly combed, which may be a factor in scalp irritation. Since curly hair is quite resistant, as is to be expected, the softening process takes a rather long time and the danger of hair damage and scalp irritation proportionately increases. Upon reaching the proper degree of straightness, the hair is rinsed and treated with a conventional neutralizer such as sodium bromate [37].

A general formula for a hair-straightening cream is:

22.00	Glyceryl monostearate, pure
1.75	Brij 35 (ICI)
1.00	Mineral oil, carnation (Sonneborn)
0.15	Preservative: M-p(OH) benzoate and P-p(OH) benzoate (3:1)
52.00–53.00	Distilled or deionized water
17.50	Ammonium thioglycolate, 50% solution = 8.75% thio acid
~4.50	Ammonium hydroxide 28%, sufficient to product pH 9.2 to 9.4
0.50–1.00	Perfume
100.00	

The Gant and Hersh patent relates to compositions and methods for straightening kinky hair or for curling hair in the event that it is straight [70].

Depilatories

These are products formulated to remove hair by physical and chemical procedures.

Physical Procedures

Wax–rosin compositions are used as depilatories. A blend of rosin (3 parts) and beeswax (1 part) is applied warm to the hair area and allowed to cool and harden. A quick pull on the hardened mix removes hair that is enmeshed.

Chemical Procedures

Calcium thioglycollate is the preferred chemical used in chemical depilatories, Calcium hydroxide is one of the alkaline materials, which in a saturated solution has a pH of 12.4, used to maintain the required alkaline pH. The more soluble strontium hydroxide is often used in combination with calcium hydroxide. The solubility of calcium thioglycollate at a pH of 12 is 0.35 mol/l. This is barely enough for active depilation, and if excess calcium thioglycollate is used the typical formulation is a cream since a more uniform suspension results from the use of an emulsion base.

The requirements of a depilatory may be briefly stated as follows:

1. It should be nonirritating and innocuous
2. It should be efficient in action, removing hair within a 10-min period
3. It should be without odor or have the pleasant odor of the perfume

The activity of the depilatory substance depends upon the following factors:

1. The nature of the depilating or denaturing agent
2. The duration of the action
3. The pH of the cosmetic medium
4. The temperature of the reaction

The depilatory most widely used today is based on the use of calcium thioglycollate in a strongly alkaline medium having a pH of approximately 12.3. In addition to the calcium thioglycollate, there is present a small amount of a wetting agent, a considerable amount of a filler such as calcium carbonate, an adequate amount of slaked lime to bring the pH up to the required value, and an alkali stable perfume.

The commercial products are sold either as a thin or very thick paste, packaged in a jar or collapsible tube.

The advantage of such preparations compared with the use of a razor is that instead of merely cutting the hair at the skin surface, it removes the hair at the mouth of the hair follicle with the result that a second treatment is necessary only after a 2- or 3-week period has elapsed; whereas if simple shaving is employed, a dark growth will be visible within a few days.

The method of use of a depilatory is as follows: The area to be treated is washed with soap and water and dried. The depilatory paste is then applied and uniformly spread over the area to be treated. The depilatory is allowed to remain at the site for the suggested time, which is usually somewhere between 5 and 10 min. At the end of this period, the cream is removed by means of a spatula or tongue depressor, the remaining cream removed or tissued off, and the skin surface washed free of all remaining cream and removed hair.

A typical formula for a depilatory follows:

Base Cream

Product G-2135	1.75 g,	wetting agent and emulsifier (ICI-U.S.)
Distilled water	60.10 g,	solvent
Calcium carbonate	32.00 g,	filler

Cetyl alcohol	6.15 g, bodying agent for cream
Preservative	q.s.
	———
	100.00

Parts by
Weight

5.4	Calcium thioglycollate	}	Depilating agent
15.0	Distilled water	}	
6.8	Calcium hydroxide	}	Alkali mixture for pH control
3.4	Strontium hydroxide · 8H$_2$O	}	
69.0	Base cream, above		
0.4	Perfume		
	———		
100.0			

There are several precautions:

1. Be sure there is no metal contamination from iron or copper or other heavy metals. Pure stainless steel is satisfactory.
2. There should be no unnecessary exposure to air. Oxidation should be avoided to maintain strength.
3. There should be proper pH control at approximately 12.3, which should be checked before packaging.
4. The proper package must be used. Pure tin is best, wax-lined lead tubes, glass, or polyethylene packaging is satisfactory.

Other depilating agents used are:

1. Sodium sulfide: A very active agent which is hazardous to use because of its action on the stratum corneum. A 2% solution has a pH pf 12.0
2. Strontium sulfide: This is effective in 3 to 5 min. It is a satisfactory agent, but the disadvantage of use is its strong odor of sulfide.

Perfumes used for depilatories have to be specially formulated to overcome the intense product odor and be stable in the high pH.

Bath Products [38–40]

The modern science of cosmetics combines the hygienic cleansing of the body with its biological care in that it recommends the use of intensive washing with soaps or detergents and active substances.

When bathing the body, dirt, perspiration, and some germs are initially removed from the surface of the skin by the water and soap. An additional effect arises in that the water causes the outermost layers of the epidermis to swell slightly so that they are shed more rapidly. In addition, if the bath contains suitable additives, it also can show effects which extend beyond the hygienic aim and contribute cosmetically to producing "well cared-for" skin.

Bath Oils

The subjective response to the use of any bath oil is generally favorable. This is to be expected since bathing itself imparts a feeling of softness to the stratum corneum, and even the slightest deposition of an emollient adds to this effect. The addition of fragrances and fragrant oils to baths was recorded during the Roman era.

Bath oils fall into one of two major categories: (1) spreading or floating bath oils, and (2) dispersible bath oils.

The spreading or floating bath oil is usually an anhydrous system which is oleaginous. The hydrophobic system, being of lighter density than water, will float to the surface of the bath water. The spreading action is facilitated through the inclusion of small amounts (usually about 5%) of an oil-soluble surfactant. The lowering of the surface tension of bath water by the surfactant permits the oil to form a continuous film rather than exist as individual droplets. Ideal surfactants should demonstrate a reasonably high HLB (approximately 9) yet remain soluble in the oleaginous composition.

From a functional standpoint, the floating layer of perfumed oil affords a pleasant fragrance and a desirable feel to the skin. When the individual emerges from a tub of water, the oleaginous film will adhere and coat the body surface and impart a hydrophobic barrier to the keratin.

The oleaginous composition may consist of one or more of the commonly accepted hydrophobic components. Natural vegetable oils, animal-derived lipids such as lanolin and mink oil, low melting point fatty alcohols such as oleyl and hexadecyl, low melting point synthetic glycerides such as glyceryl monooleate and glyceryl monolaurate, and mineral oils.

Dispersible bath oils contain a sufficient concentration of surface-active agents to (1) solubilize the oleaginous components through micellar solubilization, or (2) disperse oleaginous components as the internal phase of an oil-in-water emulsion. The addition of either system to bath water results in the oil being dispersed throughout the bath water, producing a somewhat homogeneous distribution of oil. While the spreading bath oil maintains its oleaginous character in water, the dispersible bath oil loses some of its oleaginous character as the result of oil-in-water dispersibility.

The perfume is generally employed in concentrations of from 3 to 20%, obviously depending on the nature and quantity of surfactant.

The nature (HLB) and concentration of the surfactant component will also dictate whether the dispersible bath oil will turn milky on addition to water (blooming bath oil), whether it will remain clear on dispersing in water, or whether the bath oil will generate any recognizable amount of foam (foaming bath oil). Surfactants with lower HLBs, such as polyoxyethylene oleyl ether (4.9), will be oil soluble and yield oils that produce the characteristic blooming.

Functionally, dispersible bath oils leave a layer of dilute oil in a water dispersion on the skin. As the water evaporates from this, a hydrophobic residue remains, providing a moisture barrier on the skin surface.

The dispersible bath oils are usually preferred by the consumer over the floating type because they are water dispersible (washable) while the latter may leave an oil slick in the tub and may create drain clogging problems.

Bath Salts

Two types are available. The first is formulated with crystalline salts such as rock salt and Epsom salt to which color and perfume are added. These are not water softeners. The second type is the water-softening type based on sesquicarbonates, phosphates, and borates. Color and perfume are added as are small percentages of fatty acid ester for nondrying effects in some products.

Foaming Bath Oils and Foam Baths

These are based on suitable surfactants, either liquid or powder, which are blended with color, perfume, and foam stabilizers. Addition of warm or hot running water results in a fragranced foam that leaves the skin soft and fragranced.

Soaps [41–44]

Soaps may be made from natural fats or from the isolated fatty acids. The natural fats are glycerides of various fatty acids, and are used to make soap where cost is a consideration.

Methods Used for Soap Manufacture

1. Boiled process, whereby the fats, oils, and alkalis are all boiled together until saponification is complete. The soap is salted out with or without varying percentages of glycerin left in the soap. This is the most widely used process.
2. Semiboiled process, whereby exact chemical amounts of the fats, oils, and alkalis are mixed and heated to the start of the saponification, after which the reaction is allowed to complete itself.
3. Cold process, whereby the fats, oils, and a calculated quantity (less than needed for complete saponification) of alkali are placed in a crutcher and gently heated and mixed. When saponification is underway the material is poured into "soap frames" where the reaction completes itself spontaneously and the glycerin, which is the by-product, is left in the soap.

Transparent soaps are prepared with sugar, alcohol, and glycerin or a combination of these materials in the approximate proportions [76]:

Sugar: 10 to 15% of the weight of soap
Alcohol: 50% of the fatty acid present in the soap
Glycerin: 50% of the fatty acid present in the soap

Detergency or cleaning ability is usually determined by soiling cloth with a "standard smudge" or soiling mixture made according to the following formula:

Lamp or carbon black	2.0 g
Heavy mineral oil	5.0 g
Tallow	3.0 g
Carbon tetrachloride	2000.0 cc

It is well known that soaps and detergents remove the natural oil secretions (sebaceous matter) as well as the dirt, resulting in dryness and harshness of the skin surface. This drying effect of soap or detergents can partially be overcome by superfatting the soap or detergent bar. The presence of superfatting agents reduces the cleansing efficiency, thereby overcoming a complete degreasing of the skin surface. From a dermatological aspect, the superfatting of soaps should be advantageous.

Apart from the dermatological aspect, the addition of superfatting agents to soap serve the following functions:

1. Reduces the tendency of milled toilet soap to split or crack in use.
2. Process of milling and plodding is often made easier because the soap is rendered more plastic.
3. The lather produced is usually smoother and closer in texture.
4. The soap bar or tablet has a superior velvety feel.

Superfatting agents are:

Lecithin: Softens and moisturizes the skin
Amine oxides: Improve the soap texture and lathering properties
Fatty acid amides: Same properties as amine oxides
Lanolin: Improves the texture of soap
Casein: Improves the lather and skin feel
Cocoa butter: Emollient and moisturizing

Deodorant Soaps

Some 50% of the soap market is claimed for deodorant soaps. These are usually quality soap bars containing either Irgasan DP-300 (Triclosan, Ciba Geigy) or

3,4,4-trichlorocarbanilde (Triclocarban, Monsanto). U.S. Patent 3,687,855 (assigned to Alfred Halpern, Synergistics, Inc.), August 1972, is for a germicidal detergent bar containing a polyvinyl pyrrolidone–iodine complex.

Shampoos

Present day shampoos usually contain a primary detergent which can be a fatty alcohol sulfate, ether sulfate, sarcosinate, or many other anionics. The primary detergent can also be an amphoteric or a nonionic. The secondary or auxiliary detergent is usually an alkanolamide which controls the viscosity of the product and increases the quality and volume of the foam [46–52].

Soap Shampoo

Soap shampoos are sensitive to and do not cleanse well in "hard" water areas. Soap shampoos are primarily aqueous solutions of soft soap combined with preservatives, sequestrants, color, and perfume.

Soapless Shampoo

Soapless shampoos are primarily based on aqueous solutions of sulfonated oils such as sulfonated castor oil and sulfonated olive oil combined with preservatives, color, and perfume.

Low pH Shampoos

Unlike many of the properties ascribed to shampoos, such as foaming, cleansing, lustre, and manageability, pH is an intangible. pH cannot be observed or demonstrated during product usage, and the effects of low pH are not obvious to the user. It has been reported that mild aqueous acids cause an antiswelling action on the cuticle scales of the hair. As the cuticle tightens, the hair gains lustre because light is more efficiently reflected from the surface of the hair shaft. In the absence of external charges, hair shows its highest strength and resiliency at pH 4.0 to 6.0. Some surfactants do not perform well in an acid medium. Fatty alcohol sulfates hydrolyze rapidly and decompose below pH 4.0. Alkanolamides generally will react in a similar manner. However, combinations of triethanolamine and sodium and ammonium lauryl sulfates with amine oxides may be prepared and adjusted to low pH with good stability.

An example of a clear, stable shampoo formulation adjusted to a pH of 4.0 is

6.0	Cocamidopropylamine oxide (30%)
40.0	Ammonium lauryl sulfate (30%)

54.0   Water
q.s.   Preservative, color, perfume, acid, etc.

Amphoteric Shampoos

Amphoteric shampoos are based on the use of imidazoline, betaine, and sulfobetaine surfactants. Amphoteric surfactants are generally assumed to be less irritating to the eyes of the user if accidently rubbed into the eye. Amphoteric surfactants are usually low foam materials, and it is customary to combine them with anionic surfactants. Amphoteric base shampoos are difficult to thicken and difficult to opacify. Amphoterics can be blended with equal parts of high foaming anionic lauryl sulfates or the sulfosuccinate half esters. Amine oxides may also be used instead of the more irritating alkanolamides. The keratin substantivity of the betaine amphoterics is greatest in the pH range of 5.5 to 6.5.

The following shampoo formula has been adjusted to a pH of 5.8 and illustrates the use of amphoterics:

17.0   Ammonium lauryl sulfate
6.0   Monomate CPA-40 (sulfosuccinate half ester)
2.0   Monateric ISA-35 (isostearic imidazoline)
q.s.   Perfume, preservative, color, water

Dandruff Shampoos

Dandruff can be thought of as the product of hyperkeratinization where the rate of keratinization has increased to the point that the scales become more visible. There are no known differences in the incidence of dandruff in men or women or between races. Dandruff shampoos contain ingredients that effectively control dandruff by allowing a normal turnover rate of epidermal cells. Ingredients used in some antidandruff shampoos are coal tar, quaternary ammonium compounds, resourcinol, salicylic acid, selenium sulfide, sulfur, undecylenic acid and derivatives, and zinc pyrithione. Selenium sulfide and zinc pyrithione (zinc omadine) are cytostatic (reduce the epidermal cell turnover rate). Zinc pyrithione is strongly adsorbed by the hair, and adsorption is related to concentration, pH, temperature, product formulation, and time of exposure. Since these are medicated shampoos, formulations should be nonirritating and nonsensitizing.

An example of an antidandruff shampoo formulation is:

25.5   Maprofix WAC (sodium lauryl sulfate)
5.0   Stearic acid T.P.
2.1   Ammonyx LO (lauramine oxide)
0.7   Sodium hydroxide pellets
0.4   Sodium chloride

 1.0   Zinc omadine powder, 98%
65.3   Water
——
100.0

Shampoo Additives

Foam builders are principally fatty acid alkanolamides.

Conditioning agents are principally amine oxides, fatty alcohols, lanolin derivatives, esters of fatty acids, silicones, and cationic materials. Polyacrylamides are used as conditioning agents [78, 83].

Opacifying agents are principally higher fatty alcohols, such as stearyl and cetyl, and glycol and glycerine esters of fatty acids.

Sequestering agents are principally acids such as citric and tartaric and salts of ethylenediaminetetraacetic acid (EDTA). These are used to prevent formation of insoluble calcium and magnesium soaps and to prevent discoloration due to iron contamination.

Antidandruff agents are materials such as sulfur, salicylic acid, tar extracts, zinc pyrithione, quaternaries, and undecylenic acid amides.

Viscosity builders are principally cellulose derivatives, alkanolamides, and carboxy–vinyl polymers.

Preservatives are usually the hydroxybenzoates, formaldehyde, imidazolidinyl urea compounds (Germall), sorbates, and 1-(3-chloroallyl) 3,5,7-triazo-1-azo-niaadamantane chloride (Dowicil).

Examples of shampoo formulations are:

Lotion Cream Shampoo

 3.0   Coconut fatty acid diethanolamide
 4.0   Ethylene glycol monostearate
 3.0   Propylene glycol
42.1   Sodium lauryl sulfate, 28% active
 0.4   Preservative
47.0   Distilled or deionized water
 0.5   Perfume
——
100.0

Clear Liquid Shampoo

 2.0   Oleyl alcohol
50.0   Triethanolamine lauryl sulfate, 65% active
 5.0   Lauric acid diethanolamide
 4.0   Myristic acid
10.0   Propylene glycol
 0.1   Citric acid
 0.1   Sequesterant Na$_2$
 0.4   Preservative
27.6   Distilled or deionized water

0.8   Perfume
—————
100.0

Viscous Liquid Shampoo

10.0	Coconut fatty acid diethanolamide
3.0	Myristic acid
2.0	Oleyl alcohol
0.4	Preservative
0.5	Polyvinylpyrrolidone K-30
60.0	Triethanolamine lauryl sulfate, 38% active
2.0	Propylene glycol
21.1	Distilled or deionized water
1.0	Perfume

100.0

Shaving Preparations [45]

Dry hair in the beard is usually quite compact and is resistant to cutting by a razor blade. Treatment with hot water and soap is usually a preliminary procedure in order to remove sebum, which is a complex mixture of lipids secreted by the sebaceous glands and absorbed by the beard, presenting a barrier to water and impeding hair softening. The softening action of the water is reversible since the water will quickly evaporate if the wash procedure is not followed by the use of a shaving soap or shaving cream which provides a wet, lubricating blanket of water that reduces friction between the razor blade and the skin [77].

Shave Soap

Shaving soaps or shaving sticks are hard soaps, basically sodium salts of primarily saturated fatty acids, that are usually applied with a shaving brush.

Shave Cream—Lather

Shave creams are formulated as lather shave and brushless shave. Lather shave creams are soft soaps formulated with potassium hydroxide salts of saturated and unsaturated fatty acids, with a small percentage of sodium hydroxide to regulate viscosity, body, and consistency. Added to the basic product are humectants such as glycerin and propylene glycol, emollients such as lanolin and lanolin derivatives, boric acid in small percentage to neutralize any excess alkali, and menthol for its cooling effect. The water content of a lather shave is about 35 to 40%, and the pH of the lather is about 10.

Shave Cream—Brushless

Brushless shave creams are similar to vanishing creams, with an excess of free fatty acid, usually stearic, emulsified into a nonalkaline base of suitable viscosity or consistency. Application is similar to a cream, with the fingers, and the softening effect is not as rapid as that obtained with a lather shave. The water content is higher than that of a lather shave and the pH is in the range of 6.0 to 6.5. Additives are similar to those used in lather shave formulations.

   All formulations contain preservatives and are usually fragranced to cover the base odor of the soap and additives.

Aerosol Shave Creams

Aerosol shave creams in pressurized containers are convenient to use because the lather is easily obtained in ready-to-use form from the can and is easily rinsed off the face. Formulations are either aqueous soap solutions or soap emulsions that are pressurized with a mixture of propane and isobutane. The product is expelled from the container through a valve as a liquid which expands to a foam as the propellant is released and expands. Additives are similar to those used in lather shave formulas.

Self-Heating Shave Creams

Self-heating shave creams are formed from an exothermic reaction that occurs when two components, one containing an oxidizing agent and the other a reductant, are expelled together from a dual compartment container.

Gel Shave Creams

Gel shave creams are formulations that are based on an emollient aqueous gel that is soap based and contains a volatile solvent such as pentane. When expelled as a soft gel from a pressure container, the gel is rubbed on the face and the solvent is released. The released solvent expands as it volatilizes, and the gel becomes a foam.

   Typical formulas for shaving creams follow:

	Lather Type
Stearic acid	30.8%
Coconut fatty acids	7.7
Glycerine	10.0
Lanolin	1.0
Potassium hydroxide, 100%	7.0
Sodium hydroxide, 100%	0.57
Boric acid	1.0
Sodium silicate	0.5

Water	41.52
Perfume	q.s.
Fatty acid base	20% Coconut fatty acids
	80% Stearic acid
Alkali composition	100% Potassium hydroxide or 90% KOH + 10% NaOH
Free fatty acid	3–5%
Water content	42–46%

Brushless Type

Stearic acid	17.0%
Mineral oil	2.0
Lanolin	1.0
Glycerine	4.0
Glyceryl monostearate	4.0
Triethanolamine	1.2
Water	70.8
Preservative	q.s.
Perfume	q.s.
Free fatty acids	12–15%
Soap content	3–4%
Alkali used	100% Triethanolamine
Water content	~10%

Oral Products [53, 54]

Toothpaste

Toothpastes can be considered to be toothpowders suspended in a suitable binder. They are generally packaged in a collapsible tube.

The Council on Dental Therapeutics of the American Dental Association has established definite criteria for approval of dentifrices. Some of the requirements are based on the degree of hardness, degree of abrasiveness, and particle size of the polishing agent. The use of medicinals or drugs which should be used in the mouth under the supervision of a dentist are not approved by the council for over-the-counter purchase by the consumer.

The principle ingredients in toothpastes or powders are:

1. Polishing agents: calcium carbonate, magnesium carbonate, di- or tricalcium phosphate, talc, etc.
2. Detergents: soap, anionic surface-active agents.
3. Binders (excipients) and humectants: glycerin, propylene glycol, sorbitol, mucilages or gums, etc.
4. Sweeteners: saccharin, sorbitol
5. Preservatives: benzoic acid, hydroxy-benzoates
6. Flavors: essential oils, etc.
7. Water

Other additives in toothpastes are cellulose derivatives, such as methyl cellulose, used to maintain satisfactory consistency, and anti cari agents such as stannous fluoride and sodium monofluorophosphate.

Two qualitative formulas for toothpaste are:

Saccharin sodium
Sodium carboxymethylcellulose
Sorbitol 70%
Dicalcium phosphate dihydrate
Sodium lauryl sulfate
Methyl *p*-hydroxybenzoate
Propyl *p*-hydroxybenzoate
Deionized water
Flavor

Sorbitol 70%
Titanium dioxide
Sodium carboxymethylcellulose
Veegum regular
Saccharin sodium
Sodium pyrophosphate decahydrate
Diatomaceous earth
Aluminum silicate anhydrous
Deionized water
Sodium lauryl sulfate
Methyl *p*-hydroxybenzoate
Propyl *p*-hydroxybenzoate
Flavor

A prototype formula (by weight) for toothpaste follows:

32.200	Glycerin
21.075–21.625	Distilled or deionized water
0.100	Sodium benzoate, USP
0.125	Saccharin sodium, USP
15.000	Calcium carbonate, USP
27.000	Calcium Phosphate, dibasic (Mallinckrodt)
0.400	Sodium monoflurophosphate (Pfaltz & Bauer)
1.000	Carrageenan TP-4 (Marine Colloid)
2.000	Sodium *N*-lauroyl sarcosinate
1.000–0.500	Flavor
0.100–0.050	Special toothpaste flavor
100.000	

The Pader patent is concerned with a dentifrice composition containing as the essential polishing and cleansing ingredient a synthetic, amorphous, porous silica xerogel [81].

A formula example given in the patent is

12.0	Silica Xerogel V
5.0	Silica Aerogel 1
1.5	Hydroxyethylcellulose
0.2	Saccharin
34.76	Glycerin
39.00	Water
0.41	Stannous fluoride
0.03	FD&C Blue #1, 1% solution
1.1	Flavor
6.0	Sodium lauryl sulfate, 21% in glycerin

A prototype formula for mouthwash follows:

0.025 g	Cetyl pyridinium chloride
20.000 g	Sorbitol solution, 70%
0.200 g	Disodium phosphate
67.025 cc	Distilled or deionized water
12.500 cc	Alcohol*
0.250 cc	Flavor
q.s.	Color solution[†]

*Special denatured alcohol-flavor mixture.
[†]Permitted FD&C color solutions.

Wipes

Disposable packages of cleansers, polishing agents, hand lotions, and many other functional products consist primarily of liquid formulations applied to a paper or nonwoven substrate. The package can consist of one individual unit of application or can be a multiunit package.

The Williams patent [79] related to a fresh, lightly perfumed aqueous alcohol product that could be used as a refreshing facial and hand wipe. An approximate formula would be:

0.1%	Quaternary solution
0.1	Propylene glycol
8.0	Denatured alcohol
0.1	Perfume
91.7	Water
q.s.	Preservative

The Evans patent related to hygiene cleansing and was intended to be used as a wet adjunct to toilet paper [80]. An approximate formula would be:

5.0%	Propylene glycol
1.0	Mineral oil

5.0   Emulsifier
0.3   Preservative mixture
88.7   Water

The formula could be prepackaged as a wet cleansing application or could be dispensed directly onto toilet paper.

Preservatives and Antioxidants [55–58]

Changes and deterioration in cosmetic products can be brought about by changes in acidity or alkalinity, or as a result of hydrolysis, oxidation, heat, light, and bacterial or fungal contamination.

The source of bacterial or fungal contamination of cosmetic products can be the raw materials, water obtained from dirty ion exchangers, contaminated equipment, and dirty bottles and cap liners. Cosmetic products that are properly prepared will still require the addition of preservatives to insure adequate shelf life. In addition to all other quality assurance tests run by the laboratory, suitable microbiological tests should also be run on the finished product.

The ideal preservative or mixture of preservatives should:

1.   Be active at low concentration
2.   Be physically and chemically compatible and remain effective over a wide range of pH
3.   Be effective against a wide range of microorganisms
4.   Be readily soluble and stable to heat and storage
5.   Be odorless, colorless, nontoxic, and nonirritating in the percentage used

Preservatives frequently used in cosmetic products are:

p-Hydroxybenzoates (parabens)	Most effective in acid pH
Dehydroacetic acid	Active against fungi
Sorbic acid and sorbates	Active against fungi, pH dependent
Imidazolidinylurea (Germall 115)	Synergistic with other preservatives
2-Bromo-2-nitro-1,3-propanediol (Bronopol)	Active against gram negative bacteria; releases formaldehyde
Formaldehyde	Strong antibacterial
1-(3-Chloroallyl)-3,5,7-triaza-1-azoniaadamantane chloride (Dowicil 200	Formaldehyde donor

Nonionic emulsifiers are commonly used in cosmetic formulations. When used in a concentration below 1% they do not significantly interfere with preservative action. In higher concentrations the nonionic emulsifier can form aggregates which can extract preservatives out of the oil–water phases, leaving a low concentration of preservative to distribute itself between the oil and water phases.

Lipid soluble preservatives may migrate into plastic container walls and into plastic cap liners.

Some components of a cosmetic formulation can enhance preservative action. The ingredients of many perfume oils have antimicrobial properties.

Synergisms of preservative mixtures result in increased activity. Mixtures of parabens are more effective than individual parabens, and mixtures of parabens and Germall 115 are significantly more active and widen the range of microorganisms against which the cosmetic product is protected.

Selection of a preservative or preservative system requires:

1. The preservative system must be compatible with the components of the product, including the color and fragrance
2. Microbial challenge testing of the finished cosmetic is necessary (see Table 1)
3. The antimicrobial should be relatively nontoxic, nonirritating, and non-sensitizing
4. The antimicrobial should be water soluble, have a limited tendency to migrate into the nonaqueous phase, and be stable in storage [60–62]

Antioxidants [59]

An antioxidant, in brief, is a substance capable of slowing the rate of oxidation of autoxidizable materials. It should:

1. Be soluble in both water and oil
2. Be odorless, nontoxic, nonallergenic, and effective in low concentration
3. Be stable under conditions of use

Phenolic-type antioxidants are usually soluble in oil and generally insoluble in water. Propyl gallate has a solubility of about 1.8% in water. The addition of propyl gallate requires a chelating agent to prevent reactions with iron contaminants which will cause color changes. Calcium disodium ethylene-diaminetetraacetic acid is slightly soluble in oil. The most widely used antioxidants are butylated hydroxyanisole (BHA), butylated hydroxytoluene (BHT), propyl gallate (PG) and nordihydroguaiaretic acid (NDGA). A widely used antioxidant mixture contains 20% BHA, 6% PG, 4% citric acid, and 70% propylene glycol.

Terminology, Toxicity, Test Methods, Etc.

Acute toxicity refers to the capacity of a material to produce serious injury or death following the ingestion of a single dose or single exposure.

TABLE 1 Methods of Challenge Test Proposed by USP XVIII, CTFA, and SCC of Great Britain

	USP XVIII	CTFA	SCC of Great Britain
Organisms used	*Candida albicans* ATCC 10231 *Aspergillus niger* ATCC 16404 *Escherichia coli* ATCC 8739 *Staphylococcus aureus* ATCC 6538 *Pseudomonas aeruginosa* ATCC 9027	*S. aureus* ATCC 6538 *E. coli* *P. aeruginosa* 15442 or 13388 *C. albicans* *A. niger* 9642 *P. luteum* 9644 *B. subtilis*	Vigorous strains of factory or lab product contaminants *E. coli, A. aerogenes, Proteus morganii, P. aeruginosa, P. fluorescens, S. aureus, S. epidermidis, Streptococcus faecalis,* fungi, and yeast
Inoculum	0.1 ml/20 ml; 1.25×10^5 to 5×10^5 cells/ml of product	10^6 cells/ml of product (20 ml)	Check growth in unpreserved product, 0.1 ml/30 ml; 10^5–10^7 cells/ml of product
Control	Viability of cells in nonpreserved product; 30–32°C	Cells in nonpreserved product: 32–37°C for bacteria; 25–30°C for fungi	Cells in unpreserved product: assay natural contaminants
Sampling	At least two observations, 7 d apart, during 28-d test period	0, 1–2, 7, 14, 28 d	Toothpaste: 0, 1, 2, 3, 4 weeks, 28°C Shampoo: 0, 1, 2, 3, 4 weeks, 22, 30°C Creams and lotions: 0, 1, 2, 4, 8, weeks, 22, 32, 37°C Eye cosmetics: 0, 1, 2, 4 weeks, 30, 37°C; surfaces, saliva
Effectiveness	Vegetative cells: 0.1% survival at 2 sampling periods 7 d apart. *C. albicans* and *A. niger*: no significant increase during 28-d test period	7 d; rechallenge	Toothpaste: decrease Shampoo: no count for 2 weeks; reinoculate Creams and lotions: low count for 3 weekly samplings Eye cosmetics: kill of *P. aeruginosa*; no fungi growth: kill of saliva bacteria

Chronic toxicity refers to the ability of a material to produce mild or serious injury or death following ingestion or exposure over a prolonged period of time.

Percutaneous toxicity refers to a material which can be absorbed through the skin into the blood stream producing systemic effects (usually chronic) on internal organs, glands, and brain.

M.L.D. 50 refers to the minimum lethal dose that will kill 50% of the animals used for any one series of tests. The number of animals involved in a test of this kind will vary from a minimum of 10 to many hundreds. In this test the size of the dose is always related to the weight of the animal.

Primary cutaneous irritant is a material which has the capacity to produce a dermatitis, i.e., erythema (reddening), burning, blistering, swelling, itching, or pain after a single exposure (application or handling) at the site of contact if allowed to act in sufficient quantity for an adequate period of time.

Sensitizer is a material which after initial application produces no visible injury or irritation. However, after intermittent or continued use it can produce reactions of allergic variety following each new exposure to the material at the same site or another part of the body.

Tests are performed to determine whether the material has an acute or chronic toxicity, and, if so, the M.L.D. 50. These tests involve feeding the material to the animals which are subsequently sacrificed for close examination of the organs, glands, and brain.

Percutaneous toxicity is determined by the application of the material to cleanly shaven necks and backs of animals over a regularly scheduled period. The condition of the skin is carefully checked and the animals are observed for any signs of systemic toxicity. The animals are sacrificed at the end of the experiment and their tissues are examined for pathological changes.

In actual practice the use of animals is of prime importance in screening materials for eventual human use. This method would quickly inform us whether the test material was a primary irritant and, if so, it would be discarded as a material for human use. Animals should also be used to determine whether materials may perform as sensitizers before similar tests are performed on humans as in the prophetic patch test of Schwartz-Peck. Even though true sensitization does not occur equally in animals and humans, this test provides some measure of protection by serving as a screen. Animals which are thus sensitized with a material makes the material suspect, and so great caution is exercised before using the material on humans. On the other hand, a negative sensitizing result on animals indicates that the material is probably not a strong sensitizer and further tests on humans offers very little risk.

Animal Toxicological Studies

Safety assessments in animal studies using models and conditions which simulate use conditions in man and carried out at exaggerated dose levels to elicit possible toxic effects help to evaluate the relative safety of the product. Progress in test techniques in appropriate animal models has been made, but the animal studies are not always predictive for humans. The final tests must be clinical trials on human subjects [63–69].

Draize Eye Test

Observations are made on the cornea, iris, and conjunctivae. Numerical scores are assigned to the observed lesions. Albino rabbits are used for eye mucosa toxicity studies.

Human Safety Testing

In all cases the initial trials on humans must be on a limited number of individuals, irrespective of the completeness of the program carried out with animals.

In human safety trials with topically applied materials, the principal concern in most cases is induced cutaneous reactions. If, however, animal studies indicate that there is a significant absorption of components of the formulation, clinical trials must provide for the monitoring of such absorption and also provide for laboratory evaluations of blood and urine which may reflect subclinical systemic toxic effects.

The cutaneous reactions which may result from application of cosmetics are of three types: primary irritation, contact sensitization, and reactions which develop following exposure to the sun. Tests for the potential of topically applied materials to induce these reactions are evaluated by means of patch tests.

Sensitization Potential

One of the most critical evaluations is the determination of sensitizing potential. Two basic procedures have been most frequently employed, and numerous variations of these have been used. One of the basic procedures, that of Schwartz and Peck, has limited value since it will detect only very strong sensitizers. Present consideration of the human patch test will primarily be limited to the repeated insult procedure and its various modifications. Two variations were originally presented, one by Shelanski and Shelanski and the other by Draize. Both of these procedures employ repeated applications made over a period of 4 or 5 weeks. They differ in that 10 sensitizing applications are made in the Draize procedure and 15 are made in the Shelanski procedure. They also differ in that challenge applications are made to the original test sites only in the Draize procedure, but are made to previously unpatched sites as well as the original test site in the Shelanski procedure.

A short review of these and other procedures employing human panelists has been presented by Kligman, who suggested the maximization procedure as an alternative. Various methods for achieving the objective of Kligman's maximization procedure have been employed, an interesting one being a Hill

Top Research procedure in which the skin is abraded prior to sample application. The test materials are applied to abraded sites under patches, which are applied to the upper arms. Reactions to the test applications are scored on each Wednesday, Friday, and Monday following applications on the preceding Monday, Wednesday, and Friday. The patches are removed by the panelists 24 h after application. Nine applications are made in three successive weeks, and the challenge application is made on Monday of the sixth week. The challenge applications are made to previously unpatched sites. Abrasions of the test sites are made just prior to sample applications on test days 1, 4, 7, and 11, or preferably, prior to each application. This more frequent scheduling of abrading can be done when up to four materials are evaluated on abraded skin only. The abrasions are in the form of two concentric circles with diameters of $\frac{1}{2}$ and $\frac{3}{8}$ in. and are inflicted by a specially constructed apparatus. The skin areas are cleansed with isopropyl alcohol just prior to abrading. A separate sterile abrader is used for each panelist. All adhesive patches and absorbent pads applied over abraded areas are sterilized by exposure to ethylene oxide prior to use. The usual grading scale is utilized.

Another procedure employed is to strip the skin by repeated application and removal of adhesive tape from the test sites prior to sample applications.

All of these procedures have value in certain applications. The one chosen should provide the required degree of sensitivity as well as data from which valid decisions can be derived.

Accelerated Shelf Testing [27]

After a cosmetic product has been carefully formulated and the manufacturing procedures worked out, its stability characteristics must be determined before marketing. To be sure it will stand up for at least 1 or 2 years in storage, it should be kept under storage conditions and observed for that length of time. This is obviously impractical. Moreover, small changes in formula or procedure, after marketing, would again require long delay, which is unacceptable.

Accelerated tests provide a means for shortening the testing time. Usual cosmetic stability problems are only concerned with shelf life under ordinary storage conditions, so that the only variable is time, which cannot be telescoped. Therefore, the conditions of storage have to be changed to severe extremes so that deterioration of the product occurs within a reasonably short time. This can be done by raising the temperature, since all chemical reactions are accelerated at elevated temperatures. To evaluate the test, the degree of speed-up or acceleration at this higher temperature, compared with ordinary storage temperature, must be determined.

According to an old rule of thumb, a rise of $10°C$ doubles the reaction rate. Although this rule has a sound theoretical basis and is valid in a number of simple cases, it is not directly applicable to more complex cases such as cosmetic products. This inability to correlate the degree of acceleration or exaggeration in tests at elevated temperatures is recognized in the industry.

The uncertainty of assessing the acceleration "exaggeration" at the higher temperatures is due to the fact that some preparations behave differently than others. By lumping them all together only a vague guess is possible. However, some preparations will yield to exact analysis. Others may not, but in the former case there will be almost no risk involved in predicting shelf life, and only the cases which are not easy to analyze will be subject to the usual guess work. There is occasionally an inherent difficulty in that the cosmetic preparation is in a different condition at higher temperatures. In such cases an accelerated test is not possible at all because the test is actually done on a different preparation than the one intended. An obvious example would be the testing of a product in which a suspending agent coagulates or in which the components become immiscible. Conversely, the stability of many seemingly complex systems may be governed by one predominant factor. Thus it is usual for diffusion at an interface to dominate the picture, and its temperature dependence can be established without undue difficulty. Therefore, a complex system of several compounds in two phases may present a simple problem on analysis.

Cosmetic Ingredient Review

The Cosmetic, Toiletry, and Fragrance Association has set up a Cosmetic Ingredient Review (CIR) financed and directed by the association. The operation is independent of the association to avoid any charge of bias and an expert panel of scientists supervises the evaluation of the safety of cosmetic ingredients. Reviews reflect information gathered from published scientific literature as well as data developed through test procedures.

To assign priority ranking, CIR began with the 2800 ingredients listed in *CTFA Cosmetic Ingredient Dictionary*, deferred about 700 items that are under review by the Food and Drug Administration or other programs, and selected from the remainder those that are used in 25 or more cosmetic product formulations. Seven criteria then were used by the panel for priority weighting of the 189 substances remaining: frequency of occurrence (number of formulations), use in high concentrations, area of normal use, frequency of application, use by particular groups (infants and elderly, for example), suggestions of biological activity, and frequency of consumer complaints about product categories containing the ingredient being reviewed.

Consumer Expenditures

Consumer expenditures for health and beauty aids for 1977 (originally published in *Product Marketing*) are presented in Tables 2 and 3.

TABLE 2 Consumer Expenditures for Cosmetics and Toiletries[a]

	Retail Sales, $ (add 000)			Percent Change, 1977 vs	
Item	1977	1976	1973	1976	1973
Oral hygiene products[b]	1,300,800	1,142,600	945,330	13.8	37.6
Dentifrices	575,250	544,830	423,890		
Toothpaste	571,680	541,030	419,800		
Toothpowder	3,570	3,800	4,090		
Denture products[b]	220,420	116,870	91,480		
Denture cleansers	121,020	52,560	38,680		
Denture adhesives	91,920	56,470	46,090		
Denture brushes	7,480	7,840	6,710		
Other oral hygiene	505,130	480,900	429,960		
Toothbrushes, nonelectric	127,170	107,660	98,050		
Electric toothbrushes	20,050	15,250	15,380		
Oral lavages	38,480	27,840	25,670		
Dental floss	5,020	4,660	3,640		
Mouthwashes, gargles	268,140	278,310	252,750		
Breath fresheners[c]	46,270	47,180	34,470		
Hair preparations[d]	1,759,910	1,066,150	1,405,820	65.1	25.2
Shampoos[e]	732,380	693,660	501,150		
Regular[f]	580,880	544,230	365,020		
Dandruff/medicated	142,000	136,740	124,530		
Color shampoos	9,500	12,690	11,600		
Hair color preparations	364,270	335,150	255,150		
Hair color rinses	46,050	49,630	46,420		
Hair tints/dyes	318,220	285,520	208,730		
Hair medications[g]	13,360	25,580	13,140		
Women's hair dressings	155,670	136,970	102,920		
Hair straighteners and other dressings[e]	35,560	32,170	26,280		
Cream rinse	105,460	85,150	68,350		
Depilatories	18,480	18,220	15,450		
Waveset preparations	14,030	14,650	13,350		
Women's hairspray	180,570	279,160	263,830		
Home permanent kits	57,230	55,720	52,340		
Men's hair dressings	82,900	83,380	93,860		
Aerosol	13,430	25,730	35,160		
Nonaerosol	69,470	57,650	58,700		
Cosmetics and accessories	1,924,900	1,620,160	1,231,640	18.8	56.3
Face creams	404,040	371,180	260,730		
Makeup preparation/accessories	1,520,860	1,248,980	970,910		
Makeup bases	77,070	75,470	70,410		
Cake	9,180	9,620	9,930		
Lotion	44,480	43,550	41,840		
Cream	23,410	22,300	18,640		
Pressed cake powder/compact[h]	100,360	162,160	140,620		
Loose face powder/powder puffs	37,930	39,490	42,410		

(*continued*)

TABLE 2 (*continued*)

Item	Retail Sales, $ (add 000)			Percent Change, 1977 vs	
	1977	1976	1973	1976	1973
Blush makeup[e]	72,920	—	—		
Brush	30,520	—	—		
Nonbrush	42,400	—	—		
Rouge	9,250	10,060	10,470		
Face lotions/astringents	70,250	63,500	48,460		
Liquid facial cleaners	85,370	81,180	64,100		
Talcum/body powder[i]	59,690	59,860	60,120		
Lipsticks	552,400	458,450	318,540		
Lip glosses[j]	135,800	—	—		
Eye makeup	319,820	298,810	215,780		
Mascara	56,110	52,430	45,380		
Eyebrow pencil	29,650	25,300	20,780		
Eyeshadow	43,620	41,390	32,420		
Other	190,440	179,690	117,200		
Shaving preparations	706,680	632,040	473,910	11.8	49.1
Aerosol shave cream	124,150	114,570	92,400		
Other shave cream	11,270	9,660	10,140		
Shaving soaps/sticks	6,040	5,790	5,690		
Aftershave lotions	179,860	159,570	129,680		
Preshave products	23,170	21,080	18,440		
Men's cologne	188,160	181,690	103,710		
Men's talcum	3,430	3,410	3,310		
Styptics	2,200	2,120	1,660		
Men's packaged toiletry sets	168,400	134,150	108,880		
Women's fragrances[k]	621,140	616,300	518,360	0.8	19.8
Perfumes	113,390	107,030	83,020		
Aerosol	14,320	14,100	12,500		
Nonaerosol	99,070	92,930	70,520		
Toilet water/cologne	426,690	427,210	363,570		
Aerosol cologne	306,500	311,840	274,880		
Nonaerosol	120,190	115,370	88,690		
Bubble bath	44,900	44,490	35,280		
Other bath products	31,530	33,190	32,640		
Purse atomizers	4,630	4,380	3,850		
Hand products	369,000	300,760	217,830	22.7	69.4
Hand lotions[l]	107,230	87,720	66,740		
Hand creams	21,440	19,770	18,910		
Nail polish/enamel	198,240	155,990	104,830		
Nail enamel removers	33,850	29,640	20,870		
Cuticle softeners	8,240	7,640	6,480		

(*continued*)

TABLE 2   (*continued*)

Item	Retail Sales, $ (add 000)			Percent Change, 1977 vs	
	1977	1976	1973	1976	1973
Personal cleanliness items	1,334,870	1,186,570	934,690	12.5	42.8
Bath soaps[m]	722,270	578,170	381,600		
Deodorant soaps	381,590	293,470	205,460		
Other bath soaps	340,680	284,700	176,140		
External personal deodorants	612,600	608,400	553,090		
Creams	27,490	27,980	24,500		
Liquid/squeeze containers	12,910	12,640	12,990		
Sticks	20,200	14,650	8,970		
Roll-ons	182,090	121,870	38,270		
Pads	1,590	1,550	1,530		
Feminine spray deodorants	14,460	17,270	30,600		
Other aerosols	352,770	411,450	434,990		
Powder	1,090	990	600		

[a]Copyright © 1978 by Charleson Publishing Co., New York.
[b]Data base revised to correct for prior over- or understatement.
[c]Excludes gums.
[d]Excludes products sold for professional use in beauty parlors and barber shops.
[e]Category revised in 1977 to show additional detail.
[f]Includes "baby" shampoos.
[g]Includes products promoted primarily for treatment of dandruff or scalp conditions.
[h]Blush makeup included in 1976 figure.
[i]Excludes aftershave powder and baby powder.
[j]New catagory added.
[k]Does not include door-to-door sales.
[l]Includes products promoted for hand and body.
[m]Excludes medicated soaps.

TABLE 3   Summary of Consumer Expenditures for Cosmetics and Toiletries[a]

Item	Retail Sales, $ (add 000)			Percent Change, 1977 vs	
	1977	1976	1973	1976	1973
Oral hygiene products[b]	1,300,800	1,142,600	945,330	13.8	37.6
Dentifrices	575,250	544,830	423,890	5.6	35.7
Denture products[b]	220,420	116,870	91,480	88.6	140.9
Other oral hygiene products[c]	505,130	480,900	429,960	5.0	17.5
Hair preparations[d-g]	1,759,910	1,066,150	1,405,820	65.1	25.2
Cosmetics and accessories[d,h]	1,924,900	1,620,160	1,231,640	18.8	56.3
Shaving preparations	706,680	632,040	473,910	11.8	49.1
Women's fragrances[i]	621,140	616,300	518,360	0.8	19.8
Hand products[j]	369,000	300,760	217,830	22.7	69.4

(*continued*)

TABLE 3 (*continued*)

Item	Retail Sales, $ (add 000)			Percent Change, 1977 vs	
	1977	1976	1973	1976	1973
Personal cleanliness items[k]	1,334,870	1,186,570	934,690	12.5	42.8
Bath soaps and detergents	722,270	578,170	381,600	24.9	89.3
Personal deodorants	612,600	608,400	553,090	0.7	10.8
Total for cosmetics and toiletries listed	8,017,300	6,564,580	5,727,580	22.1	40.0

[a]Copyright © 1978 by Charleson Publishing Co., New York.
[b]Data base revised to correct for prior over- or understatement.
[c]Excludes gums.
[d]Category revised in 1977 to show additional detail.
[e]Excludes products sold for professional use in beauty parlors and barber shops.
[f]Includes "baby" shampoos.
[g]Includes products promoted primarily for treatment of dandruff or scalp conditions.
[h]Excludes aftershave powder and baby powder.
[i]Does not include door-to-door sales.
[j]Includes products promoted for hand and body.
[k]Excludes medicated soaps.

Bibliography

Texts

Association of Official Agricultural Chemists, *Official Methods of Analysis*, 9th ed., Washington, D.C., 1960.

Balsam, M. S., and Sagarin, E., *Cosmetics—Science and Technology*, 2nd ed., Vols. 1 and 2, Wiley-Interscience, New York, 1972.

de Navarre, M. G., *The Chemistry and Manufacture of Cosmetics*, Van Nostrand, Princeton, New Jersey, 1st ed., 1 vol., 1941; 2nd ed., Vols. 1 and 2, 1962, Vols. 3 and 4, 1975.

Harry, R. G., *Modern Cosmeticology* (Vol. 1 of *Principles and Practice of Modern Cosmetics*, revised by J. B. Wilkinson), Chemical Publishing Co., New York, 1962.

Harry, R. G., *Cosmetic Materials* (Vol. 2 of *Principles and Practice of Modern Cosmetics*, revised by W. W. Myddleton), Chemical Publishing Co., New York, 1963.

Keithler, W. R., *The Formulation of Cosmetics and Cosmetic Specialties*, Drug and Cosmetic Industry, New York, 1956.

Sagarin, E. (ed.), *Cosmetics—Science and Technology*, Interscience, New York, 1957.

Wells, F. V., and Lubowe, I. I., *Cosmetics and the Skin*, Reinhold, New York, 1964.

Journals

Aerosol Age, Industry Publications, Inc., Cedar Grove, New Jersey 07009.
Archives of Dermatology, American Medical Association, Chicago 60610.

Beauty Fashion, Beauty Fashion Inc., New York 10017.
Chemical and Engineering News, American Chemical Society, Washington, D.C. 20036.
Chemical Week, McGraw-Hill, New York 10020.
Cosmetics and Toiletries, Allured Publishing Corp., Wheaton, Illinois 60187.
CTFA Cosmetic Journal, Cosmetic, Toiletry, and Fragrance Association, Washington, D.C. 20006.
Drug & Cosmetic Industry, Harcourt Brace Jovanovich, New York 10017.
Household & Personal Products Industry, Rodman Publishing Corp., Ramsay, New Jersey 07446.
Journal of the American Pharmaceutical Association, American Pharmaceutical Association, Washington, D.C. 20037.
Journal of the Society of Cosmetic Chemists, Society of Cosmetic Chemists, New York 10017.
Parfümerie Kosmetik, Postfach 10 28 69, Heidelberg, Germany.
Soap, Cosmetics, Chemical Specialties, MacNair-Dorland Co., New York 10001.
Soap, Perfumery & Cosmetics, United Trade Press Ltd., London, W1, England.

References

1. F. E. Wall, "Origin and Development of Cosmetic Science and Technology," in *Cosmetics—Science and Technology*, Interscience, New York, 1957, Chap. 2.
2. M. G. de Navarre, *International Encyclopedia of Cosmetic Material Trade Names*, Moore, New York, 1957.
3. L. A. Greenburg and D. Lester, *Handbook of Cosmetic Materials*, Interscience, New York, 1954.
4. Cosmetic, Toiletry and Fragrance Association, Inc., *Standards*, New York, 1948.
5. F. E. Wall, *The Principles and Practices of Beauty Culture*, 2nd ed., Keystone, New York, 1946.
6. Federal Food, Drug, and Cosmetic Act, as amended (F.D.C. Act, October 1976, revision), Food and Drug Administration, Washington, D.C.
7. J. F. L. Chester, "The Function of Cosmetic Components," *J. Soap, Perfum. Cosmet.*, *46*, 205–209 (April 1973).
8. P. Carter, "Basic Emulsion Technology," *Am. Perfum. Aromat.*, *1*, 233 (1960).
9. P. Carter, "Application of Emulsification Theories to Cosmetic Formulation," *J. Am. Perfum. Cosmet.*, *77*(4), 10–18 (July 1962).
10. G. Barnett, "Emollient Creams and Lotions," in *Cosmetics—Science and Technology*, Vol. 1, Wiley-Interscience, New York, 1972, Chap. 2, pp. 27–104.
10a. W. H. Mueller and R. P. Quatrale, "Antiperspirants and Deodorants," in *The Chemistry and Manufacture of Cosmetics*, Vol. 3, Van Nostrand, Princeton, New Jersey, 1975, pp. 205–228.
11. A. B. G. Lansdown, "Aluminum Compounds in the Cosmetic Industry," *J. Soap, Perfum. Cosmet.*, *47*, 209–212 (May 1974).
12. S. I. Kreps, "The Structure, Function and Formulation of Topical Sunscreens," *J. Soc. Cosmet. Chem.*, *14*(12), 625–630 (1963).
13. S. I. Kreps, "Sun Burn Protection and Sun Tan Preparations," *J. Am. Perfum. Cosmet.*, *78*, 73 (October 1963).
14. E. G. Klarman, "Suntan Preparations," in *Cosmetics—Science and Technology*, Interscience, New York, 1957, Chap. 8, pp. 189–212.

15. E. I. du Pont de Nemours & Co., *Package for Profit* (*Freon*), 1960.
16. J. R. Frauenheim, "Alternative Liquified Gas Propellants," *Aerosol Age*, *20*(12), 23–27 (1975).
17. M. A. Johnsen, "Business Strategies for the Future," *Aerosol Age*, *20*(12), 36–40 (1975).
18. C. S. Hayes, "Carbon Dioxide," *Aerosol Age*, *20*(6), 34–36 (1975).
19. P. A. Sanders, *Principles of Aerosol Technology*, Van Nostrand Reinhold, Princeton, New Jersey, 1970.
20. M. A. Johnsen, W. E. Dorland, and E. K. Dorland, *The Aerosol Handbook* Dorland, New York, 1972.
21. Chemical Specialties Manufacturers Association, *Glossary of Terms Used in the Aerosol Industry*, August 25, 1955.
22. Chemical Specialties Manufacturers Association, *Bulletin No. 211–66*, December 27, 1966.
23. W. H. Peacock, *The Application Properties of the Certified Coal Tar Colors* (Calco Technical Bulletin No. 715), Calco Chemical Division, American Cyanamid Co., December 1944.
24. D. Nickerson, *Color Measurement*, U.S. Department of Agriculture, March 1946.
25. W. H. Peacock, *The Practical Art of Color Matching* (Calco Technical Bulletin No. 573), Calco Chemical Division, American Cyanamid Co., 1948.
26. S. Zuckerman, "Colors for Foods, Drugs and Cosmetics," in *Encyclopedia of Chemical Technology*, Vol. 4, Interscience, New York, 1949, pp. 287–313.
27. H. Isacoff, "Cosmetics," in *Encyclopedia of Chemical Technology*, Vol. 6, Wiley-Interscience, New York, 1965, pp. 346–375.
28. H. Hilfer, "New Developments in Lipsticks," *Drug Cosmet. Ind.*, *65*, 518 (1949).
29. H. Bishop, "A Method for the Semiquantitative Analysis of Lipstick," *J. Soc. Cosmet. Chem.*, *5*, 2 (1954).
30. P. G. Lauffer, "Lipsticks," in *Cosmetics—Science and Technology*, Interscience New York, 1957, Chap. 13.
31. M. G. de Navarre, *The Chemistry and Manufacture of Cosmetics*, Van Nostrand, New York, 1941, p. 371.
32. H. Hilfer, "Mascara and Eye Makeup," *Drug Cosmet. Ind.*, *74*, 678 (1954).
33. P. Rutkin, "Eye Make-Up," in *The Chemistry and Manufacture of Cosmetics*, Continental Press, Orlando, Florida, 1975, pp. 709–740.
34. H. J. Wing, "Nail Preparations," in *The Chemistry and Manufacture of Cosmetics*, Continental Press, Orlando, Florida, 1975, pp. 983–1010.
35. P. Alexander, "Nail Lacquer," *J. Soap, Perfum. Cosmet.*, *48*, 153–159 (April 1975).
36. J. Peirano, *Cosmetics—Science and Technology*, Interscience, New York, 1957, p. 678.
37. D. Y. Hsiung, "Hair Straightening," in *The Chemistry and Manufacture of Cosmetics*, Van Nostrand, Princeton, New Jersey, 1975, pp. 1155–1166.
38. E. A. Taylor, "Cutaneous Adsorption of Bath Oils," *Arch. Dermatol.*, *87*, 137–139 (March 1963).
39. Anonymous, "The Formulation of Bath Products," *J. Soap, Perfum. Cosmet.*, *48*, 437–441 (October 1975).
40. A. P. R. James, "Bath Oils in the Management of Dry, Puritic Skin," *J. Am. Geriatr. Soc.*, *9*(5), 367–369 (1961).
41. E. Jungermann, "Antibacterial Soaps and Normal Skin Flora," *J. Cosmet. Toiletries*, *91*, 50–58 (July 1976).
42. D. Osteroth, "Toilet Soaps: Practical Characteristics and Test Methods," *J. Soap*

Cosmet. Chem. Spec., *53*, 29 (April 1977).

43. E. T. Webb, "Transparent Soaps," *J. Soap, Perfum. Cosmet.*, *31*, 770 (August 1958).
44. R. C. Reald, *Germicidal Detergent Bar* (Norda Briefs, No. 447), March 1973.
45. C. Jones, "The Chemistry of Shaving," *Nucleus* (*Boston*), 3–5 (June 1976).
46. J. R. Hart and E. F. Levy, "Creme Rinse Shampoos," *J. Soap Cosmet. Chem. Spec.*, *53*, (August 1977).
47. A. J. Harris, K. C. James, and M. Powell, "Assessment of Auxiliary Detergents in Shampoo Mixtures," *J. Cosmet. Perfum.*, *90*, 23 (October 1975).
48. F. V. Wells and I. I. Lubowe, "Shampoos," in *Cosmetics and the Skin*, Reinhold, New York, 1964, pp. 397–415.
49. L. R. Smith and D. C. Zajac, "The Formulation of Low pH Shampoos with Amine Oxides," *J. Household Pers. Prod. Ind.*, *13*, 34 (March 1976).
50. L. R. Smith and M. Weinstein, "Formulating Dandruff Shampoos," *J. Household Pers. Prod. Ind.*, *14*, 54 (October 1977).
51. R. L. Goldemberg, "Amphoteric Shampoos," *Drug Cosmet. Ind.*, *121*, 26 (January 1976).
52. W. R. Markland, *Low Eye Irritation Shampoo Systems* (Norda Briefs, No. 479), March 1977.
53. S. D. Gershon, H. H. Pokras, and T. H. Rider, "Dentifrices," in *Cosmetics—Science and Technology*, Interscience, New York, 1957, Chap. 15.
54. H. W. Zussman, *Proc. Sci., Sect., Toilet Goods Assoc.*, *19*, (1953).
56. F. V. Wells and I. I. Lubowe, "Preservatives and Antioxidants," in *Cosmetics and the Skin*, Reinhold, New York, 1964, pp. 586–598.
57. M. Yanagi, "Microbiology," in *The Chemistry and Manufacture of Cosmetics*, Continental Press, Orlando, Florida, 1975, pp. 67–84.
58. W. E. Rosen and P. A. Berke, "Modern Concepts of Cosmetic Preservation," *J. Soc. Cosmet. Chem.*, *24*, 663–675 (September 1973).
59. L. Chalmers, "Antioxidants," *J. Soap, Perfum. Cosmet.*, *44*, 29–38 (January 1971).
60. I. R. Gucklhorn, "Antimicrobials in Cosmetics," *TGA Cosmet. J.*, *1*, 15–32 (Fall 1969).
61. E. M. Owen, "A Method for the Evaluation of Preservative Systems in Cosmetic Formulations," *TGA Cosmet. J.*, 1(3), 12–15 (1969).
62. A. Marinaro, "Good Microbiological Practices for Aqueous Cosmetics and Toiletries," *TGA Cosmet. J.*, 1(3), 16–18 (1969).
63. R. A. Quisno, G. L. Quisno, and P. A. Majors, "Minimal Validation of Product Claims," *J. Cosmet. Perfum.*, *90*, 44 (February 1975).
64. M. J. Thomas and P. A. Majors, "Animal, Human and Microbiological Safety Testing of Cosmetics Products," *J. Soc. Cosmet. Chem.*, *24*, 140–144 (1973).
65. L. Schwartz and S. M. Peck, The Patch Test in Contact Dermatitis, *Public Health Rep. 59*, 2 (1944).
66. H. A. Shelanski and M. V. Shelanski, "A New Technique of Human Patch Test," *Proc. Sci. Sect., Toilet Goods Assoc.*, *19*, 46–49 (1953).
67. J. H. Draize, *Dermal Toxicity, Appraisal of the Safety of Chemicals in Foods, Drugs and Cosmetics*, The Staff of the Division of Pharmacology of the Federal Food and Drug Administration, The Editorial Committee of the Association of Food and Drug Officials of the United States, Austin, Texas, 1959, p. 5.
68. A. M. Kligman, "The Identification of Contact Allergens by Human Assay," *J. Invest. Dermatol.*, *47*, 369–374 (1966).
69. B. Magnusson and A. M. Kligman, "The Guinea-Pig Maximization Test," *J. Invest. Dermatol.*, *52*, 268 (1964).

Patents

70. U.S. 2,787,274 (April 2, 1957), V. A. Gant and H. I. Hersh (Hersh), "Ammonium Polysiloxanolate Hair Treatment."
71. U.S. 2,927,055 (March 1, 1960), M. Lanzet (Airkem), "Air Treatment Gels."
72. U.S. 2,976,217 (March 21, 1961), S. I. Kreps (Van Dyk Co.), "Sunscreen Agents."
73. U.S. 3,341,419 (September 12, 1967), H. J. Eiermann and I. Rappaport (Shulton), "Sunscreen Compositions."
74. U.S. 2,814,585 (November 26, 1957), E. W. Daley (Procter & Gamble), "Buffered Antiperspirant Compound."
75. U.S. 2,854,382 (September 30, 1958), M. Grad (Procter & Gamble), "Zirconyl Antiperspirants."
76. U.S. 2,820,768 (January 21, 1958), L. E. G. H. Fromont (Molenbeek St. Jean, Belgium), "Transparent Soap."
77. U.S. 3,072,536 (January 8, 1963), D. J. Pye (Dow Chemical Co.), "Acrylamide Shave Lotions."
78. U.S. 3,313,734 (April 11, 1967), E. W. Lang and H. W. McCune (Procter & Gamble), "Conditioning Shampoos."
79. U.S. 3,057,467 (October 9, 1962), R. R. Williams (Colgate Palmolive), "Disposable Applicator Wipes."
80. U.S. 3,634,259 (January 11, 1972), E. C. Evans (Kimberly Clark Corp.), "Cleansing Lotions—Hygiene."
81. U.S. 3,662,059 (May 9, 1972), W. Wiesner and M. Pader (Lever Bros.), "Dentifrice Compositions."
82. U.S. 3,148,125 (September 8, 1964), S. J. Strianse and M. Havass (Yardley), "Lipstick."
83. U.S. 3,001,949 (September 26, 1961), K. R. Hansen (Colgate Palmolive), "Acrylamide Shampoos."

HARRY ISACOFF

Cosmetics, Japan

The protracted economic recession following the 1973 oil crisis greatly discouraged equipment investments in all Japanese industries as well as consumer spending. In spite of this, however, the cosmetics industry was little

TABLE 1   Deliveries of Cosmetics (unit: ¥1,000)

Year	Value	Increase over Previous Year (%)
1970	212,308,031	15.1
1971	255,170,066	20.2
1972	301,916,057	18.3
1973	367,528,242	21.7
1974	419,539,443	14.2

affected by the business slump with total sales for 1974 amounting to more than ¥419,500* million for a rise of 14.2% over the previous year.

This is probably sufficient proof that toilet goods have become one of people's daily necessities, consumed by them regardless of economic conditions. Table 1 shows sales in value for the years 1970–1974.

Table 2 shows sales in value in 1974 classified by kinds and their comparison with the results of the previous year. The largest increase ratio (23.1%) was registered for skin care products, which accounted for 40.8% of all cosmetic products sold in 1974.

Japanese women, traditionally attentive to their skin and enthusiastic about its care, buy many skin care goods, but the sudden rise in the demand for this type of product in 1974 was because various new excellent kinds of creams and lotions were put on sale and won a favorable reception from consumers. Sales of medicated cosmetics and hair preparations went up by 20 and 8.4%,

*As of October, 1979, the value of the yen was $.004352 U.S.A.

TABLE 2   Sales of Cosmetics (classified by items) (unit: ¥1,000)

Item	Value	Rate of Increase vs Previous Year (%)	Percentage of Total
Skin care products	171,000,128	23.1	40.8
Hair preparations	91,566,181	8.4	21.8
Makeup preparations	72,077,700	−0.5	17.2
Fragrance	13,698,236	11.6	3.3
Men's toiletries	5,579,856	−28.6	1.3
Medicated cosmetics	43,523,737	20.0	10.4
Others	22,093,605	44.8	5.2
Total	419,539,443	14.2	100.0

respectively. Particularly noteworthy is the fact that sales of perfumed goods, consumed less here than in Western countries, rose by 11.6%.

On the other hand, a slight drop was reported in the sales of makeup preparations, which had markedly grown in the past. This is because demand had reached a peak. Men's toiletry goods plunged in sales for an unknown cause.

Domestic cosmetics makers are now concentrating their energies on ensuring the safety of their products. The safety of toilet goods is covered by the Pharmaceutical Law. Before producing new products, the makers are obliged to apply to the Health and Welfare Ministry for permission by submitting detailed data about the ingredients used in each product. The ministry conducts a close examination of the raw materials and, if and when their safety is confirmed, grants production and marketing permits to the makers. The makers also carry out various tests on their own to confirm the safety of their new goods before actually selling them.

Knowledge covering many scientific fields, ranging from chemistry, dermatology, and biology to toxicology and microbiology is needed for the production of cosmetics, and since most of the raw materials are chemicals, their safety must be established through detailed examinations. Tests are carried out to check their toxicity, carginogenicity, hallucinogenicity, metabolism, and mutagenicity. It is also necessary to study the results of chemical tests conducted by public laboratories, universities, and individual experts throughout the world. The Japan Cosmetic Association is planning to set up an Information Subcommittee as a lower-echelon panel of its Safety Committee to collect as much reference materials at home and abroad as possible for use by member companies.

Such information gathering can be conducted more effectively through tieups with similar overseas organizations. Cosmetics manufacturers' associations of more than 20 countries, including the United States, France, Japan, West Germany, and Britain, established the International Information Center of the Cosmetic Industry in December 1973. Member countries are required to send any important information they may obtain to its headquarters in Stockholm, and the gathered data, after being edited by the headquarters, are made available to all other member countries.

Thus the Japanese cosmetics producers are making every effort to supply better and safer goods in the hope that they will be used by consumers all over the world to make them beautiful and happy.

This material appeared in *Asahi Evening News International 1975*.

TAKEJI BABA

Cost–Capacity Relationships

Preliminary cost estimates are frequently required in the petrochemical and refining industries. Many times, these initial estimates need not be of a high order of accuracy. Several simplified techniques have been used to estimate the capital investment required for a proposed venture. With a knowledge of this capital investment and the net cash flow, one can quickly estimate the investor's rate of return and the present value of the proposal.

The difficulty of estimating the capital requirement is proportional to the desired accuracy. Peters and Timmerhaus [1] described five different types of estimates and their probable accuracies:

1. Order-of-magnitude: $\pm 30\%$. An estimate based on similar, previous cost data.
2. Study estimate: $\pm 30\%$. The cost of the major equipment items serves as the basis for this estimate.
3. Preliminary estimate: $\pm 20\%$. The estimate is based on data used in budgeting.
4. Definitive estimate: $\pm 10\%$. An estimate made just prior to the completion of drawings and specifications.
5. Detailed estimate: $\pm 5\%$. An estimate based on completed drawings and specifications.

For many situations, the least accurate estimate is sufficient to permit economic decisions.

Information about the makeup of the total capital investment is necessary—how much fixed capital and how much working capital are required? A useful rule of thumb is that the working capital initially amounts to 10 to 20% of the total capital investment. With this rule and with knowledge of the fixed capital investment, one can easily determine the rate of return for a given cash flow.

Two methods of estimating fixed capital investments are presented: (1) the order of magnitude estimate and (2) the study estimate. Investment data have been collected from various trade publications and textbooks and corrected to a CE, or Chemical Engineering, plant cost index of 199 (March 1977). The accuracy of these data is unknown.

Order-of-Magnitude Estimates of the Fixed Capital Investment

Published accounts of cost–capacity data have been used to construct logarithmic plots for chemical processes and for refining operations. These data were correlated using the expression

$$C = aS^h$$

where C = fixed capital investment
a = coefficient
S = plant size
h = size exponent

Cost data were found for 44 chemicals and for 26 refining processes. For 36 other chemicals, literature references were such that constants a and h in the above equation could not be evaluated. For these situations, h was assumed to be 0.67 and the coefficient a was calculated to fit the single, known cost–capacity citation.

Usually one would expect the slopes of the cost vs capacity plots to fall near 0.6 or 0.7. The data presented in Table 1 for chemical plants show some deviation from this expectation. However, the average size exponent h for the Table 1 values is 0.67. With this value for h, values of a were calculated for 36 other chemicals. Table 2 shows these a values.

More variation in the size exponent exists for the refining processes as shown by Table 3. This is to be expected because of varying degrees of process complexity. The average of the h values in Table 3 is 0.59.

The data of Tables 1, 2, and 3 thus allow one to estimate the fixed capital investment, at a CE plant cost index of 199, for a given size plant with a probable accuracy of $\pm 30\%$.

Study Estimates of the Fixed Capital Investment

The cost of the major equipment items serves as the basis for this estimation method. The "equipment ratio technique" is another name for this method.

The development of fixed capital requirements by this method involves:

Devising a detailed process flow diagram
Sizing the equipment items
Evaluating the installed cost of each item, which generally involves the use of some source of cost data such as Chilton [2], Peters and Timmerhaus [1], Page [3], Guthrie [4], and Mills [5]

Mathematical representations of equipment cost as a function of capacity were presented by Drayer [6] for over 125 equipment items. One of three equation forms was used to correlate the cost data:

$$Y = aX^n \tag{1}$$
$$Y = c + aX^n \tag{2}$$
$$Y = b + dX + eX^2 + fX^3 \tag{3}$$

TABLE 1 Cost Data Correlations: Chemicals, I ($C = aS^h$; C in millions of dollars, S in thousands of metric tons per year; Chemical Engineering Cost Index = 199)

Compound	Size Exponent, h	Coefficient, a
Acetic acid	0.68	0.890
Acetone	0.45	1.794
Acetylene	0.65	4.260
Acrylonitrile	0.60	3.566
Ammonia	0.58	2.362
Ammonium nitrate	0.65	0.145
Ammonium sulfate	0.73	0.066
Butadiene	0.68	1.112
Caprolactam	0.90	3.038
Carbon black	0.70	1.367
Cement	0.78	0.373
Chlorine	0.45	2.645
Cyclohexane	0.50	0.560
Ethanol	0.73	4.636
Ethylene	0.83	0.408
Ethylene glycol	0.75	1.450
Ethylene oxide	0.78	0.585
Formaldehyde	0.55	2.464
Hydrochloric acid	0.68	0.744
Hydrofluoric acid	0.70	0.887
Hydrogen	0.70	0.870
Hydrogen peroxide	0.75	0.493
Isoprene	0.55	2.464
Isopropyl alcohol	0.65	0.524
Maleic anhydride	0.82	0.866
Methanol	0.60	1.637
Oxo alcohols	0.75	1.492
Phenol	0.75	0.973
Phosphoric acid	0.60	0.589
Phthalic anhydride	0.70	0.841
Polyester fiber	0.65	2.713
Polyethylene, low density	0.65	2.611
Polypropylene	0.70	2.981
Polystyrene	0.55	1.973
Polyvinyl chloride	0.66	1.084
Propylene	0.70	0.286
Styrene	0.67	1.463
Sulfur	0.65	1.219
Sulfuric acid	0.65	0.347
Terephthalic acid	0.70	2.778
Urea	0.70	0.288
Vinyl acetate	0.65	0.188
Vinyl chloride	0.80	0.715
p-Xylene	0.55	1.782

TABLE 2   Cost Data Correlations: Chemicals, II $(C = aS^{0.67}$; C in millions of dollars, S in thousands of metric tons per year; Chemical Engineering Cost Index = 199)

Compound	Coefficient, a
Acetaldehyde	0.883
Acrylic acid	3.513
Alkylbenzene	1.141
Amino acids	7.664
Aniline	0.621
Asphalt	0.049
Beet sugar	0.357
Benzene	0.914
Butyraldehydes	1.352
Chloroacetic acid	1.075
Chloroprene	0.970
Coke	1.389
Cumene	1.208
Detergents	2.943
Dimethyl terephthalate	3.218
Dioctyl phthalates	2.271
Epichlorohydrin	1.546
Ethylhexanol	0.926
Hydrogen cyanide	2.824
Hydroxylamine	1.872
Iron oxide pigment	4.135
Lime	0.586
Magnesia (from seawater)	2.285
Melamine	1.536
Methionine	9.739
Methyl ethyl ketone	1.741
Methyl mercaptan	1.437
Nonene	0.534
Nylon 6	4.907
Polybutadiene	2.822
Propylene oxide	0.579
Proteins	3.122
Soda ash	1.466
Soybeans	0.387
Toluene diisocyanate	1.468
Uranium oxide	69.696

where Y = installed equipment cost at a known cost index
          X = equipment size or capacity
a, b, c, d, e, f, n = constants

Shown in Table 4 are values of the correlation constants for several equipment items. The calculated, installed costs are for a CE cost index of 199.

All of the mathematical expressions have been tested to insure that no unexplainable maxima or minima exist within the intended range of application. As with most empirical expressions, extrapolations made using these expressions beyond their intended ranges of applicability could give meaningless results.

Use of these equations is straightforward. One merely substitutes the value of X in the equation and solves for Y, which then in turn can be adjusted to the year in question by use of the appropriate cost indices. For example,

$$Y \text{ at } (CE = 300) = Y \text{ at } (CE = 199)\frac{300}{199} \qquad (4)$$

TABLE 3 Cost Data Correlations: Processes ($C = aS^h$; C in millions of dollars; Chemical Engineering Cost Index = 199)

Process	Size of Units		Size Exponent, h	Coefficient, a
Alkylation	M	bbl/d	0.60	2.945
Coal gasification	MM	SCF/d	0.64	28.80
Coking: Delayed	M	bbl/d	0.38	5.587
Fluid	M	bbl/d	0.42	2.670
Cracking: Catalytic	M	bbl/d	0.55	4.007
Thermal	M	bbl/d	0.70	0.620
Distillation: Atmospheric	M	bbl/d	0.90	0.270
Vacuum	M	bbl/d	0.70	0.469
Gas oil desulfurization	M	bbl/d	0.51	0.558
Gas processing	MM	SCF/d	0.42	0.637
Hydrocracking	M	bbl/d	0.52	5.638
Hydrotreating	M	bbl/d	0.65	0.502
Isomerization	M	bbl/d	0.65	1.868
Lube blending	M	bbl/d	0.52	1.728
Lube plant	M	bbl/d	0.29	24.269
Polymerization (gasolines)	M	bbl/d	0.58	0.691
Propane deasphalting	M	bbl/d	0.60	0.810
Propane dewaxing	M	bbl/d	0.45	1.176
Refinery	M	bbl/d	0.99	1.575
Reforming, catalytic	M	bbl/d	0.61	3.873
Solvent dewaxing	M	bbl/d	0.68	2.068
Steam and power	MM	W	0.48	3.300
Substitute natural gas	MM	SCF/d	0.87	1.243
Sweetening	M	bbl/d	0.65	0.195
Town gas	MM	SCF/d	0.40	3.147
Visbreaking	M	bbl/d	0.60	1.002

TABLE 4 Equipment Cost Correlations (Chemical Engineering Cost Index = 199)

Item	Description	P	S	W	Form of Correlation Equation*	a	b	c	d	e	f	n	X (Units of)	Y (Units of)	Range of Usage in Values of X
Demineralized water systems	—				1	2,582						0.85	gal/min	\$	4–250
Dryers and kilns	Drum dryers				1	5,841						0.42	ft²	\$	10–400
	Rotary dryers and kilns:														
	Hot air heated, steel				1	168						0.87	ft²	\$	100–2,000
	Flue gas heated, direct				1	0.60						1.62	ft²	\$	100–4,000
	Flue gas heated, indirect				1	207						0.95	ft²	\$	100–2,000
	Hot air heated, stainless steel				1	171						1.19	ft²	\$	80–1,000
Dryers, Spray					1	16,877						0.96	ft (diameter)	\$	9–20
Ejectors	Condensing:														
P = mmHg abs.	Two-stage:	10	14	1.15	2	42		2,534				1.13	lb dry air/h	\$	12–58
$S = \dfrac{\text{lb steam/h}}{\text{lb dry air/h}}$		20	11	0.95	2	524.7		927				0.52	lb dry air/h	\$	16–80
$W = \dfrac{\text{gal/min water}}{\text{lb dry air/h}}$		25	6.5	0.35	3		2,223		36.8	-0.274	0.00122		lb dry air/h	\$	10–175
		40	5.7	0.30	3		2,342		15.5	0.091	-0.00043		lb dry air/h	\$	10–200
		50	4.8	0.25	3		1,946		28.7	-0.140	0.00051		lb dry air/h	\$	15–200
		75	4.3	0.20	3		1,650		33.5	-0.249	0.00100		lb dry air/h	\$	15–200
		100	3.2	0.17	3		1,732		14.6	0.001	0.00003		lb dry air/h	\$	15–200
	Three-stage:	7.5	10	1.2	3		4,326		117.6	-0.714	0.00377		lb dry air/h	\$	10–100
		10	9	0.7	2	74.1		4,446				1.00	lb dry air/h	\$	10–100
		20	7	0.6	2	67.0		4,199				1.00	lb dry air/h	\$	10–100
		25	5.5	0.4	3		4,010		49.1	0.214	-0.00165		lb dry air/h	\$	10–100
		40	4.5	0.3	3		4,252		35.2	-0.043	0.00019		lb dry air/h	\$	20–200
		50	4.0	0.25	3		4,228		15.1	0.102	-0.00021		lb dry air/h	\$	20–200
	Four-stage:	1	55	9.0	2	950.5		4,715				0.84	lb dry air/h	\$	2–14
		2	21	3.5	2	593.4		4,285				0.78	lb dry air/h	\$	4.2–34
		3	15	2.4	2	344.6		4,563				0.81	lb dry air/h	\$	7–60
		4	12	1.7	2	177.6		4,576				0.95	lb dry air/h	\$	10–70
	Noncondensing:														
	Single-stage:	100	6.2	—	1	142.1						0.45	lb dry air/h	\$	14–300
		150	3.4	—	1	107.6						0.46	lb dry air/h	\$	25–550
		200	2.1	—	1	84.4						0.46	lb dry air/h	\$	42–900
		250	1.6	—	1	76.5						0.45	lb dry air/h	\$	55–1,200

Equipment	Description	Extra params	n	Cost coeff	Alt coeff	Exponent		Unit	Capacity range
	Two-stage:	300, 1.2, —	1	66.5		0.45	$	Lb dry air/h	75–1,600
		5, 145, —	1	1,165		0.40	$	Lb dry air/h	1.7–11
		10, 45, —	1	754.7		0.38	$	Lb dry air/h	6–35
Evaporators	Long tube vertical, all steel		1	201.3		0.73	$	ft^2	100–25,000
	Standard horizontal tube, cast iron body, Cu tubes, or all steel		1	1,299		0.53	$	ft^2	100–6,000
	Standard vertical tube or basket, C.I. body, Cu tubes, or all steel		1	1,515		0.53	$	ft^2	100–6,000
	Long tube vertical: C.I. body, Cu tubes		1	220.3		0.80	$	ft^2	100–25,000
	All Cu		1	1,400		0.60	$	ft^2	200–2,000
	Standard vertical tube or basket: Pb-lined body, Pb tubes		1	2.1		1.47	$	ft^2	150–3,000
	Forced circulation: C.I. body, Cu tubes		2	57.1	44,670	1.03	$	ft^2	15–6,000
	Ni C.I. body, Ni tubes		2	203.7	40,459	0.97	$	ft^2	15–6,000
Fans, blowers and com-pressors	Fans and blowers		1	2,146		0.40	$	hp	0.1–1
	Fans and blowers		2	1,339	774	0.60	$	hp	1–100
	Fans and blowers		1	832		0.72	$	hp	100–5,000
	Compressors: Centrifugal		1	1130		0.76	$	hp	100–23,000
	Reciprocating: One- and two-stage		1	899		0.84	$	hp	7.5–4,800
	Multistage		1	3,933		0.68	$	hp	10–1,000
Filters, liquid	Filter presses: Aluminum		1	476.6		0.60	$	ft^2	15–1,500
	Bronze or pressure leaf filters, 4-in. spacing		1	943		0.58	$	ft^2	15–1,500
	Cast iron or wood		1	308.9		0.58	$	ft^2	15–1,500
	Lead or pressure leaf filters, 2-in. spacing		1	632.4		0.62	$	ft^2	15–1,500
	Stainless steel		1	1,763		0.58	$	ft^2	15–1,500
	Vacuum filters continuous		1	3,879		0.43	$	ft^2	40–1,200
Furnaces	Tubular: Cr-Ni steel tubes		1	61,846		0.70	$	MM Btu/h	10–80
	Cr steel tubes		1	23,259		0.73	$	MM Btu/h	10–75
	Steel tubes		2	11,494	8,495	0.77	$	MM Btu/h	0.5–75
	Dowtherm units		1	40,395		0.36	$	MM Btu/h	0.5–3
Gas holders			1	223.1		0.59	$	ft^3	$1,000–10^6$
Gas producers		—, —	1	48,314		0.42	$	t coal/d	27–140

(continued)

TABLE 4 (*continued*)

Item	Description	Form of Correlation Equation*	Correlation Constants* a	b	c	d	e	f	n	Units of X	Units of Y	Range of Usage in Values of X
Heat exchangers	Drip coolers:											
	Steel	1	618.3						0.51	ft²	$	40–1,500
	Stainless steel	1	930						0.51	ft²	$	40–1,500
	Jacketed pipe:											
	Glass-lined	1	892						0.61	ft²	$	20–1,500
	Stainless steel	1	753						0.61	ft²	$	20–1,500
	Shell and tube:											
	Stainless-clad shell, stainless tubes	1	1,480						0.57	ft²	$	150–4,000
	Steel shell, Cu or brass tubes	1	338.3						0.58	ft²	$	20–6,000
	Steel shell, cupro-nickel tubes	1	512.2						0.59	ft²	$	20–6,000
	Steel shell, stainless tubes	1	583.0						0.59	ft²	$	20–6,000
	Steel shell, steel tubes	1	282.4						0.57	ft²	$	10–20,000
	Shell and tube calandrias:											
	Steel shell, stainless tubes	1	821						0.58	ft²	$	20–6,000
	Steel shell, steel tubes	1	407						0.59	ft²	$	20–4,000
	Waste heat boilers	1	38.1						0.91	ft²	$	1,500–15,000
Motors, electric, and motoreducers	Motors, electric	2	78.4		928				0.94	hp	$	0.9–250
	Motoreducers	2	372.1		1,324				0.64	hp	$	1–25
Pipelines, cross-country		1	1,821						1.47	in. (diameter)	$/mi	7–33
Pumps, liquid	Centrifugal, iron	3	1,814	1,631		1,166	−608.8	82.28		hp	$	0.1–1
	Centrifugal, iron	2	250.1		456.5				0.26	hp	$	1–10
	Centrifugal, iron	2	315.1		2,235				0.78	hp	$	10–100
	Centrifugal, iron	1	1,348						0.78	hp	$	100–1,000
	Centrifugal, stainless steel	2	997		1,607				0.40	hp	$	0.1–20
	Rotary, iron	2	5,598		843				0.40	hp	$	0.1–5
	Triplex reciprocating, stainless steel	1							0.60	hp	$	0.7–500
Refrigeration units	Triplex reciprocating, steel	1	2,488						0.60	hp	$	1–50
	Mechanical units:											
	10°C	1	1,823						0.68	t	$	5–500
	0°C	1	2,561						0.69	t	$	5–500
	−20°C	1	6,286						0.68	t	$	5–500
	−40°C	1	12,491						0.68	t	$	5–500
	−60°C	1	31,031						0.68	t	$	5–500

Equipment	No.	Coefficient	Exponent	Cost	Unit of capacity	Capacity range
Separators						
Steam jet units:						
20°C	2	94.3	1.03	$	t	20–1,000
10°C	2	143.2	1.03	$	t	20–1,000
Centrifugal:						
ATM suspended basket:						
Stainless steel	2	2,460	0.65	$	hp	1.5–55
Steel	2	980	0.78	$	hp	1.5–35
Bird solid bowl:						
Stainless steel	2	970	1.01	$	hp	5–200
Steel	2	760	0.96	$	hp	7.5–200
Sharples super D, stainless	2	2,832	0.88	$	hp	10–60
Cyclone:						
Stainless steel	2	16.8	1.80	$	in. (diameter)	6–120
Steel	2	10.5	1.42	$	in. (diameter)	3–30
Steel	1	4.9	1.73	$	in. (diameter)	30–100
Size reduction equipment						
Ball mills	1	2,614	0.61	$	hp	10–400
Gyratory crushers	1	3,277	0.68	$	hp	20–300
Jaw crushers or rotary crushers	1	2,410	0.59	$	hp	1–100
Roll crushers or Mikro-Pulverizers	1	2,527	0.68	$	hp	2.5–65
Rotary cutters	1	7,370	0.61	$	hp	8–50
Swing hammer mills	1	752	0.75	$	hp	7–400
Tanks						
Steel pressure:						
Chlorine	1	660	0.43	$	gal	2,000–10,000
Cylindrical	1	211.9	0.54	$	gal	100–10^5
Spheres						
25 lb/in.^{2}	1	38.4	0.67	$	gal	10,000–65 (10^4)
50 lb/in.^{2}	1	50.4	0.66	$	gal	10,000–55 (10^4)
100 lb/in.^{2}	1	104.7	0.62	$	gal	10,000–40 (10^4)
Spheriods, 15 lb/in.^{2}	1	12.4	0.74	$	gal	50,000–85 (10^4)
Storage:						
Aluminum	1	71.2	0.62	$	gal	100–50,000
Copper	1	64.5	0.61	$	gal	100–30,000
Glass-lined steel	1	441.7	0.44	$	gal	100–5,000
Lead-lined or rubber-lined steel	1	93.0	0.56	$	gal	100–10^6
Lithcote-lined steel	1	70.2	0.55	$	gal	5,000–40,000
Monel or Type 316 stainless	1	303.0	0.54	$	gal	100–4 (10^5)
Silver-lined steel	1	144.6	0.72	$	gal	100–40,000
Type 304 stainless or Monel-clad steel	1	220.0	0.54	$	gal	100–75 (10^4)
Stainless-clad steel or nickel-clad steel	1	131.8	0.57	$	gal	100–25 (10^4)
Steel	1	65.4	0.53	$	gal	100–10^6
Thickeners						
Single compartment	3	−338.7, 1247, −2.51, 0.0469	0.53	$	ft (diameter)	13–100
Two compartment	1	1,550	1.06	$	ft (diameter)	14–65

(continued)

TABLE 4 (continued)

Item	Description	Form of Correlation Equation*	Correlation Constants*							Units of		Range of Usage in Values of X
			a	b	c	d	e	f	n	X	Y	
Towers	Bubble plate tower:											
	Copper	3		187.1		27.7	0.017	0.0009		in. (diameter)	$/plate	10–150
	Stainless steel	3		223.5		41.1	0.171	0.0014		in. (diameter)	$/plate	10–150
	Packed tower:											
	Copper	3		83.7		17.7	0.122	0.0002		in. (diameter)	$/ft height	8–350
	Glass-lined steel	3		776		−15.1	1.510	−0.0103		in. (diameter)	$/ft height	18–55
	Haveg	3		166.5		8.7	0.156	−0.0003		in. (diameter)	$/ft height	20–125
	Silver-lined steel	3		886		−6.3	1.808	−0.0123		in. (diameter)	$/ft height	20–60
	Stainless steel	3		89.6		25.0	0.198	0.0003		in. (diameter)	$/ft height	3–150
	Steel	3		65.7		9.5	0.073	−0.00007		in. (diameter)	$/ft height	5–200
Vessels, agitated	Autoclaves:											
	Glass-lined	1	1,326						0.44	gal	$	100–4,000
	Stainless steel	1	8,342						0.25	gal	$	100–2,000
	Steel	1	2,200						0.33	gal	$	300–3,000
	Tanks:											
	Stainless steel	1	537.8						0.47	gal	$	100–25,000
	Steel	1	365.6						0.44	gal	$	100–25,000

*Correlation equations:　1: $Y = aX^n$
　　　　　　　　　　　　2: $Y = c + aX^n$
　　　　　　　　　　　　3: $Y = b + dX + eX^2 + fX^3$

Rate of Return Calculations

The profitability of a venture is frequently determined by an estimation of the investor's rate of return. This rate of return is defined as that rate of interest which discounts the sum of the cash flows to a present value of zero. This discounted cash flow technique takes into consideration the time value of money. The technique is discussed at length by several authors, e.g., Peters and Timmerhaus [1] and Stermole [7].

The basic mathematical expression for calculating the rate of return is

$$\text{P.V.} = 0 = \text{T.I.} - \frac{\text{N.C.F.}}{(1+i)^1} - \frac{\text{N.C.F.}}{(1+i)^2} - \cdots - \frac{\text{N.C.F.}}{(1+i)^9} - \frac{\text{N.C.F.} + \text{W.C.}}{(1+i)^{10}} \quad (5)$$

where P.V. = present value, here set equal to zero
  T.I. = total investment (working capital plus fixed capital)
  N.C.F. = net cash flow, here assumed constant for each year
  W.C. = working capital
  i = investor's rate of return, a decimal

Implied in Eq. (5) is (1) a constant net cash flow, (2) recovery of the working capital in the tenth year, and (3) a 10-year plant life.

Equation (5) can be rewritten as

$$\text{T.I.} = \left[\sum_{n=1}^{10} \frac{1}{(1+i)^n} \right] \text{N.C.F.} + \frac{1}{(1+i)^{10}} \text{W.C.} \quad (6)$$

Values of $\sum_{n=1}^{10} \frac{1}{(1+i)^n}$ and $\frac{1}{(1+i)^{10}}$ can be found in present value tables.

Present Value Calculations

The profitability of a venture is sometimes discussed in terms of its present value. In general terms, the present value of a future sum of money is that amount of money which must be invested today at a given interest rate in order to yield that future sum. The present value of an investment opportunity can be calculated using

$$\text{P.V. at } i\% = -\text{T.I.} + \text{N.C.F.} \sum_{n=1}^{10} \frac{1}{(1+i)^n} + \frac{\text{W.C.}}{(1+i)^{10}} \quad (7)$$

If the present value is some positive sum, the actual rate of return is greater than the value of i chosen to evaluate the discount factors, i.e.,

$$\sum_{n=1}^{10} \frac{1}{(1+i)^n} \quad \text{and} \quad \frac{1}{(1+i)^{10}}$$

For various values of i, Eq. (7) becomes as follows for $n = 10$ years:

$i = 0.10$:

$$\text{P.V. at } 10\% = -\text{T.I.} + 6.144567(\text{N.C.F.}) + 0.385543(\text{W.C.}) \quad (7a)$$

$i = 0.15$:

$$\text{P.V. at } 15\% = -\text{T.I.} + 5.01768(\text{N.C.F.}) + 0.247184(\text{W.C.}) \quad (7b)$$

$i = 0.20$:

$$\text{P.V. at } 20\% = -\text{T.I.} + 4.192472(\text{N.C.F.}) + 0.161505(\text{W.C.}) \quad (7c)$$

$i = 0.25$:

$$\text{P.V. at } 25\% = -\text{T.I.} + 3.570503(\text{N.C.F.}) + 0.107374(\text{W.C.}) \quad (7d)$$

$i = 0.30$

$$\text{P.V. at } 30\% = -\text{T.I.} + 3.091539(\text{N.C.F.}) + 0.072538(\text{W.C.}) \quad (7e)$$

Nomographic solutions of Equations (7a) through (7e) have been presented [6].

Sample Problems

Four example problems are presented to illustrate the utility of the estimation techniques.

Example 1. Estimation of Capital Investments. Estimate the fixed and the total capital investments required for a 200,000 t/yr styrene plant (CE index = 199). *Solution*: From Table 1, $h = 0.67$ and $a = 1.463$. Therefore,

$$C = 1.463(200)^{0.67} = 50.9 \text{ MM \$} = \text{fixed capital investment}$$

If the fixed capital investment is assumed to be 85% of the total investment, then

Total capital investment	= 59.9 MM$
Fixed capital investment	= 50.9 MM$
Working capital investment	= 9.0 MM$

Example 2.   Estimation of Net Cash Flow. A 20% rate of return is desired on the investment of Example 1. Determine the net cash flow required. *Solution*:   From Eq. (6),

$$\text{N.C.F.} = \frac{\text{T.I.} - \dfrac{1}{(1+i)^{10}}\text{W.C.}}{\displaystyle\sum_{n=1}^{10}\frac{1}{(1+i)^{10}}} = \frac{59.9 - 0.161505(9)}{4.192472} = 13.9 \text{ MM\$}$$

Example 3.   Rate of Return and Present Value. A 50M bbl/d refinery realizes an annual net profit of 10 MM\$. If the life of this refinery is 10 yr and if the annual net profit remains constant, does this operation yield an investor's rate of return of 20%? *Solution*:   From Table 3,

Fixed capital investment, C	$= 1.575S^{0.99} = 75.7$ MM\$
Total capital investment	$= 75.7/0.85 = 89.1$ MM\$
Working capital	$= 89.1 - 75.7 = 13.4$ MM\$
Net cash flow	$=$ net profit $+$ depreciation
Depreciation, straight-line	$= \dfrac{\text{fixed capital investment}}{\text{life}}$
	$= 75.7/10 = 7.57$ MM\$

Using Equation (7c),

$$\text{P.V. at } 20\% = -89.1 + 4.192472(10 + 7.57) + 0.161505(13.4)$$

$$= -13.3 \text{ MM\$}$$

So, the rate of return is *less* than 20%. In fact, for a 20% rate of return, the present value at 20% would be zero, and the net cash flow would be about 20.7 MM\$. Therefore, the net profit per year should be $20.7 - 7.57 \approx 13.1$ MM\$.

Example 4.   Errors in Capital Estimates. Consider the styrene plant of Example 1 and suppose that the fixed capital estimate was 30% low. What would be the difference in the required selling price if both were to yield a 20% rate of return? Take a total production cost of 27 MM\$ for each case.

Item	Case A	Case B
Fixed capital investment	50.9 MM\$	72.7 MM\$
Total capital investment at F.C.I./0.85	59.9 MM\$	85.5 MM\$
Working capital	9.0 MM\$	12.8 MM\$
Net cash flow	13.9 MM\$	19.9 MM\$
Depreciation at 10% F.C.I.	5.09 MM\$	7.27 MM\$
Net profit = N.C.F. − depreciation	8.81 MM\$	12.63 MM\$
Gross profit = net profit/0.52	16.94 MM\$	24.29 MM\$
Production cost (P.C.)	27.0 MM\$	27.0 MM\$
Total revenue = gross profit + P.C.	43.94 MM\$	51.29 MM\$
Required selling price $= \dfrac{\text{total revenue}}{\text{production rate}}$	\$0.100/lb	\$0.116/lb

Hence the error in product selling price caused by this error in fixed capital is 1.6¢/lb

With the data and techniques presented, one has a ready source of cost data and a means for rapidly estimating venture profitability. Even though the cost data are of unknown accuracy, useful cost estimates can be made.

This is an adaptation of an article by the author that appeared in *Petro/Chem Engineer*. The original article has much material which would be helpful to the reader. Part 1 of the article appeared in May 1970, pp 10–38. This adaptation is used with the permission of *Petro/Chem Engineer*.

References

1. M. S. Peters and K. D. Timmerhaus, *Plant Design and Economics for Chemical Engineers*, 2nd ed., McGraw-Hill, New York, 1968, p. 97.
2. C. H. Chilton (ed.), *Cost Engineering in the Process Industries*, McGraw-Hill, New York, 1960.
3. J. S. Page, *Estimator's Manual of Equipment and Installation Costs*, Gulf Publishing Co., Houston, 1963.
4. K. M. Guthrie, *Chem. Eng.*, *77*(13), 140–156 (June 15, 1970).
5. H. E. Mills, *Chem. Eng.*, *71*(6), 133–156 (March 16, 1964).
6. D. E. Drayer, *Petro/Chem Eng.*, *42*(5), 39–42 (1970).
7. F. J. Stermole, *Engineering Economy and Investment Decision Methods*, Investment Evaluations Corp., Golden, Colorado, 1974.

DENNIS E. DRAYER

Cost, Capital Estimation

The preparation of reliable capital cost estimates is an important responsibility of the project manager. A clear and specific scope definition is essential in fixing the basis for any estimate. A collection of numbers is meaningless if its basis is not clear or is inconsistent. In any decisions between alternate process arrangements, site locations, plant capacities, etc., the factors which identify the significant differences ultimately become expressed as costs.

This article is not a complete estimating manual due to the volume of special detail required; however, it is a guide to needs, techniques, and sources of information.

Estimate Scope

The scope of the estimate is the basic definition of a project in which all key parameters are identified. The cost that is developed from the scope evaluation is intended to represent the expenditure required to accomplish the project on the basis of the scope.

To obtain costs for alternative schemes of the same basic project requires specific identification of the variables and the degree to which they might change. These variables may include one or several of factors such as

1. Site location
2. Process flow scheme
3. Provision for utilities (purchased or constructed)
4. Indoor versus outdoor construction
5. Production levels

As the project narrows toward a decision, the number of alternates becomes less; however, sometimes the choices of interest to management may be, for example, several estimates involving items 1 and 5 in combination, or 2 and 5, or just 1 with all other factors fixed.

The more relevant details that are defined during the planning stage, the more meaningful an estimate should be. Time given to planning that anticipates questions by the engineers and estimators is usually well spent. Obviously, the best planning draws on the experience of those who can see through the project development and can recognize the need to answer certain cost-related questions. The detail that can be accommodated and converted into costs is a function of the time available to complete the estimate. The quality and/or accuracy of an estimate is often related to the time allowed to assemble reliable costs and to convert them in an orderly manner to a total estimate.

For some projects it is well to recognize that a 1-h factored estimate may be as good as one made in 24 h, since the 24 h still may not allow enough time to obtain worthwhile numbers. The next estimate accuracy improvement step might require 1 week. These extended situations often develop for complex projects.

Updating Cost Estimates and Projections

A single completed cost estimate is very seldom the last estimate made for the project. As the project progresses, estimates are needed for changes in scope or other extra costs approved for the project, including conditions of unusual problems or complexity during construction. Also, cost control is best exercised by using known costs from job records and adding to these cost projections (estimates looking forward) of the work required to complete the project. This cost projection is actually the estimate of work to be done to bring the project to completion. It is almost meaningless if arrived at by merely subtracting the known costs from the original estimate.

Figure 1 represents the summary of an updated detailed estimate about at the midpoint of project construction. The final cost came to within $2000 of the updated projection.

Classification of Estimates by Accuracy

The American Society of Cost Engineers has published definitions of various classes of estimates which evolved from industry practice [1]:

> *Estimate*—An evaluation of all the costs of the elements of a project or effort as defined by an agreed-upon scope. Three specific types based on degree of definition of a Process Industry Plant are:
>   1. *Order of Magnitude Estimate*—An approximate estimate made without detailed engineering data. Some examples would be: an estimate from cost–capacity curves, an estimate using scale up or down factors, and an approximate ratio estimate. It is normally expected that an estimate of this type would be accurate within $+50\%$ or -30%.
>   2. *Budget Estimate*—Budget in this case applies to the Owner's Budget and not to the Budget as a project control document. A budget estimate is prepared with the use of flow sheets, layouts, and equipment details. It is normally expected that an estimate of this type would be accurate within $+40\%$ or -15%.
>   3. *Definitive Estimate*—As the name implies, this is an estimate prepared from very defined engineering data. The engineering data includes, as a minimum, fairly complete plot plans and evaluations, piping and instrument diagrams, one line electrical diagrams equipment data sheets and quotations, structural sketches, soil data and sketches of major foundations, building sketches, and a complete set of specifications. This category of estimate covers all types from the minimum described above to the maximum definitive type which would be made from "Approved for Construction" drawings and specifications. It is expected that an estimate of this type would be accurate within $+15\%$ and -5%.

Industry terminology varies in different organizations. The order-of-magnitude estimate is also known as a factored, ratio, or approximate estimate. The budget estimate is often referred to as a preliminary estimate. The definitive estimate is identified as a detailed, firm, or take-off estimate. In this latter category the differences are small and relate to the degree of detail available for estimating or establishing costs. There are some differences between the accuracies given by the above definitions and some parts of industry practice. For the three classifications listed, the ranges of accuracy referenced to industry use are respectively (1) $\pm30\%$, (2) $\pm20\%$, and (3) ±5 to 10%.

ESTIMATE SUMMARY

PROJECT Solids Drying, Chemical ABC					TYPE OF ESTIMATE	Sheet 1 of 1
						REVISION # 1
LOCATION						JOB NO 831-X
					P ☐	DATE 7/10/70
STRUCTURE					D ☒	ESTIMATOR
						CHECKED BY

ACCT. NO.	DESCRIPTION	LABOR	CONTRACT/ SUBCONTRACT	MATERIAL	TOTAL
	Equipment: Package System, Drying			87,650	88,543
	Other Process Equipment			37,560	40,000
*	Structural Steel			9,350	14,000
*	Piping			5,000	10,000
	Foundations			4,000	4,000
*	Labor for Erection of above			24,225	29,000
	Instruments, erected			3,945	7,300
	Insulation, in place			9,000	9,000
	Electrical, L. & M.			10,600	10,600
	Painting, L. & M.			850	850
	Safety and Fire Protection			650	650
	Hoist			400	300
	TOTAL DIRECT COSTS (Battery Limits)			193,230	213,643
	Indirects for Battery Limits			26,500	26,500
	Sub-Total			219,730	240,143
*	Estimated based up increased complexity of detail, no firm bids				
	Contingency			16,000	---
	Sub-Total			235,730	240,143
	Outside Battery Limits			42,000	42,000
	GRAND TOTAL ESTIMATED COST			$ 277,730	$ 282,143

FIG. 1. Updated cost projection.

Estimate Preparation

Sources of Information for Order-of-Magnitude Estimates

Order-of-magnitude estimates are usually prepared to furnish an idea of the cost of a unit, system, plant, or facility. Often they are used for the broad-brush screening, sifting, and selection of concepts that precede deeper interest in a particular project. Normally, they are prepared with a minimum of time invested and use the information available from cost references, cost factors, cost index ratios, and generalized percentages. When several different schemes for the same project are being compared, this type of estimate is valuable in establishing the relative positions of the alternates.

Budget or Preliminary Estimate

Budget or preliminary estimates are usually prepared to establish the preliminary magnitude of a project's cost. Although as the scope is refined the cost figures will probably change, these estimates do reflect a reasonable view of the project.

Preliminary estimates are usually sufficiently accurate to allow their use with manufacturing or operating costs to determine a first look at project return on investment. Management normally accepts this type of estimate when compiling probable budget items for yearly forecasts and planning. Only in rare, urgent situations are they used to authorize release of funds for purchase and construction of a project. Under such circumstances a definitive or detailed estimate is prepared as soon as enough firm information becomes available.

Costs for the budget estimate are usually accumulated from individual equipment cost record tabulations, charts, or plotted curves of preliminary estimates or quotations from a manufacturer. Labor costs for installation may be obtained from accumulated records for similar equipment, percentage tabulations referenced to equipment costs, and experienced analysis by professional labor estimators.

Definitive or Detailed Estimates

Definitive or detailed estimates are the most accurate since they utilize the maximum amount of known or firmly established costs for equipment through firm quotations and purchase orders, plus reliable individual estimates and price book figures for the balance. In addition, layout, structural, and a majority of the piping drawings are either completed or essentially so, and these serve as a basis for actual line-by-line material take-offs. These take-offs are priced for material and labor of assembly, fabrication, and installation. Standard price book costs for welds, flanged connections, setting of concrete, etc. have been developed, and these add uniformity and reliability to the assembled figures.

Computers are used for compiling much of the estimate once the basic take-offs and listing of materials and equipment have been entered into the estimating memory of the program. From this, all excavation yardages, concrete quantities, form and reinforcing bar quantities along with the piping and its welding can be tabulated on standard printout forms [2–4]. Often these are updated by field records to compare estimated quantities against actual. Using good drawings, many contractors expect to be within 3 to 5% of their estimated figures.

Almost without exception, the owner or contractor preparing a bid will seek advice and call sources of information to establish the validity of various costs. Often the costs are a function of particular assembly, prefabrication standards for pipe and structural steel, and other methods peculiar to the circumstance. These exceptions must be recognized if the resultant assembled costs are to reflect the project requirements. The safety aspects of the construction procedures influence costs; however, there should be no compromise on this point.

Good estimates can be made better by having the various field foremen, supervisors, purchasing representative, expeditors, and inspectors review them.

The project schedule is an important factor in the efficient integration of all equipment and labor into the project. The actual availability of key equipment items often controls certain segments of the schedule for foundations, piping, electrical, and building errection.

Estimate Accounting Components

If too many diverse phases of the project are lumped into one cost estimate number, the effects of the components cannot be properly monitored for cost control. Good practice suggests separate cost estimates for

1. Project battery limits
2. Project outside battery limits
3. Project utilities (other than pipe lines)
4. Project special requirements (not in 1, 2, or 3)

Figure 2 indicates a simple cost accounting scheme tied to the classification of the project's major cost components listed above.

Cost Accounting Codes

The order-of-magnitude and budget estimates normally do not identify cost components in category or element of cost but lump equipment, piping, etc. as groups. The definitive estimate is usually assembled by cost code. The contractors and most industry engineering groups have detailed cost code identification systems for (1) feedback collection of actual field cost information and (2) collection of estimate figures and factors for preparing detailed cost estimates.

Bauman [5] suggests one form of cost code, while another is suggested in Ludwig [45]. The contractor's codes reflect considerably more subbreakdowns of the work steps such as excavation, forming, reinforcing bar, concrete, setting pumps, piping to pump, piping from pump, set electric motor, and wire motor.

In general, it is convenient to break the process areas into subcost codes to avoid intermixing unrelated costs. Well-assembled costs often have real value in later analysis. In addition, the preparation of asset records is more useful with costs separated by areas such as reaction, gas purification, drying, finishing, and

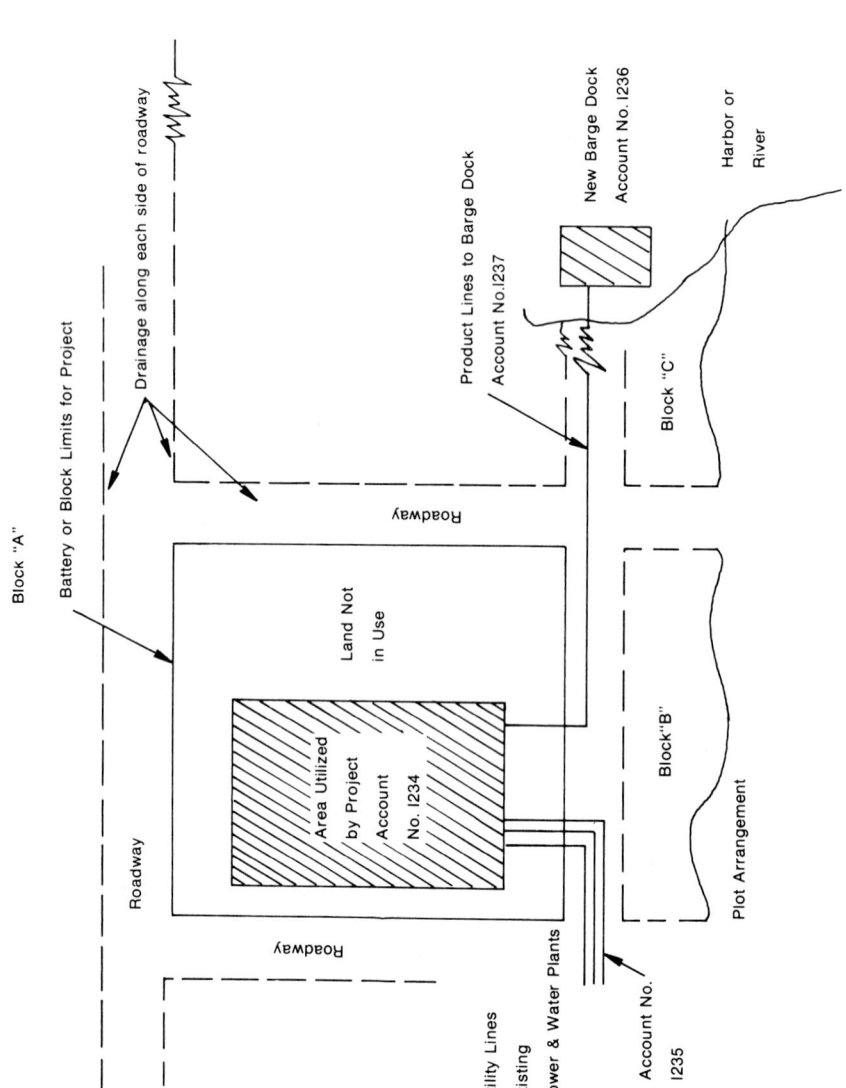

FIG. 2. Project cost estimate and cost control segments.

storage. The process system costs are then broken down into the usual categories of buildings, roads, substructures (foundations), above grade structures, process equipment, piping, miscellaneous mechanical, instrumentation, electrical, etc.

Components of a Capital Estimate

The capital cost is only one of the many components which ultimately make up the total financial obligation of a company when it decides to undertake a specific project. Its importance is defined by the magnitude of the dollars that must be committed before any return on the investment can commence. The costs that go into designing, purchasing materials, and erecting the plant are all spent before the next phase, operations for production, can begin. If the capital costs have been poorly spent, the profit and payout of the total investment may be in jeopardy.

The two primary capital cost components are termed direct and indirect.

Direct Costs

Direct costs are those which result in the permanent installation of the project's physical components and are comprised of the factors described below.

Physical Facilities Related to the Process Plant. All costs associated with the purchase of each item, including the equipment itself, delivery charges to site, taxes, insurance, and any other costs incident to a complete transaction for

1. Mechanical equipment
2. Instruments and controls
3. Electrical equipment, switchgear, etc.
4. Pipe (process, utility, instruments)
5. Insulation for pipe and equipment
6. Structures, supports for equipment, piping, etc.
7. Building for all permanent uses: office, warehouse, process maintenance, safety, fire protection, laboratory, utility, etc.

Labor. All labor for installation and erection of the items listed in the preceding paragraph.

Site Development, Auxiliary Facilities. All work required as labor and materials to establish the site; grade to elevations; establish roads, drainage, fencing, parking areas; off-site facilities such as railroad sidings, truck stations, docks; and general landscaping.

Land Purchase. New projects usually have direct land costs from purchases or allocated costs on an area basis from earlier purchases.

Office and General Facilities. Costs of office furnishings for administration and all buildings and office-related functions; shop equipment; safety (including fire protection) and first aid equipment; mobile equipment such as cars, trucks, fire trucks, lift fork trucks, ambulance; change room and locker facilities; lunch room facilities; laboratory facilities; and storage bins for parts in inventory.

Spare Parts. Spare equipment is sometimes, but not always, installed as an "installed spare"; other parts are purchased and placed in inventory to be used when a breakdown occurs.

Service Utilities. All steam, air, water, power, gas, fuel oil, waste disposal, or other facilities required to complete the needs for an operating plant. These may be on-site as independent systems or may be connected to off-site new or existing service plants or systems.

Indirect Costs

Indirect costs are primarily for *nonpermanent* facilities required to execute the project, all engineering and construction supervision, and support personnel and services required for execution of the project. They are not directly applied construction labor but include other services which are not directly process or utility systems oriented.

Engineering. All engineering functions including specifications, reproduction of drawings, estimating, etc., and their costs are charged to the project as indirect costs. If the course or scope of the project changes and some of this work is useless, it is usually segregated and charged as an ongoing operating expense of the company, rather than capitalized. Every unused study or design consideration does not come into the expense category. Only initially useful work intended to serve the project scope that has to be scrapped and new work initiated are treated as expenses.

Purchasing and Procurement. This includes the expediting, inspection, etc. as well as actual purchasing.

Construction. The set-up and operation of the temporary facilities, roads, fences, and other physical arrangements to allow the construction forces to work, plus costs for tools; rental (or owned) equipment such as cranes, welding machines, and others; safety supplies; medical costs for any treatments; personnel assigned to various construction functions such as time keeping, cost records, and accounting; purchasing; warehousing; security guards; local and state building permits and licenses; taxes, insurance; overhead fringe benefits costs for employees; and interest on money borrowed to conduct the project.

Contractor's Fee or Profit.   These costs are charged as indirect costs.

Contingency Fund.   Amounts allocated to the contingency fund are treated as indirect costs.

Collection and Assembly of Reliable Cost Data

Experienced estimators maintain well-organized files of cost data referenced to recent quotations for specific equipment. To be useful, this requires continuous updating. Labor costs are often quoted using a piece price list for different labor operations. These require careful take-offs and tabulations to obtain a particular total labor figure.

Experience indicates that within certain accuracy percentages, labor costs for a particular work function such as welding, assembly, insulation, rigging, etc. can be priced as a percent of the materials costs or of some other correlated reference base.

The computer is being used as a time and labor-saving tool for storing cost factors, data, and mathematical equations for developing details which otherwise might be too time consuming to be practical.

Updating Cost Data

As the costs of equipment, labor, and other components of the project's cost increase due to inflation or economic conditions, specific data often becomes obsolete. It can be made usable for many estimating purposes by using published cost indexes. With the ratio of indexes for year A and year B (latest), the old original cost times (Index B/Index A) gives an acceptable figure for the current year B cost. This assumes the same capacity of equipment. When the cost data is over 10 yr old, the accuracy of this technique falls off. Size or capacity changes can be adjusted by using specific equipment exponents referenced to capacity.

Cost Indexes

Several cost indexes are in common use. These record the evaluated cost base specific to the cost factors integrated into the index for a given date. Data are available in different forms and are referenced to the first of the year or to the average for the year. Besides showing the changing trend in costs and economics of a particular industry, indexes serve as a reference base for

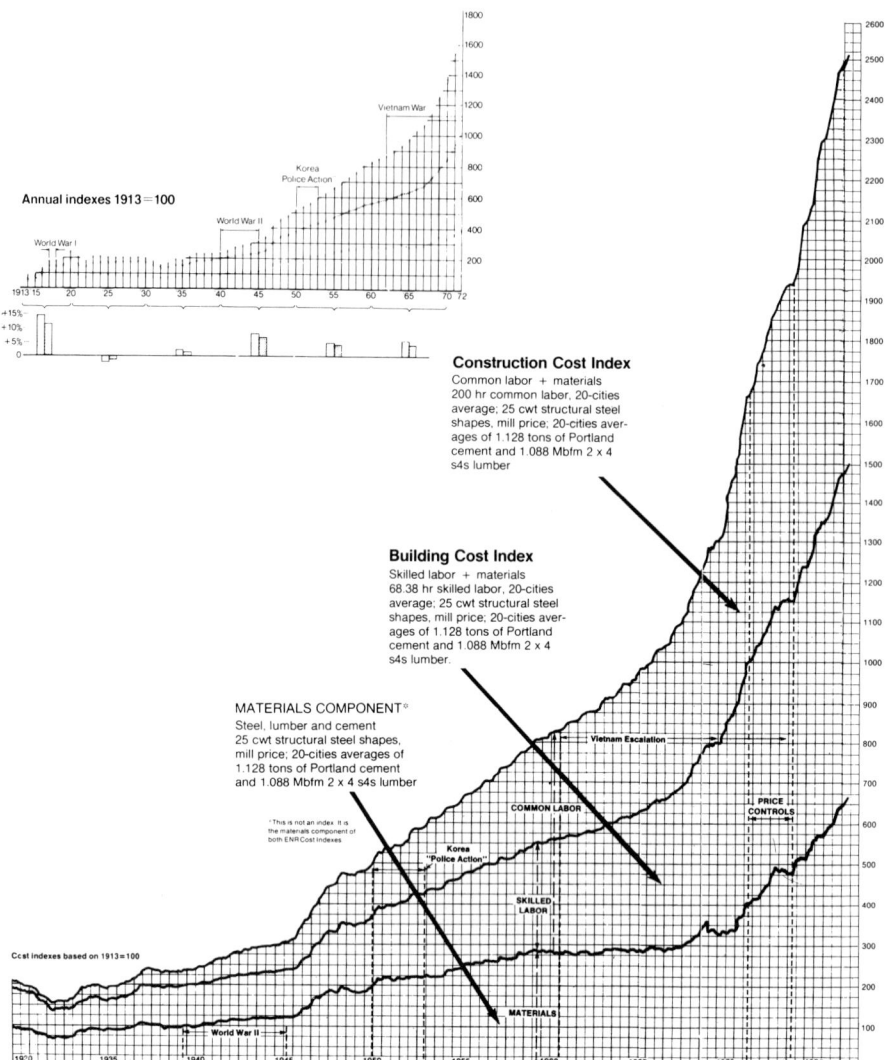

FIG. 3. *Engineering News Record,* Construction and Building Cost Indexes and materials components. The inset shows the annual rate of change compounded. (Reprinted by special permission from *Engineering News Record,* March 24, 1977, copyright © by McGraw-Hill, Inc., New York.)

updating costs as noted in the preceding section. Although there are many indexes for different business purposes, the ones of most use to the process industry are summarized here.

Engineering News-Record (ENR) Construction Cost Index

The *Engineering News-Record* (ENR) Construction Cost Index [6] is intended to reflect *construction* trends least influenced by local conditions. The material

TABLE 1   Weighted Components of the *Chemical Engineering* Plant Cost Index
(1957–1959 = 100)[a]

1.	Equipment, machinery, and supports	61%
2.	Erection and installation labor	22
3.	Buildings, materials, and labor	7
4.	Engineering and supervision manpower	10
	Total	100%

[a]Source: C. H. Chilton and T. H. Arnold, *Chemical Engineering* (February 18, 1963). Reprinted by special permission. Copyright © by McGraw-Hill, Inc., New York.

components of the cost are based on steel, lumber, and cement. The ENR index is divided into two parts—construction and building costs. The Construction Index uses common labor and the Building Cost Index uses skilled labor; however, both are directed to the general construction industry and do not specifically reflect process plant costs. Figure 3 represents the charting of this index to the reference year of 1913.

Chemical Engineering Plant Cost Index

The *Chemical Engineering* Plant Cost Index [7] originated in February 1963 and uses the period 1957–1959 = 100 as reference for costs related to construction in the chemical industry. The four major components, weighted, are shown in Table 1.

The breakdown of the 61% for equipment, machinery, and supports component is shown in Table 2.

Figure 4 presents this index graphically, and Table 3 summarizes the yearly average index. This index is constructed to reflect costs of a "typical" chemical plant.

TABLE 2   Breakdown of Item 1 in Table 1[a]

Fabricated equipment	37%
Process machinery	14
Pipe, valves, and fittings	20
Process instruments and controls	7
Pumps and compressors	7
Electrical equipment and materials	5
Structural supports, insulation, and paint	10
Total	100%

[a]Source: C. H. Chilton and T. H. Arnold, *Chemical Engineering* (February 18, 1963). Reprinted by special permission. Copyright © by McGraw-Hill, Inc., New York.

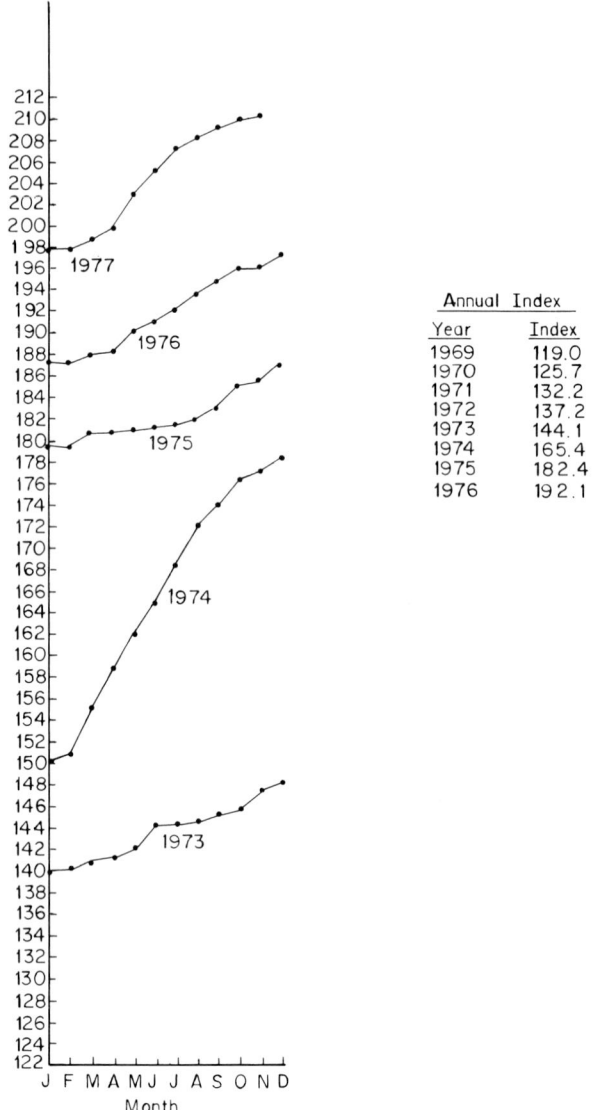

Annual	Index
Year	Index
1969	119.0
1970	125.7
1971	132.2
1972	137.2
1973	144.1
1974	165.4
1975	182.4
1976	192.1

FIG. 4. *Chemical Engineering* Plant Cost Index (1959 = 100). (Reprinted by special permission from *Chemical Engineering*, copyright © by McGraw-Hill, Inc., New York.)

Marshall and Stevens Equipment Cost Index

The Marshall and Stevens Equipment Cost Index [8] has an all-industry equipment index and a process industry equipment index. The first index is based on equipment for 47 different industrial, commercial, and housing industries. The second index is for eight process industries. The year 1926 is referenced as 100. These indexes include equipment cost and/or appraised value plus installation labor but exclude buildings and engineering costs [9]. See Fig. 5.

TABLE 3   Tabular Summary of Yearly Cost Indexes[a]

	1950	1951	1952	1953	1954	1955	1956	1957	1958	1959	1960	1961	1962	1963	1964	1965	1966	1967	1968
Chemical Engineering Trend of Plant Costs since 1950 (base period: 1957–1959 = 100)																			
Plant Cost Index	73.9	80.4	81.3	84.7	86.1	88.3	93.9	98.5	99.7	101.8	102.0	101.5	102.0	102.4	103.3	104.2	107.2	109.7	113.6
Equipment, machinery, and supports	69.8	77.6	77.8	80.9	82.3	85.1	92.7	98.5	99.6	101.9	101.7	100.2	100.6	100.5	101.2	102.1	105.3	107.7	111.5
Erection and installation labor	80.5	85.1	87.4	91.6	93.5	93.5	95.8	98.6	100.0	101.4	103.7	105.1	105.6	107.2	108.5	109.5	112.5	115.8	120.9
Buildings, materials, and labor	82.2	88.1	88.5	91.4	93.1	95.0	98.0	99.1	99.5	101.4	101.5	100.8	101.4	102.1	103.3	104.5	107.9	110.3	115.7
Engineering and supervision manpower	78.2	82.2	84.8	88.3	87.9	92.0	94.2	98.2	99.3	102.5	101.3	101.7	102.6	103.4	104.2	105.6	106.9	107.9	108.6
Trend of Equipment Costs since 1950 (history of the equipment component and major subcomponents of Chemical Engineering Plant Cost Index. Base period: 1957–1959 = 100)																			
Equipment, machinery, and supports	69.8	77.6	77.8	80.9	82.3	85.1	92.7	98.5	99.6	101.9	101.7	100.2	100.6	100.5	101.2	102.1	105.3	107.7	111.5
Fabricated equipment	71.5	78.4	79.0	81.3	81.4	84.2	92.5	99.5	99.6	100.9	101.2	100.1	101.0	101.7	102.7	103.4	104.8	106.2	109.9
Process machinery	69.4	76.5	77.5	80.6	82.8	85.3	92.2	98.1	100.1	101.8	101.8	101.1	101.9	102.0	102.5	103.6	106.1	108.7	112.1
Pipe, valves, and fittings	64.7	73.1	73.8	78.0	79.5	85.2	94.8	97.9	98.8	103.3	104.1	101.1	100.6	100.7	101.6	103.0	109.6	113.0	117.4
Process instruments, and controls	71.9	79.8	80.0	82.9	85.1	86.7	91.2	96.7	100.4	102.9	105.4	105.9	105.9	105.7	105.6	106.5	110.0	115.2	120.9
Pumps and compressors	65.4	73.9	73.4	77.5	79.5	81.7	90.0	97.5	100.0	102.5	101.7	100.8	101.1	100.1	101.0	103.4	107.7	111.3	115.2
Electrical equipment and materials	68.3	79.7	79.3	82.0	83.0	84.3	93.5	98.4	100.6	101.0	95.7	92.3	89.4	87.6	85.5	84.1	86.4	90.1	91.4
Structural supports, insulation, and paint	77.6	82.0	83.0	86.0	88.6	90.5	92.5	98.0	100.4	101.6	101.9	99.8	99.2	97.3	98.3	98.8	101.0	102.1	105.7
Marshall and Stevens Annual Indexes of Comparative Equipment Costs, 1950 to 1968 (base period: 1926 = 100)																			
Average of all	167.9	180.3	180.5	182.5	184.6	190.6	208.8	225.1	229.2	234.5	237.7	237.2	238.5	239.2	241.8	244.9	252.5	262.9	273.1
Process industries																			
Cement	161.6	172.7	172.8	174.6	177.6	182.6	199.4	216.4	222.8	228.7	232.1	231.1	231.8	232.5	235.9	239.3	249.6	258.1	268.3
Chemical	169.6	180.7	181.1	183.1	186.2	191.5	209.1	226.5	232.3	236.5	239.2	237.7	238.0	238.7	241.1	243.8	246.1	261.8	271.6
Clay products	156.6	167.7	167.8	169.5	172.4	177.3	193.8	210.2	216.8	222.2	225.7	224.6	225.5	225.8	229.2	232.6	239.0	250.7	260.7
Glass	159.7	170.8	171.0	173.0	176.0	180.9	197.5	213.8	219.3	223.2	225.3	224.4	224.7	225.4	227.6	230.1	240.6	271.1	256.3
Paint	162.9	174.0	174.4	176.3	179.3	184.3	201.2	217.6	223.2	226.9	229.5	230.0	231.5	232.1	235.0	238.1	243.9	255.7	267.4
Paper	163.2	174.3	174.7	176.6	179.6	184.6	201.5	218.2	223.8	227.8	229.9	229.0	229.3	229.9	232.3	234.8	247.5	252.1	261.6
Petroleum products	166.0	177.1	177.6	179.7	182.8	188.0	205.4	222.2	228.0	231.8	234.3	235.0	238.2	238.8	241.8	244.9	253.9	263.4	275.2
Rubber	168.4	179.5	180.0	182.1	185.2	190.5	207.9	224.9	230.8	234.6	237.3	237.9	239.2	240.0	243.0	246.2	251.2	264.7	276.5
Related industries																			
Electrical power equipment	171.2	182.3	182.8	185.0	188.0	193.3	211.0	229.2	235.2	239.0	241.0	236.3	235.6	234.7	236.8	239.4	246.5	257.3	264.9
Mining, milling	170.7	181.4	181.9	184.1	187.1	192.6	210.4	227.9	233.8	237.1	240.6	239.2	239.5	240.1	242.6	245.3	253.0	263.5	273.2
Refrigerating	185.2	200.1	200.7	202.8	204.8	211.6	234.3	254.2	260.8	265.1	268.2	268.8	270.4	271.2	274.6	278.2	287.1	299.1	312.5
Steam power	158.4	169.9	170.5	172.6	175.5	180.4	197.0	213.0	218.6	222.9	224.7	225.3	226.6	227.2	230.1	233.0	240.4	250.6	261.8

[a]Source: R. B. Norden, *Chemical Engineering* (May 5, 1969). Reprinted by special permission. Copyright © by McGraw-Hill, Inc., New York.

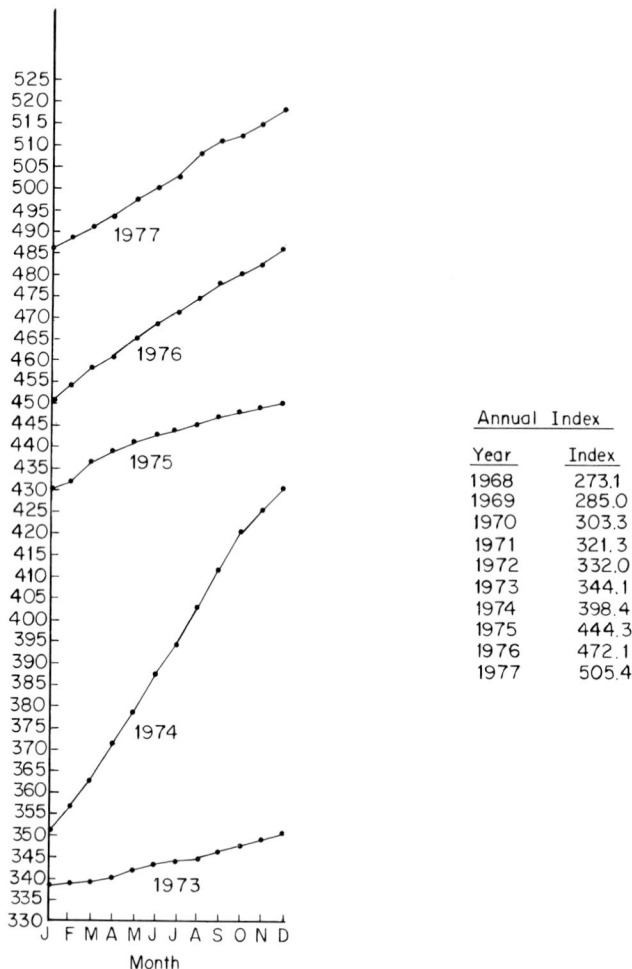

FIG. 5. Marshall and Stevens Equipment Cost Index (1926 = 100). (Reprinted by special permission from *Chemical Engineering*, copyright © by McGraw-Hill, Inc., New York.)

These costs are not directly comparable with the equipment component of the *Chemical Engineering* Plant Cost Index, since the CE Index excludes labor and the overall CE Index includes buildings and their labor. The M&S Index does, however, follow the same general pattern as the CE Index.

Nelson Refinery Construction Cost Index

The Nelson Refinery Construction Cost Index [10, 11] primarily reflects refinery costs and uses the year 1946 as a reference base. Two Nelson indexes are now published. The original Nelson Index is known as the Nelson Refinery (Inflation) Construction Cost Index and is the one most commonly referred to;

the second index is termed a True Cost Index (since 1966) and reflects the result of dividing the inflation index by productivity [11].

The Nelson indexes are composed of prices of materials and equipment (40%) and wages of labor (60%). The 40% materials–equipment portion is composed of (1) iron and steel, 20%; (2) nonmetallic building materials, 8%; and (3) miscellaneous fabricated equipment, 12%.

The labor segment of 60% is composed of (1) skilled labor, 39%; and (2) common labor, 21%. Table 4 compares selected Nelson Index figures. Figure 6 represents the yearly index values.

U.S. Department of Labor Monthly Statistics

Monthly labor and materials indexes for various industries are published by the U.S. Department of Labor. Reference should be made to the detailed reports to determine their use for a particular situation.

Many other specialized indexes are prepared by companies and service organizations; however, their application is usually too narrow for general use in process plants.

Although the trend is ever upward for composite costs, Fig. 7 illustrates how the downward movement of some specific items has helped keep the total index down. In many cases the cost improvements have been the result of improvements in process technology and/or larger size or volume plants using fewer pieces of larger equipment.

TABLE 4   Illustrations of Actual Cost Indexes and the Nelson (Inflation) Refinery Construction Cost Index[a]

	Productivity in 1966	Actual Cost Indexes Based on 1946		
		1956	1961	1966
Original Nelson Inflation (refineries)	0	195.3	232.7	273.0
Average U.S. refineries[b]	2.653	102.0	101.8	102.9
Large catalytic crackers	6.3[c]	58	45	45[c]
Alkylation plants	2.3[d]	176	151	119
Bubbleplate fractionators	1.36	166.7	182.6	200.1
Tubestill heaters	2.97	119	95	91.8
Synthetic ammonia plants	13.12	51.2	37.6	20.8

[a]Source: W. L. Nelson, *Trans. Am. Assoc. Cost Eng.*, p. 45–1 (1968). Selected data reprinted by permission. Note: Since August 1, 1966, the original Nelson Inflation Index has been corrected for the productivity attained in the design and construction of refineries.
[b]Now published as the Nelson Construction (True Cost) Cost Index.
[c]Estimate.
[d]Cost increased until 1953 and then began declining rapidly. (Approximation.)

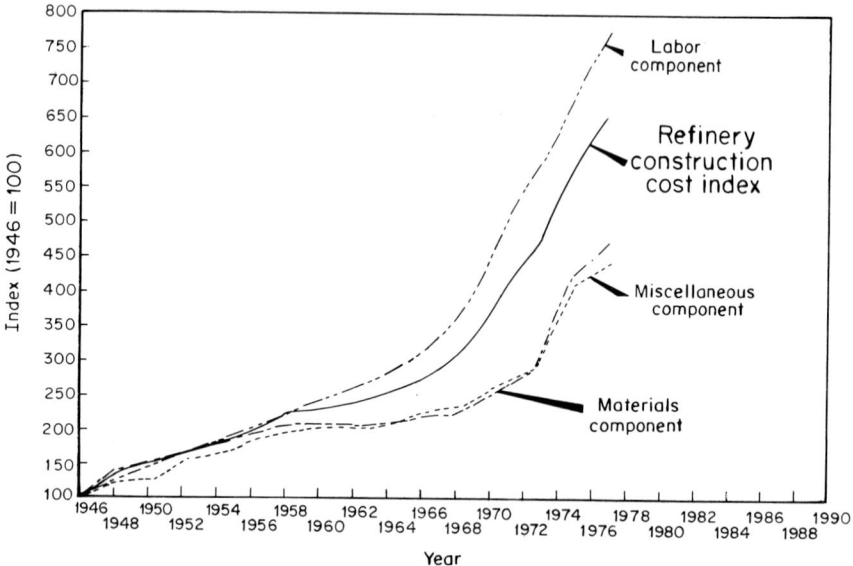

FIG. 6. Nelson Refinery (Inflation) Construction Cost Index. [Reprinted by permission from W. L. Nelson, *Oil Gas J.*, *71*, 193 (January 30, 1978).]

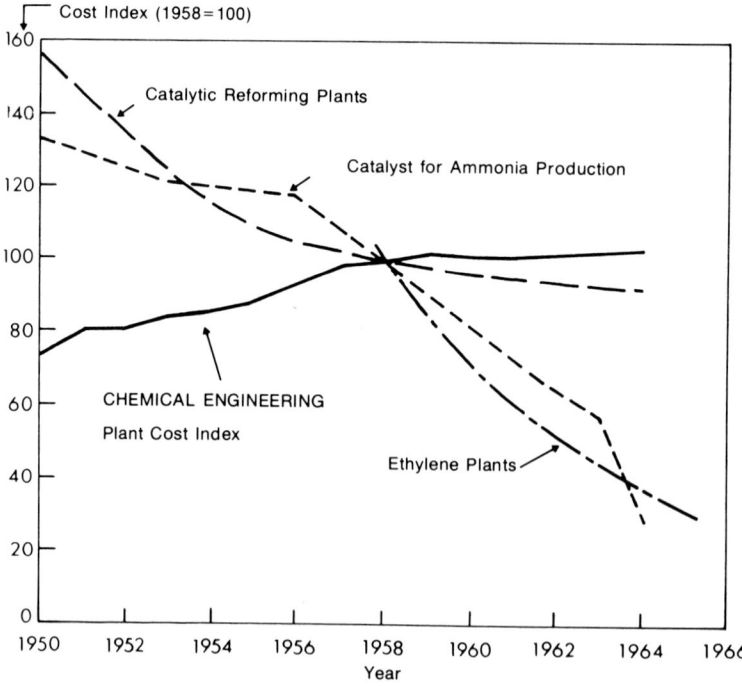

FIG. 7. Composite Index trend and the movement of individual segments in the industry. [Reprinted by permission from R. B. Norden, *Trans. Am. Assoc. Cost Eng.*, p. 46 (1968).]

Application of Indexes

For evaluating costs at different time references:

$$C_2 = C_1(I_2/I_1)$$

where I_1 = index value for time (year) represented by 1
        I_2 = index value for time (year) represented by 2
        C_1 = actual cost at time represented by 1
        C_2 = estimated cost at time represented by 2

When costs are rising rapidly, monthly index values are useful in attempting to reflect a specific point in time.

Example 1.   Updating Cost of Ethylene Glycol Plant. The XYZ Chemical Company constructed an ethylene glycol plant which was completed in June 1969 at a cost of $6,500,000.

1. If the same size unit were to be built starting January 1971, what would be the approximate cost, exclusive of escalation to the probable completion date of June 1972? Using the *Chemical Engineering* Cost Index:

$$
\begin{aligned}
I_1 \ (\text{June 1969}) \quad &= 118.2 \\
I_2 \ (\text{January 1971}) &= 128.2 \\
C_1 \qquad\qquad &= \$6,500,000 \\
C_2 \qquad\qquad &= 6,500,000 \ (128.2/118.2) \\
&= \$7,050,000
\end{aligned}
$$

2. If inflation increases at 6% a year, what is a reasonable cost to anticipate if the plant is completed in June 1972? Escalation amounts to 9% rate for the 18 months from January 1971 through June 1972.

$$
\begin{aligned}
\text{Expected increase} &= 0.09 \ (7,050,000) \\
&= \$634,500 \\
\text{Escalated plant cost to June 30, 1972} &= \$7,050,000 + 634,500 \\
\text{Total} &= \$7,684,500 \ (\text{use } \$7,690,000)
\end{aligned}
$$

It is important not to convey a sense of false accuracy in the numbers generated by these techniques. There would be nothing wrong with using $7,700,000.

3. If the capacity considered for this plant is to be different than the original, the factor scaling or other applicable technique to be described later would be applied to the figure generated in Step 2 *before* adding the escalation of 9%.

The same concepts apply to adjusting the anticipated costs of equipment by using the applicable materials cost portions of any of the indexes described previously, but be careful in using the ENR Index for this purpose as the results could be too high.

Plant and Equipment Capacity Factoring

Factoring or use of exponents on cost-related capacity data is one technique for developing order-of-magnitude, preliminary, and budget estimates. The accuracy of the results depends on the applicability of the new project equipment to the available data. The estimator needs to apply a judgment to the information and identify some range of accuracy. Certainly, the "age" of the reference data is important (such as an old purchase order), as capacity factors and cost indexes cannot properly recognize all the cost influences on information older than about 10 to 12 yr.

Unfortunately, good cost information has a time value, and by the time a new, updated compilation is made and published, it is 1 to 2 yr old. Engineering journals publish cost articles throughout the year; therefore it is helpful to assemble this type of data for reference.

Factoring—Lang and Six-Tenths Rules

Lang [12, 13] and Williams [14] present correlations useful for capacity scaling of equipment of like design, materials of construction, pressure rating of design and characteristics, and for process systems and entire processing plants. The basic concept is:

$$C_2 = C_1(W_2/W_1)^f$$

where C_1 = cost of equipment or facility of capacity, 1
C_2 = cost of equipment or facility of different capacity, 2
W_1 = capacity of equipment or facility, 1
W_2 = capacity of equipment or facility, 2
f = exponent correlation factor; when no experience or published factors are available, use 0.6

Capacity increases of over 10 times may lead to errors of serious magnitude, unless actual confirmation data are available.

Exponent values usually vary from 0.2 to 0.8 but can be more extreme for unique situations. Some tests of reasonable validity of the exponent for a particular case are recommended, as all published values are not consistent, or the exponent should be referenced to sufficient data to be representative. The literature summarizes exponents which vary considerably for the same equivalent description of equipment and for process plants [15–21]. Unfortunately, this places the burden of selecting the factor on the estimator. Judgment should be exercised in selecting an exponent that suits the situation. Table 5 presents some comparison exponents selected from the literature.

Figures 8, 9, 10, and 11 from Guthrie [22] represent a few typical cost charts for correlating basic equipment and the costs for installation as a process system or module.

Table 6 summarizes additional exponent values from Ref. 15. Due to the

TABLE 5 Selected Cost–Capacity Equipment Exponent Ranges for Cost Estimating

Equipment Description	Capacity or Size	Exponent f Range
Blowers: centrifugal/motor	To 10 hp	0.15–0.40
	15–500 hp	0.60–0.90
Blower: axial/motor	To 5,000 ft^3/min	0.12–0.35
	To 15,000 ft^3/min	0.30–0.40
	To 50,000 ft^3/min	0.50–0.75
Boilers: package	To 4,000 lb/h	0.90–0.98
	To 15,000 lb/h	0.65–0.80
	To 50,000 lb/h	0.40–0.75
waste heat	To 5,000 lb/h	0.88–0.93
	To 25,000 lb/h	0.60–0.75
Compressors: centrifugal		
Motor drive	To 100 hp	0.90–1.25
Steam turbine drive	To 2,000 hp	0.30–0.70
Turbo; 3500 rev/min		
Process service	500–2,000 hp	0.60–0.80
Compressors: reciprocating		
Motor drive, air	To 500 hp	0.80–0.95
	To 5,000 hp	0.65–0.80
Motor drive, process	To 100 hp	0.70–0.90
	To 1,500 hp	0.80–0.95
Gas engine drive		0.80–0.90
Compressors: rotary		
Motor drive	To 500 hp	0.60–0.70
Dust collectors: bag filter (only)		
Shaking type	To 10,000 ft^3/min	0.50–0.65
Light duty	To 40,000 ft^3/min	0.90–1.0
Heavy duty	To 20,000 ft^3/min	0.80–0.95
Dust collectors: centrifugal		
Impeller; includes hopper only		
Dry type	To 20,000 ft^3/min	0.60–0.70
	To 75,000 ft^3/min	0.80–1.0
Wet type	To 20,000 ft^3/min	0.70–0.80
	To 75,000 ft^3/min	0.80–1.0
Dust collectors: cyclone		
No frame and mounting		
Light duty	To 10,000 ft^3/min	0.4–0.6
	To 20,000 ft^3/min	0.80–1.0
Heavy duty	To 10,000 ft^3/min	0.60–0.80
	To 40,000 ft^3/min	0.90–1.0
Electrostatic precipitator		
Automatic		
Range of all types	To 100,000 ft^3/min	0.60–0.90
	To 200,000 ft^3/min	0.65–0.90
	To 500,000 ft^3/min	0.30–0.85
Scrubber, water spray		
Light	To 20,000 ft^3/min	0.60–0.85

(*continued*)

TABLE 5   *(continued)*

Equipment Description	Capacity or Size	Exponent f Range
Heavy	To 30,000 ft³/min	0.60–0.85
	To 50,000 ft³/min	0.80–0.95
Ejectors, range, single, CI	To 2,000 ft³/min	0.45–0.80
Ejectors, four stage, CI		
Barometric type, air	To 80 ft³/min	0.40–0.60
Surface type, air	To 80 ft³/min	0.45–0.65
Filters, plate and frame		
CI, aluminum, PVC, rubber	To 650 ft²	0.55–0.70
stainless	To 450 ft²	0.80–0.90
Heat exchangers, kettle type		
U-tube, 150 lb/in.², CS	To 10,000 ft²	0.45–0.60
	To 20,000 ft²	0.60–0.70
	To 4,000 ft²	0.60–0.75
Floating heat, 150 lb/in.², CS	To 600 ft²	0.30–0.50
	To 2,000 ft²	0.55–0.75
300 lb/in.², CS	To 600 ft²	0.35–0.55
	To 2,000 ft²	0.50–0.75
Heat exchangers, shell and tube		
Fixed T. S., CS, 150 lb/in.²,		
CS tubes	To 400 ft²	0.50–0.60
	To 2,000 ft²	0.65–0.85
SS tubes	To 300 ft²	0.60–0.75
Floating head, 150 lb/in.²,		
Admiralty tubes	To 400 ft²	0.60–0.75
	To 2,000 ft²	0.80–0.95
CS tubes	To 400 ft²	0.55–0.70
	To 2,000 ft²	0.75–0.90
U-tube, remove bundle,		
CS shell, 150 lb/in.²,		
Admiralty tubes	To 1,000 ft²	0.75–0.90
CS tubes	To 1,000 ft²	0.65–0.85
SS tubes	To 1,000 ft²	0.80–1.0
Kettles, cast iron, jacketed,		
100 lb/in.²	To 800 gal	0.30–0.40
	To 2,500 gal	0.40–0.60
Glass lined, jacketed	To 800 gal	0.35–0.45
	To 2,500 gal	0.55–0.75
Mixer; agitator		
Propeller or turbine type		
with motor, SS	To 10 hp	0.75–0.85
	To 20 hp	0.50–0.70
	To 30 hp	0.20–0.35
Side entering, w/motor		
Gear drive, CS	To 25 hp	0.40–0.60
Gear drive, SS	To 25 hp	0.40–0.65

(continued)

TABLE 5   (*continued*)

Equipment Description	Capacity or Size	Exponent f Range
Motors: induction, 440 V,		
explosionproof	To 20 hp	0.60–0.85
1800 rev/min, 60 cycle, 3-phase	To 200 hp	0.80–0.95
Open drip-proof	To 20 hp	0.60–0.75
	To 200 hp	0.80–1.0
TEFC	To 20 hp	0.55–0.65
	To 200	0.85–1.0
Motors: induction, 2300 V,		
explosionproof, open drip-proof,		
TEFC	250–500 hp	0.65–0.85
Vessels		
Flat head		
Carbon steel	To 1,500 gal	0.65–0.85
CS glass lined	To 1,000 gal	0.40–0.60
CS jacketed	To 1,500 gal	0.65–0.80
SS	To 1,000 gal	0.65–0.85

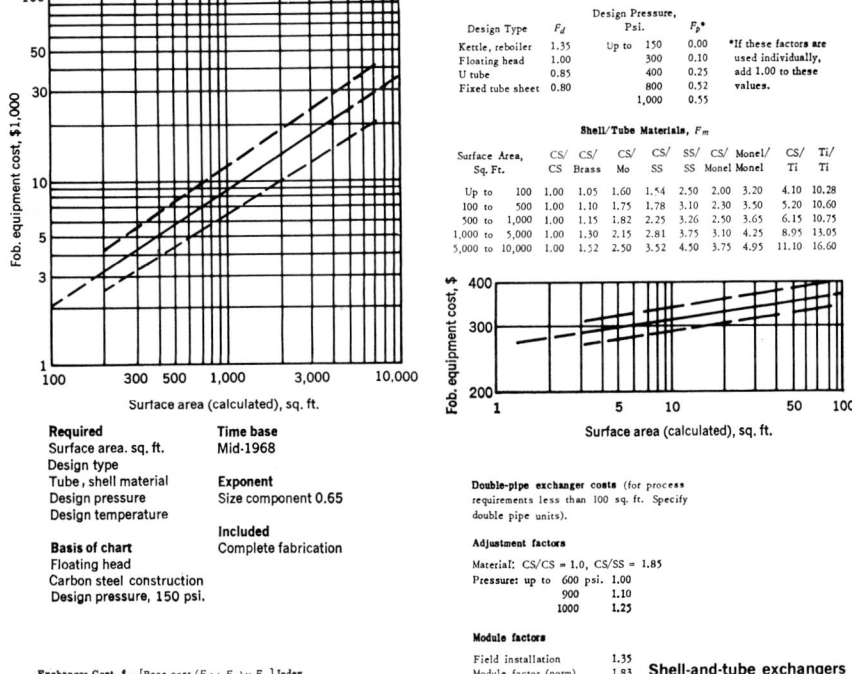

FIG. 8.   Estimating heat exchanger costs. [Reprinted by special permission from K. M. Guthrie, *Chem. Eng.*, *76*, 114 (March 24, 1969), copyright © by McGraw-Hill, Inc., New York.]

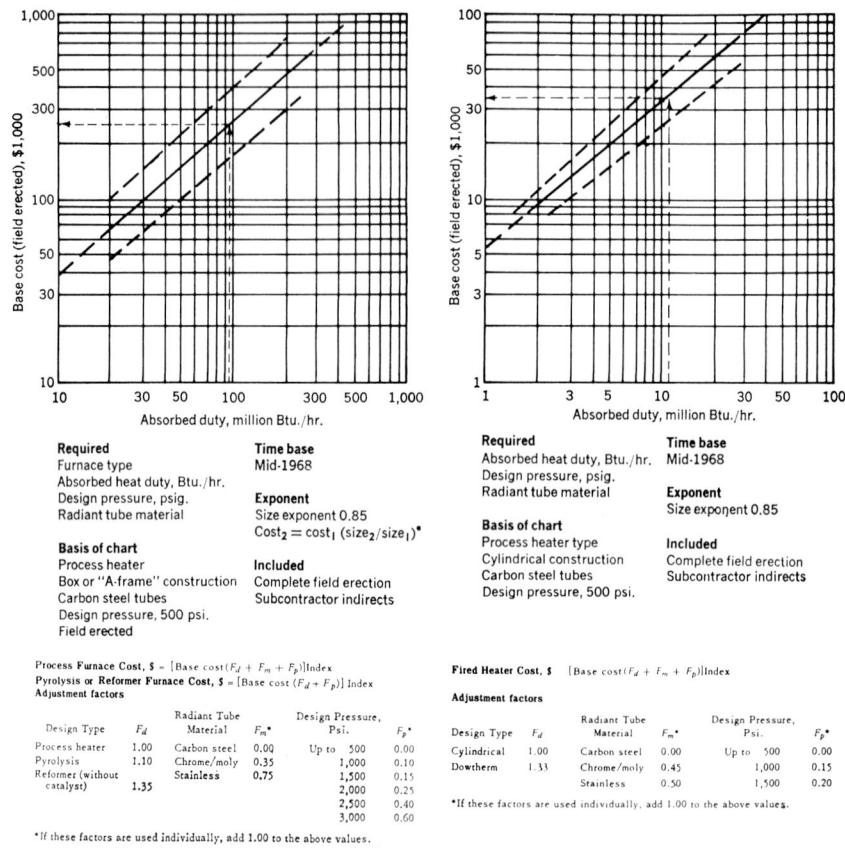

a. Process Furnaces

Required	Time base
Furnace type	Mid-1968
Absorbed heat duty, Btu./hr.	
Design pressure, psig.	
Radiant tube material	**Exponent**
	Size exponent 0.85
	$Cost_2 = cost_1 \ (size_2/size_1)^*$
Basis of chart	
Process heater	**Included**
Box or "A-frame" construction	Complete field erection
Carbon steel tubes	Subcontractor indirects
Design pressure, 500 psi.	
Field erected	

Process Furnace Cost, $ = [Base cost$(F_d + F_m + F_p)$]Index
Pyrolysis or Reformer Furnace Cost, $ = [Base cost $(F_d + F_p)$] Index
Adjustment factors

Design Type	F_d	Radiant Tube Material	F_m^*	Design Pressure Psi.	F_p^*
Process heater	1.00	Carbon steel	0.00	Up to 500	0.00
Pyrolysis	1.10	Chrome/moly	0.35	1,000	0.10
Reformer (without catalyst)	1.35	Stainless	0.75	1,500	0.15
				2,000	0.25
				2,500	0.40
				3,000	0.60

*If these factors are used individually, add 1.00 to the above values.

b. Direct Fired Heaters

Required	Time base
Absorbed heat duty, Btu./hr.	Mid-1968
Design pressure, psig.	
Radiant tube material	**Exponent**
	Size exponent 0.85
Basis of chart	
Process heater type	**Included**
Cylindrical construction	Complete field erection
Carbon steel tubes	Subcontractor indirects
Design pressure, 500 psi.	

Fired Heater Cost, $ [Base cost$(F_d + F_m + F_p)$]Index

Adjustment factors

Design Type	F_d	Radiant Tube Material	F_m^*	Design Pressure, Psi.	F_p^*
Cylindrical	1.00	Carbon steel	0.00	Up to 500	0.00
Dowtherm	1.33	Chrome/moly	0.45	1,000	0.15
		Stainless	0.50	1,500	0.20

*If these factors are used individually, add 1.00 to the above values.

FIG. 9. Estimating furnace costs: (left) process furnaces, (right) direct fired heaters. [Reprinted by special permission from K. M. Guthrie, *Chem. Eng.*, *76*, 114 (March 24, 1969), copyright © by McGraw-Hill, Inc., New York.]

variety of items and types of equipment and the way the data are presented, it becomes difficult to match exact comparison information from the published references.

Guthrie [22] presents an excellent estimating system which lends itself to rapid estimates and provides data for somewhat detailed estimates. Table 7 summarizes Guthrie's exponents; however, reference to the original presentation is recommended in order to obtain additional data and costs. Tables 8, 9, and 10 present estimating unit cost information for various equipment and site development requirements.

The referenced information is not all that is available on the subject of exponent estimating. When selecting data, it is important to examine the basis of the data before applying it to a particular situation. Interesting collections of costs were published in the 1950s; however, these are not referenced here due to the inaccuracy of scale-up and the variations and changes that have taken place in industrial equipment design and standardization during the 1960s.

Installation Costs—Overall Factors

To complete a short-cut estimate referenced primarily to equipment, overall multipliers have been developed from historical data [22–24] which allow conversion to an approximate installed cost. Table 11 summarizes suggested multipliers which are used as

$$C_e = FP$$

where C_e = installed equipment costs of complete unit as described by notes to Table 11
  F = factor for multiplication, overall
  P = quoted or estimated price of equipment item(s)

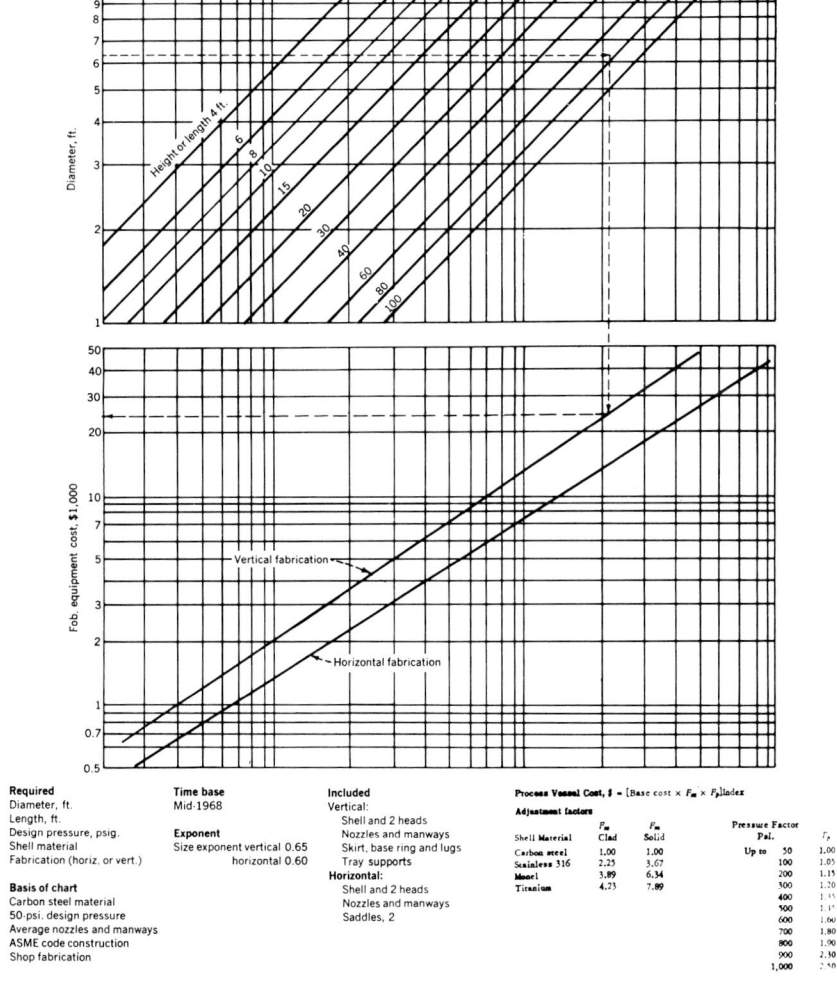

Required	Time base	Included	Process Vessel Cost, $ = [Base cost × F_m × F_p]Index

FIG. 10.   Estimating pressure vessel costs. [Reprinted by special permission from K. M. Guthrie, *Chem. Eng.*, *76*, 114 (March 24, 1969), copyright © by McGraw-Hill, Inc., New York.]

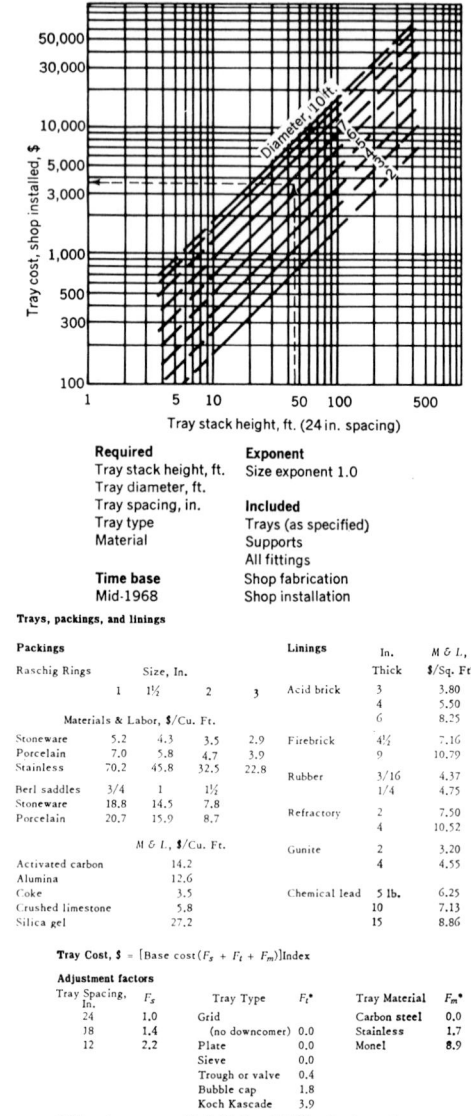

Required
Tray stack height, ft.
Tray diameter, ft.
Tray spacing, in.
Tray type
Material

Time base
Mid-1968

Exponent
Size exponent 1.0

Included
Trays (as specified)
Supports
All fittings
Shop fabrication
Shop installation

Trays, packings, and linings

Packings

Raschig Rings	Size, In.				Linings	In. Thick	M & L, $/Sq. Ft.
	1	1½	2	3	Acid brick	3	3.80
						4	5.50
	Materials & Labor, $/Cu. Ft.					6	8.25
Stoneware	5.2	4.3	3.5	2.9	Firebrick	4½	7.16
Porcelain	7.0	5.8	4.7	3.9		9	10.79
Stainless	70.2	45.8	32.5	22.8	Rubber	3/16	4.37
Berl saddles	3/4	1	1½			1/4	4.75
Stoneware	18.8	14.5	7.8		Refractory	2	7.50
Porcelain	20.7	15.9	8.7			4	10.52
	M & L, $/Cu. Ft.				Gunite	2	3.20
Activated carbon	14.2					4	4.55
Alumina	12.6						
Coke	3.5				Chemical lead	5 lb.	6.25
Crushed limestone	5.8					10	7.13
Silica gel	27.2					15	8.86

Tray Cost, $ $= [\text{Base cost}(F_s + F_t + F_m)]\text{Index}$

Adjustment factors

Tray Spacing, In.	F_s	Tray Type	F_t*	Tray Material	F_m*
24	1.0	Grid		Carbon steel	0.0
18	1.4	(no downcomer)	0.0	Stainless	1.7
12	2.2	Plate	0.0	Monel	8.9
		Sieve	0.0		
		Trough or valve	0.4		
		Bubble cap	1.8		
		Koch Kascade	3.9		

*If these factors are used individually, add 1.00 to the above values.

FIG. 11. Estimating tower packings and tray costs. [Reprinted by special permission from K. M. Guthrie, *Chem. Eng.*, *76*, 114 (March 24, 1969), copyright © by McGraw-Hill, Inc., New York.]

TABLE 6 Scale-up Exponents for Major Types of Process Equipment[a]

Heat Exchangers				
Material: Shell/Tube	Length of Tube (ft)	Pressure (lb/in.^{2} gauge)	Size Range (ft^2)	Exponent
Shell and tube exchangers				
CS/CS	20	150	<1500	0.47
	20	150	>1500	0.78
	20	4,800	200–5000	0.24
	5–12	≤600	10–3000	0.60
All aluminum	5–20	—	100–300	0.49
Admiralty.; SS heads or shell	5–20	—	100–4500	0.58
CS/304 SS	20	—	20–100	0.36
	20	—	100–3000	0.62
CS/316 SS	20	—	10–200	0.36
	20	—	200–4000	0.73
	5–20	6000	100–700	0.80
All SS	20	—	25–500	0.39
	20	—	500–7000	0.73
CS/NI	20	—	70–900	0.70
CS/Hastelloy C or Carpenter. 20	20	—	100–1000	0.84
CS/graphite	7–16	—	250–4000	0.75
CS/TI	7–20	—	200–2000	1.11
Tank and bayonet heaters				
CS/CS	—	≤160	20–150	0.10
	—	≤160	150–1000	0.53
SS/SS	—	≤160	80–1000	0.81
(finned tube)	—	≤160	50–600	0.75
Double pipe exchangers				
CS/CS	—	—	30–1500	0.70
Air-cooled exchangers				
CS	24	—	300–1500	0.69
SS	24	—	300–1500	0.93

Miscellaneous Equipment			
Equipment Type	Material	Size Range	Exponent
Agitators	CS or SS	1–30 hp	0.54
Twin-cone blenders	304 SS	60–600 ft^3	0.57
Compressors (including driver and gear)	—	70–5000 hp	0.38
Steam ejectors	—	—	0.54
Single-screw extruders	Stainless	2–6 in. diameter	1.09
	Stainless	6–12 in. diameter	2.60
Double-screw extruders	Stainless	50–160 mm diameter	1.30

(*continued*)

TABLE 6   (*continued*)

Miscellaneous Equipment			
Equipment Type	Material	Size Range	Exponent
Pumps	CS	0.1–25 hp[b]	0.04
	CS	25–400 hp[b]	0.65
	CS	5–4M hp[c]	0.31
	Stainless	0.25–5 hp[b]	0.11
	Stainless	5–100 hp[b]	0.40

		Vessels		
Type	Material	Pressure (lb/in.^{2}gauge)	Size Range (gal) (M = 1000)	Exponent
Storage tank	CS	atm	200–1MM	0.46
Vacuum	CS	—	100–10M	0.35
Pressure	CS	15–35	250–15M	0.47
	CS	35–120	50–10M	0.49
	CS	35–120	10M–600M	0.65
	CS	120–160	50–10M	0.58
	CS	200–300	100–1MM	0.44
	CS	300–4000	50–10M	0.57
Jacketed	CS	50–65	100–15M	0.52
Storage	Rubber or plastic lined	atm	100–15M	0.57
Pressure	Rubber or plastic lined	100–175	200–30M	0.50
Agitated	Glass lined	atm	60–6M	0.43
	Glass lined	40–100	80–8M	0.34
Storage	Aluminum	atm	50–100M	0.61
Cone-bottom bin	Aluminum	atm	400–10M	0.39
	Aluminum	atm	10M–25M	0.81
Storage	304 SS	atm	100–500M	0.50
Pressure	304 SS	20–40	100–800	0.34
	304 SS	40–150	250–20M	0.55
Vacuum	304 SS	—	250–2M	0.42
	304 SS	—	2M–15M	0.61
Pressure	304 SS	150–600	250–10M	0.39
Storage	316 SS	atm	250–50M	0.54
Pressure	316 SS	30–260	30–800	0.41
	316 SS	30–260	800–30M	0.71
	316 SS	260–400	400–10M	0.92
	316 SS	600–800	400–10M	1.46
	Nickel	25–100	10–100	0.30
	Nickel	25–100	100–5M	0.54
	Hastelloy C or Carp. 20	30	400–10M	0.29
	Ti lined	175	3–2M	0.19
	Ti lined	175	2M–50M	0.95
	Ti lined	500	1$\frac{1}{2}$M–6M	0.78

(*continued*)

TABLE 6   (*continued*)

Type	Material	No. of Trays	Diameter Range (in.)	Exponent
Towers				
Sieve tray	CS	10–20	60–120	1.40
	CS	21–45	30–60	0.28
	CS	21–45	60–200	0.71
	CS	46–75	24–60	0.30
	CS	46–75	60–200	0.62
	304 SS	10–35	30–120	0.65
	304 SS	36–75	30–120	0.38
	316 SS	10–80	18–60	0.34
	316 SS	10–80	60–200	0.72
Packed	Stainless	—	5–36	0.35
	Stainless	—	36–100	0.85
Bubble cap	CS	—	36–100	0.86
	Stainless	—	36–100	0.52

[a]Source: J. D. Chase, *Chem. Eng.*, (April 6, 1970). Reprinted by special permission. Copyright © by McGraw-Hill, Inc., New York, and Celanese Chemical Co.
[b]$\Delta P \leq 125$ lb/in.^{2}.
[c]$\Delta P \leq 250$ lb/in.^{2}.

The total installed plant or modification cost can be approximated by adding the individual installed equipment costs:

$$\Sigma\, C_e = \Sigma\, [(F_1 P_1) + (F_2 P_2) + (F_3 P_3)] + \ldots]$$

where Σ = sum of subscripts: 1, 2, 3, 4, . . . refer to individual equipment items listed on flow sheet. Guthrie [22] suggests that for a "normal" chemical plant, the multiplier is 3.48 times bare equipment FOB prices to obtain an erected total plant cost. This is too general for many applications, but it is reasonable to assume that the range of multipliers from simple to complex plants might be 2.5 to 5.0, with 4.0 being a good average.

Probably the easiest to obtain and most readily available data are the delivered equipment cost. Tables 12a and 12b summarize the percent of such equipment item class costs that can be attributed to the various direct and indirect cost elements. Note that the tabular values are percentages of total plant costs and can be used with delivered equipment cost to obtain an overall installed cost for that equipment system. Some judgment should be exercised to adjust any obvious percentage that would appear to give too high or too low results for the type and complexity of plant under consideration.

TABLE 7 Scale-up Exponents for Selected Equipment[a]

Equipment	Scale-up Basis	Exponent f
Boilers, packaged complete, to 250 lb/in.^{2}gauge	Steam capacity: 1000 lb/h	0.70
Boilers, field erected complete, to 400 lb/in.^{2}gauge	Steam capacity: 1000 lb/h	0.80
Cooling towers, field erected complete/pumps, basin cooling range, 15°F	gal/min	0.60
Coolers, air; field erected complete/motor; supports	ft^2 area; (calculated (area/15.5)	0.80
Heat exchangers	See Fig. 8	
Heaters, direct fired 500 lb/in.^{2}gauge, CS tubes field erected	Absorbed heat, Btu/h. Also see Fig. 9	0.85
Heaters, process furnaces 500 lb/in.^{2}gauge, CS tubes	Absorbed heat, Btu/h. Also see Fig. 9	0.85
Process gas compressors /drivers; to 1000 lb/in.^{2}gauge; applies to centrifugal and reciprocating types	Brake hp	0.82
Pumps, centrifugal with driver; complete	(gal/min)(lb/in.^{2})	0.52
Pumps, reciprocating with driver; complete	(gal/min)(lb/in.^{2})	0.70
Pressure vessels, 50 lb/in.^{2}gauge	See Fig. 10	
Towers, packings or trays	See Fig. 11	
Tanks, storage; to 40,000 gal, tank only		
vertical API conical	gal	0.30
light gauge	gal	0.28
Tanks, storage; above 40,000 gal; vertical cone roof, field erected	gal	0.63
Refrigeration, mechanical, erected; complete, with centrifugal compressor	Tons refrigeration	0.70

[a]Selected values reprinted by special permission from K. M. Guthrie, *Chem. Eng.*, 114 (March 24, 1969). Copyright © by McGraw-Hill, Inc., New York.

TABLE 8 Estimating Unit Costs for Chemical Plant Equipment[a]

	Unit	Unit Cost ($)[b]	Size Exponent	Field Installation Factor,[c] M and L	L/M Ratio
Agitators					
Propellers	hp	350	0.50	1.62	0.27
Turbine	hp	750	0.30	1.62	0.27
Air compressors (capacity)					
125 lb/in.^{2} gauge (capacity)	ft^3/min	2900	0.28	1.60	0.27
Air conditioners					
Window vent	Each	300	—	1.12	0.12
Floor-mounted	Each	200	—	1.12	0.12
Rooftop 10 ton	Each	3800	—	1.20	0.20
20	Each	6500	—	1.20	0.20
30	Each	8100	—	1.20	0.20
Air dryers (capacity)	ft^3/min	200	0.56	1.74	0.37
Bagging machines (capacity)					
Weight	Bags/min	3300	0.80	1.45	0.11
Volume	Bags/min	1000	0.80	1.45	0.11
Blenders (capacity)	ft^3	850	0.52	1.61	0.27
Blowers and fans (capacity)	ft^3	7	0.68	1.59	0.25
Boilers (industrial)					
15 lb/in.^{2} gauge	lb/h	400	0.50	1.50	0.26
150	lb/h	440	0.50	1.50	0.26
300	lb/h	500	0.50	1.50	0.26
600	lb/h	560	0.50	1.50	0.26
Centrifuges					
Horizontal basket	Diameter, in.	140	1.25	1.57	0.23
Vertical basket	Diameter, in.	310	1.00	1.57	0.23
Solid bowl (SS)	hp	1900	0.73	1.60	0.27
Sharples (SS)	hp	5200	0.68	1.60	0.27
Conveyors (length)					
Belt:[d] 18 in. wide	ft	450	0.65	1.69	0.33
24	ft	540	0.65	1.69	0.33
36	ft	620	0.65	1.64	0.28
42	ft	700	0.65	1.64	0.28
48	ft	750	0.65	1.64	0.28
Bucket (height)					
30 tons/h (8 in. × 5 in.)	ft	220	0.65	1.84	0.44
75 tons/h (14 in. × 7 in.)	ft	400	0.83	1.84	0.44
120 tons/h (15 in. × 8 in.)	ft	500	0.83	1.84	0.44
Roller, 12 in. wide	ft	7	0.90	1.69	0.33
15	ft	8	0.90	1.69	0.33
18	ft	9	0.90	1.65	0.29
20	ft	10	0.90	1.65	0.29

(*continued*)

TABLE 8   (*continued*)

	Unit	Unit Cost ($)[b]	Size Exponent	Field Installation Factor,[c] M and L	L/M Ratio
Screw, 6 in. diameter	ft	230	0.90	1.59	0.25
12	ft	270	0.80	1.59	0.25
14	ft	290	0.75	1.59	0.25
16	ft	300	0.60	1.59	0.25
Vibrating, 12 in. wide	ft	80	0.80	1.64	0.28
18	ft	110	0.80	1.64	0.28
24	ft	120	0.90	1.60	0.26
36	ft	150	0.90	1.60	0.26
Cranes (capacity)					
Span: 10 ft	tons	1800	0.60		
20	tons	2400	0.60		
30	tons	3800	0.60	Field erected costs	
40	tons	4800	0.60		
50	tons	6300	0.60		
100	tons	8500	0.60		
Crushers (capacity)					
Cone	tons/h	750	0.85	1.57	0.23
Gyratory	tons/h	55	1.20	1.57	0.23
Jaw	tons/h	85	1.20	1.57	0.25
Pulverizers	lb/h	520	0.35	1.59	0.25
Crystallizers (capacity)					
Growth	tons/d	5500	0.65	1.75	0.38
Forced circulation	tons/d	7900	0.55	1.75	0.38
Batch	gal	170	0.70	1.60	0.26
Dryers (area)					
Drum	ft^2	3000	0.45	1.74	0.36
Pan	ft^2	1900	0.38	1.74	0.36
Rotary vacuum	ft^2	3100	0.45	1.74	0.36
Ductwork (shop fabricated and field erected)					
Aluminum	Linear ft	5.42	0.55	Incl.	0.87
Galvanized	Linear ft	8.00	0.55	Incl.	0.84
Stainless	Linear ft	15.12	0.55	Incl.	0.44
Dust collectors (capacity)					
Cyclones	ft^3/min	3	0.80	1.69	0.32
Cloth filter	ft^3/min	25	0.68	1.69	0.32
Precipitators	ft^3/min	390	0.75	1.69	0.32
Ejectors (capacity)					
4 in.Hg suction	lb/h	2000	0.79	1.10	0.10
6	lb/h	200	0.67	1.10	0.10
10	lb/h	200	0.55	1.10	0.10

(*continued*)

TABLE 8 (*continued*)

	Unit	Unit Cost ($)[b]	Size Exponent	Field Installation Factor,[c] M and L	L/M Ratio
4-stage barometric					
2.5 mmHg suction	lb/h	2500	0.45	1.12	0.12
5.0	lb/h	1400	0.48	1.12	0.12
10.0	lb/h	900	0.53	1.12	0.12
20.0	lb/h	700	0.54	1.12	0.12
5-stage barometric					
0.5-mmHg suction	lb/h	4200	0.50	1.15	0.15
0.8	lb/h	3200	0.50	1.15	0.15
1.0	lb/h	2800	0.48	1.15	0.15
1.4	lb/h	2500	0.49	1.15	0.15
Elevators (height)					
Freight 3,000 lb	ft	3600	0.32		
5,000	ft	4000	0.32	Field erected costs	
10,000	ft	5400	0.32		
Passenger 3,500 lb	ft	3900	0.48		
Evaporators					
Forced circulation	ft^2	6000	0.70	1.90	0.35
Vertical tube	ft^2	1200	0.53	1.90	0.35
Horizontal tube	ft^2	800	0.53	1.90	0.35
Jacketed vessel (glasslined)	gal	1000	0.50	1.74	0.37
Filters (effective area)					
Plates and press	ft^2	330	0.58	1.79	0.42
Pressure leaf-wet	ft^2	410	0.58	1.79	0.42
dry	ft^2	1500	0.53	1.79	0.42
Rotary drum	ft^2	1400	0.63	1.60	0.27
Rotary disk	ft^2	1000	0.78	1.60	0.27
Flakers (effective area)					
Drum	ft^2	1300	0.64	1.59	0.25
Generator sets (portable)					
10 kW	Each	1500	—	—	—
15	Each	2000	—	—	—
25	Each	3000	—	—	—
50	Each	5000	—	—	—
100	Each	7000	—	—	—
Hoppers (capacity)					
Conical	ft^3	1.0	0.68	1.04	0.04
Silos	ft^3	0.9	0.90	1.10	0.10
Hydraulic presses (plate area)					
100 lb/in.^{2}gauge	ft^2	2500	0.955	1.74	0.36
300	ft^2	3600	0.95	1.74	0.36
500	ft^2	5000	0.95	1.74	0.36
1,000	ft^2	6200	0.95	1.74	0.36

(*continued*)

TABLE 8   (*continued*)

	Unit	Unit Cost ($)[b]	Size Exponent	Field Installation Factor,[c] M and L	L/M Ratio
Mills (capacity)					
Ball	tons/h	550	0.65	1.70	0.34
Roller	tons/h	5000	0.65	1.70	0.34
Hammer	tons/h	500	0.85	1.70	0.34
Screens (surface)					
vibrating single	ft²	900	0.58	1.32	0.18
double	ft²	1100	0.58	1.32	0.18
Stacks (height)					
24 in. (CS)	Linear ft	25.83	1.00	1.24	0.16
36 in. (CS)	Linear ft	58.20	1.00	1.24	0.16
48 in. (CS)	Linear ft	78.25	1.00	1.24	0.16
Tank heaters (area)					
Steam coil[e]	ft²	94.12	0.32	1.25	0.25
Immersion	kW	18.75	0.85	1.20	0.20
Weigh scales					
Portable beam	Each	250	—	—	—
dial	Each	1500	—	—	—
Truck 20 tons	Each	4000	—	1.08	0.08
50	Each	7200	—	1.08	0.08
75	Each	8500	—	1.08	0.08

[a]Source: K. M. Guthrie, *Chemical Engineering* (March 24, 1969). Reprinted by special permission. Copyright © by McGraw-Hill, Inc., New York.
[b]All unit costs are based on mid-1968. These are not general unit costs.
[c]Field installation included equipment foundations, electrical, paint, and field labor (no indirects).
[d]For enclosed conveyors walkway, multiply by 2.10.
[e]Stainless factor 2.4.

TABLE 9   Estimating Costs for Site Development[a]

	Unit	Field Installation, $ M and L			L/M Ratio
		Minimum	Normal	Maximum	
Dewatering and drainage					
Pumping system (rented)	d	25	32	40	—
Wellpoint dewatering system	month	6500	7500	8500	—
Drainage trench	Linear ft	0.75	0.85	0.95	—

(*continued*)

TABLE 9 (*continued*)

	Unit	Field Installation, $ M and L			L/M Ratio
		Minimum	Normal	Maximum	
Fencing					
Complete fence (light)	Linear ft	1.34	1.88	2.42	0.32
Complete fence (heavy)	Linear ft	1.51	2.13	2.75	0.32
Chain link	Linear ft	5.48	5.93	6.38	0.48
Gates, 6 ft					
Light	Each	55.50	67.50	79.50	0.15
Heavy	Each	69.25	82.50	95.75	0.15
Chain	Each	105.65	128.00	150.65	0.15
Corner posts	Each	31.50	32.00	32.50	0.19
Fire protection					
Pumps					
Firehouse	Allowance	100,000	150,000	200,000	—
Firetrucks (2)					
Land surveys and fees					
General surveys and fees % total cost		4.0	9.0	14.0	—
Soil tests	Each	300	400	500	—
Landscaping					
General	yd^2	1.50	1.70	1.90	0.90
Piling					
Wood (untreated)	Linear ft	1.70	2.15	2.60	2.20
Wood (creosoted)	Linear ft	2.15	2.60	2.90	2.50
Concrete:					
Precast	Linear ft	6.75	7.00	7.25	0.50
Cast in place	Linear ft	4.75	6.62	8.50	1.25
Steel pipe (concrete filled)	Linear ft	7.50	9.50	11.50	0.25
Steel section	Linear ft	7.40	8.50	9.50	0.27
Sheet piling,					
Steel	ft^2	1.45	2.60	3.75	0.20
Wood	ft^2	1.25	1.75	2.25	0.35
Pile driver setup	Each	6800	7500	8200	—
Roads, walkways, paving					
4-in.-thick reinforced, 6-in. subbase	yd^2	6.35	7.87	8.39	1.75
6-in.-thick reinforced 6-in. subbase	yd^2	7.61	9.37	10.13	1.75
2-in.-thick asphalt top, existing base	yd^2	2.37	3.12	3.87	0.22
2-in.-thick asphalt top, 4-in. subbase	yd^2	3.58	4.68	5.78	0.22
3-in.-thick asphalt top, 12-in. subbase	yd^2	6.37	7.62	8.87	0.22
Gravel surface					

(*continued*)

TABLE 9 (*continued*)

	Unit	Field Installation, $ M and L			L/M Ratio
		Minimum	Normal	Maximum	
2-in.-thick gravel	yd²	0.33	0.58	0.83	0.52
4-in.-thick gravel	yd²	0.55	0.87	1.19	0.52
6-in.-thick gravel	yd²	1.00	1.38	1.76	0.52
Parking lots					
Black-top surface	yd²	5.30	6.25	7.54	0.45
Sewer facilities					
Asbestos cement pipe					
(general)	Linear ft	4.55	4.85	5.15	0.38
Concrete pipe (reinforced)					
18 in. diameter	Linear ft	5.65	5.80	5.95	0.38
36	Linear ft	14.75	15.96	17.23	0.39
72	Linear ft	50.33	52.18	54.03	0.40
Vitrified clay piping					
18 in. diameter	Linear ft	7.55	7.80	8.45	0.86
24	Linear ft	15.15	16.10	17.05	0.88
36	Linear ft	33.95	36.20	38.95	0.89
Septic tank (45,000 gal)	Each	—	7,500	—	0.05
Site clearing, excavation and grading					
Site preparation					
Machine cuts	yd³	0.50	0.56	0.63	0.30
Clearing and grubbing	yd²	0.13	0.15	0.18	—
General grading	yd²	0.63	0.44	0.48	—
Final leveling	yd²	0.25	0.31	0.38	—
Foundation excavation					
Machine excavation	yd³	1.50	1.63	1.75	0.58
Machine plus hand trim	yd³	2.50	3.44	3.75	0.90
Hand work	yd³	7.56	10.00	12.50	—
Trench excavation					
Machine 3½ ft deep × 2 ft wide	Linear ft	0.38	0.44	0.50	0.50
Machine 4 ft. × 3 ft	Linear ft	0.56	0.63	0.68	0.50
Machine 4½ ft × 4 ft	Linear ft	1.12	1.13	1.25	0.50
Machine 5 ft × 5 ft	Linear ft	1.38	1.50	1.63	0.50
Hand labor	yd³	8.75	10.12	13.75	—
Trench shoring					
Sheeting	ft²	1.25	1.52	1.75	1.12
Trench and foundation backfill					
Machine plus hand trim	yd³	1.44	1.56	1.68	0.90
Hand labor only	yd³	5.79	6.25	6.75	—
Miscellaneous materials					
Sand	yd³	3.05	4.80	5.55	—
Gravel	yd³	1.50	2.25	3.00	—
Dirt fill	yd³	1.30	2.15	3.00	—
Crushed stone	yd³	2.55	4.37	5.19	—

[a]Source: K. M. Guthrie, *Chemical Engineering* (March 24, 1969). Reprinted by special permission. Copyright © by McGraw-Hill, Inc., New York.

TABLE 10   Estimating Cost for Miscellaneous Services and Facilities[a]

	Unit	Field Installation, $ M and L			L/M Ratio
		Minimum	Normal	Maximum	
Air systems					
Instrument air					
Compression facilities, air dryer, air receiver, and distribution	$M	18.75	43.75	62.50	0.80
Plant air					
Compression facilities, air receiver, and distribution	$M	12.50	31.25	50.85	0.80
Blowdown and flare					
For general purposes (including flare lines, blowdown drum, and disposal pit)	$M	81.75	102.58	187.52	0.45
Cooling tower and CW distribution					
Use 1.15 design factor on estimated throughput.					
Cooling tower costs		See original source[a]			
Distribution systems for general purposes	gal/min	12.58	36.50	43.25	0.85
River intake installation for general purposes	gal/min	8.22	16.25	24.37	1.12
Fireloop and hydrants					
For general purposes	$M	12.58	22.50	40.24	0.80
Fuel systems					
Fuel oil (includes pumps, storage, piping, controls, and distribution)	$M	6.25	25.12	43.75	0.85
Fuel gas (includes receiver, piping, controls, and distribution)	$M	12.50	37.52	62.58	0.85
General water systems					
Treated water					
Filtered and softened	gal	0.15	0.23	0.30	
Distilled	gal	0.65	0.92	1.20	
Drinking and service water, General facilities	$M	2.50	5.40	7.58	
Power generation and distribution					
Use 1.10 design factor on estimated consumption.					
Generating facilities	kW	See original source[a]			

(*continued*)

TABLE 10   (*continued*)

	Unit	Field Installation, $ M and L			L/M Ratio
		Minimum	Normal	Maximum	
Electrical distribution for general purposes	kW	87.5	93.75	98.75	0.75
Main transformer stations					
Three phase, 60 cycle					
Capacity 3,000 kVA	kVA	33.0	37.0	44.0	
5,000	kVA	20.0	23.0	26.0	
10,000	kVA	13.0	14.0	16.0	
20,000	kVA	10.0	12.0	13.0	
Secondary transformer stations					
4,200/575 V, 600 kVA	kVA	30.1	33.8	42.3	
1,000	kVA	20.1	25.2	31.5	
1,500	kVA	15.6	19.5	24.3	
2,000	kVA	14.8	18.5	23.2	
13,200/575 V, 600 kVA	kVA	28.2	35.3	44.2	
1,000	kVA	21.2	26.5	33.1	
1,500	kVA	16.6	20.8	26.2	
2,000	kVA	15.4	19.3	24.1	
Receiving, shipping, storage					
Automotive					
Forklift trucks					
3,000 lb	Each		7,800		
5,000	Each		11,000		
10,000	Each		16,200		
Pallet truck					
Hydraulic 4,000 lb	Each		930		
Electric 4,000 lb	Each		3,600		
Payloaders					
2yd^3 (gas)	Each		21,000		
4yd^3 (gas)	Each		33,700		
2yd^3 (diesel)	Each		22,900		
4yd^3 (diesel)	Each		36,500		
Tank trailers					
Carbon steel	Each		14,800		
Aluminum	Each		21,600		
Stainless	Each		36,500		
Tractors					
Gasoline	Each		12,500		
Diesel	Each		27,500		
Tractor shovel					
2-yd^3 bucket	Each		26,300		
3	Each		28,700		
4	Each		35,600		
Automotive shipping facilities, One outlet per 2,000 bbl			9,800		0.85

(*continued*)

TABLE 10   (*continued*)

	Unit	Field Installation, $ M and L Minimum	Normal	Maximum	L/M Ratio
Docks and wharves					
Light construction					
2-in. deck	ft²	5.15	5.63	6.25	0.45
3	ft²	6.25	6.87	7.50	0.45
Medium construction					
3-in.	ft²	8.75	9.38	10.12	0.45
4	ft²	10.15	11.25	12.50	0.45
Heavy construction					
4-in.	ft²	12.50	15.62	18.75	0.45
Concrete	ft²	17.50	21.25	25.25	0.45
Dredging					
General operations	yd³	4.32	10.81	17.28	
Tankage					
General		See original source[a]			
Railroad					
Straight track					
(railroad siding)	Linear ft		26.25		0.58
Turnout	Each		2,800		0.10
Bumper	Each		790		0.12
Blinker and gate	Each		9,300		0.15
Grade and ballast	Linear ft		6.25		0.95
Locomotives (battery)					
9 ton	Each		35,000		
12	Each		41,800		
Locomotives (diesel)					
1½ tons	Each		11,000		
3	Each		14,000		
Tank car (10,000 gal)	Each		10,800		
Railroad shipping facilities, one outlet per 2,000 bbl	Each		4,800		0.85
Steam generation and distribution					
Use 1.10 design factor on estimated consumption					
Package boilers (up to 150,000 lb/h)	lb/h	See original source[a]			
Field erected (above 150,000 lb/h)	lb/h	See original source[a]			
Steam distribution for general purposes	lb/h	0.94	1.52	1.68	0.85
Yard lighting and communications					
For general purposes	$M	18.75	52.25	93.75	0.75
Yard transfer lines and pumps					
For general purposes	$M	17.25	31.25	56.25	0.65

[a]Source: K. M. Guthrie, *Chemical Engineering* (March 24, 1969). Reprinted by special permission. Copyright © by McGraw-Hill, Inc., New York.

TABLE 11 Installed Cost Multipliers for Plant Equipment[a]

Equipment	Multiplier, F[b]		
	A	B	C
Blender	2.0		
Blowers and fans, including motor	2.5		
Centrifuges, process	2.0		
Compressors, centrifugal			
Motor driven, less motor	2.0	2.5	3.21[c]
Steam turbine, including turbine	2.0		
Compressors, reciprocating			
Steam and gas drive	2.3		
Motor driven, less motor	2.3		
Ejectors (vacuum units)	2.5		
Furnaces, package unit	2.0	2.0	
Heat exchangers	4.8	3.5	3.39
Instruments	4.1	4.0	
Motors, electric	8.5		
Pumps, centrifugal			
Motor driven, less motor	7.0	4.0	3.48[c]
Steam turbine, including turbine	6.5		
Pumps, positive displacement, less motor	5.0		
Reactors[d]	—		
Refrigeration, package units	2.5		
Tanks			
Process	4.1	4.0	
Storage	3.5		
Fabricated and field erected,			
50,000 gal	2.0		
Towers (distillation, absorption columns)	4.0	4.0	
Miscellaneous major equipment	—	2.5	
Process vessels, vertical			4.34
Process vessels, horizontal			3.29

[a]Sources: Column *A*, from W. F. Wroth, *Chem. Eng.*, *67*, 204 (October 17, 1960); Column *B* from J. T. Gallagher, *Chem. Eng.*, *74*, 89 (December 18, 1967); and W. E. Hand, *Pet. Refiner*, p. 133 (September 1958) (data date estimated to be 1957); Column *C* from K. M. Guthrie, *Chem. Eng.*, *76* (March 24, 1969) (data mid-1968). Data reprinted by special permission. Copyright © by McGraw-Hill, Inc., New York.

[b]These factors include cost of site development, buildings, electrical installations, carpentry, painting, contractor's fee and rentals, foundations, structures, piping, insulation, engineering, overhead, and supervision.

[c]Includes driver.

[d]Factor as approximate equivalent type of equipment.

TABLE 12a Development of Capital Estimates for Different Type Processes at Different Sites[a]

Type Plant, Predominate Processing →	New Plant at New Site			New Plant at Existing Site			Expanded Plant at Existing Site		
	Fluids	Fluids-Solids	Solids	Fluids[b]	Fluids-Solids	Solids	Fluids	Fluids-Solids	Solids
Equipment	53	54	50	64	57	53	66	65	69
Fabricated equipment	21	22	30[c]	41	35	35	45	10	18
Process machinery	21	23	30[c]	9	19	31	5	33	51
Pipe, valves, fittings	18	24	9	21	19	11	23	22	5
Process instruments and controls	7	5	5	8	8	6	12	12	6
Pumps, and compressors	13	10	10[c]	6	7	2	5	3	1
Electric equipment and materials	5	6	6	5	6	7	3	7	9
Structural support, insulation, paint	15	10	10[c]	10	6	8	7	13	10
	100	100	100	100	100	100	100	100	100
Buildings, materials, and labor	13	14	24	4	10	9	2	2	7
Erection and installation labor	22	24	18	23	22	26	22	25	17
Engineering and supervision	12	8	8	9	11	12	10	8	7
Total costs, percent of total	100	100	100	100	100	100	100	100	100

[a]Source: Cost File No. 81, *Chemical Engineering* (September 30, 1963). Reprinted by special permission. Copyright © by McGraw-Hill, Inc. New York. (Also see T. H. Arnold and C. H. Chilton, *Chemical Engineering* (February 18, 1963). Note: Data from project battery limit costs ranging from $1 to $10 million.

[b]Average for chemicals and petroleum refineries, this situation only.

[c]Estimated by this author, data not available.

TABLE 12b   Percent Distribution for Costs of Typical Process Plants[a]

	Type Process		
	Fluids	Fluids-Solids	Solids
Direct costs			
Total principal items	21	26.5	30
Installation of total principal items, including painting, insulation	9	10	12
Process piping, installed	15.5	11	4.5
Instrumentation, installed	4	2.5	2.5
Electrical, installed	2	2	2
Process buildings	2	2	1.5
Other direct costs			
Utilities, installed	10	11	6
General services	2.5	3	6
Buildings, general	3	6.5	8.5
Receiving, shipping, storage facilities	4.5	6	6
Average total direct costs	73.5	80.5	79
Indirect costs			
Engineering and overhead	11.5	9	9.5
Contingencies	15	10.5	11.5
Grand total costs	100.0%	100.0%	100.0%

[a]Source: J. E. Haselbarth and J. M. Berk, Cost File No. 31, *Chem. Eng.*, *70*, 158 (May 16, 1960). Reprinted by special permission. Copyright © by McGraw-Hill, Inc., New York. Data presented is average of high and low values. Notes: *General services* includes plant site development, fire protection, waste disposal, fences, drainage; *Utilities, installed*, includes plant power, steam, cooling water and electrical costs, other than process wiring, starters, panels, etc., utilities piping and buildings; *buildings, general*, includes offices, warehouses, shops, garage, etc.

Correcting Equipment Costs for Construction Materials

The materials of construction for corrosion and mechanical considerations are important in the determinations to arrive at proper equipment costs for any estimating system. For short-cut or quick estimating, multipliers have been used to convert costs based on carbon steel construction to comparable costs using some other material, such as 316 stainless steel (Table 13).

Example 2.   Determining Approximate Cost. A heat exchanger with carbon steel shell and tubes costs $5,620 by a current quotation. If the same square feet of heat exchange surface is needed in all stainless 316 construction, determine an approximate cost.

Referring to Table 13, the multiplier for a stainless 316 heat exchanger is 3.0. Approximate estimating cost of unit is (3)(5620) or $16,860.

Example 3.   Finding the Current Price. If the heat exchanger cost data in carbon steel construction had been from a quotation dated January 1969, what would be a current price for September 1971?

First, the 1969 cost of $5620 would be updated using the *Chemical Engineering* Index. If the *Fabricated Equipment Indexes* were available for 1969, they would be preferable; but, if not, use the main index.

January 1969, Index = 117
September 1971, Index extrapolated = 134
Forecast cost for exchanger = 5620(134/117) = $6450

TABLE 13   Cost Ratio of Nonferrous Material Cost to Carbon Steel Costs for Major Equipment[a]

Material of Construction[b]	Fabricated Equipment		Heat Exchanger	Centrifugal Pumps	Air Coolers[c]
	Clad	Alloy			
Cast steel				1.0	
				1.3*	
Stainless, Type 410	2.1	2.1		1.5	
405	2.25	2.25			
304	2.75	2.50		1.8	
316	3.0	3.0		2.0	
310	3.25	3.25			
Bronze				1.5	
Monel	6.5	4.0		2.5	
				3.2*	
CS shell/Al-brass tubes			1.25		
CS shell/Monel tubes			2.25		2.2*
			3.0*		
Monel shell and tubes			3.6		
			4.2*		
Stainless 304 shell and tubes			3.2		
			3.8*		
Stainless 316 shell and tubes			3.0		
CS shell/stainless tubes			2.8*		1.85*
Alloy-20				2.1*	
Hastelloy-C				2.89*	
Nickel				3.5*	
Titanium				8.89*	
Aluminum		1.4*			0.50*
Rubber-lined		1.48*			
Glass-lined		4.25*			
Pressure vessels, see Fig. 10					

[a]Sources: All figures are from J. Clerk, *Chemical Engineering* (February 18, 1963), except those from Guthrie, *Chemical Engineering* (March 24, 1969), as noted by the symbol *. Reprinted by special permission. Copyright © by McGraw-Hill, Inc., New York. Ratio data from Clerk are approximately 1963; data of Guthrie are 1968.
[b]Carbon steel = 1.0
[c]Tubes, shell CS.

This cost is in carbon steel. To convert to 316 stainless:

$$\text{Stainless estimated cost} = 6450(3.0) = \$19,350 \text{ for September 1971}$$

Installed Plant Costs Reflecting Nonferrous Construction

Different techniques have been used to account for the costs of field fabrication and installation of the plant components where significant quantities of nonferrous materials are involved. Usually, the common metal component is stainless steel 304 or 316; however, nickel, aluminum, and other metals have their special applications. For plants involving considerable glass equipment and piping or special plastic coatings or linings such as Teflon or Kynar, the multiplier should be specifically developed for that type of construction, as the generalities of the usual factors will be too inaccurate.

Accountings by Gallagher [24] and Clerk [25] use Fig. 12 and Table 13. The procedure is to determine from Table 13 the nonferrous cost ratio, R_m, for the materials. Then refer to Fig. 12, reading R_{ci}, the ratio of installed capital cost to major equipment cost. This is expressed for each unit of equipment as:

$$C_i = R_m R_{ci} C_{cs}$$

where C_i = installed capital cost for individual item
        C_{cs} = cost of equipment, fabricated of carbon steel

To accumulate a complete plant, sum the appropriate C_i values which correspond to the materials of construction. Obviously, all materials in a plant

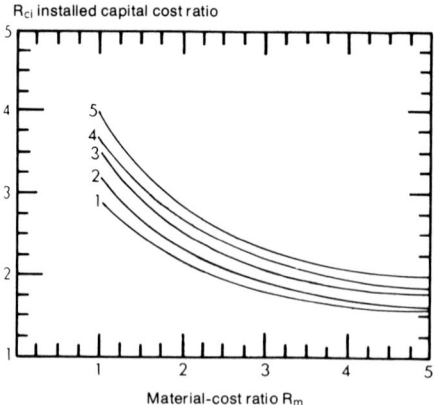

R_{ci} installed capital cost ratio

Material-cost ratio R_m

(Cost of equipment made of alloy material ÷ Cost of equivalent carbon-steel equipment)

FIG. 12.   Installed cost ratio of nonferrous material cost to carbon steel cost for major equipment. [Reprinted by special permission from J. T. Gallagher, *Chem. Eng.*, *74*, 89 (December 18, 1967), copyright © by McGraw-Hill, Inc., New York.]

TABLE 14 Ranges of Equipment-Type Distribution for "Average" Chemical/ Petrochemical Plants (Dollars)[a]

	A (%)	B (%)	
Equipment class or type	—	Liquid plant	Liquid–solid plant
Furnaces	14	—	—
Exchangers	18	15–25	15–25
Process vessels, vertical, including distillation	15	20–30	8–12
Process vessels, horizontal	8	18–28[b]	16–20[b]
Pumps and drivers	7	8–16	—
Compressors	30	10–20[c]	40–50[c]
Storage tanks	8		
Total equipment	100		

[a]Sources: Column A from K. M. Guthrie [22]; Column B from J. D. Chase [15]. Reprinted by special permission. Copyright © by McGraw-Hill, Inc., New York.
[b]May include storage tanks.
[c]Referred to as machinery.

will not be nonferrous construction, so the appropriate recognition can be made by this technique. Note that C_i accounts for erection, field costs, and home office costs, and therefore attempts to lump all associated cost factors of a system to arrive at a total, including directly associated piping. If the cost of the equipment is known in the proper materials of construction, the ratio factor, R_m, is not used except to obtain R_{ci} from Fig. 12.

Guthrie [26] proposed a short-cut technique for arriving at total plant costs. His method also recognizes the carbon steel and nonferrous construction components and is rapid to use in developing preliminary costs.

Ranges of Equipment Type Distribution for "Average" Chemical/Petrochemical Plants

Table 14 shows the ranges of equipment-type distribution for both liquid plants and liquid–solid plants. The average used by Guthrie [22] for the cost module concept is also tabulated for comparison.

Direct Labor Costs

Estimates of the direct labor cost for installation of plant equipment, structures, etc. are usually made using man-hour estimates for the work of the various crafts involved. Experienced estimators maintain detailed tabulations of labor required for segments of work and the appropriate factors to adapt these tabulations to a specific situation.

When such detail is not available, the general mark-up factors based on equipment cost are used to obtain an installed cost as suggested in the previous paragraph.

The data of Guthrie [22] and Page [27–31] are complete in supplying individual equipment and systems installation costs. For the overall direct labor costs for an average petrochemical plant, multiply the sum of the delivered equipment costs times 0.58 [22]. When considering the installation of mechanical equipment only or an isolated mechanical system (not a plant), you can use a multiplier of 0.34 [22]. For plants with obvious differences in the types and relative amounts of labor compared to a petrochemical plant, the ratios should be adjusted.

Other guidelines regarding installed costs for a new project in the $2–5 million range suggest (overall):

1. Engineering 10–15%
2. Materials 65–45%
3. Labor 25–40%

Plant Capacity Exponents and Cost Charts

Estimates of total process plants can be obtained by using historical capacity versus cost charts, or capacity ratio exponents. In such cases the process details are not broken down according to a flow sheet but are taken as overall, all-inclusive figures. It is important to understand the content of the data in order that costs are determined for a defined scope. Extra costs may have to be added to satisfy the needed total information for the type of project under study.

The exponent concept is similar to that described for equipment in that it relates changes in capacity using a known cost.

$$C_{T_2} = C_{T_1}(W_2/W_1)^f$$

where C_{T_2} = complete plant cost at capacity, 2
        C_{T_1} = complete plant cost at capacity, 1
        W_2 = capacity of plant, condition 2
        W_1 = capacity of plant, condition 1
        f = exponent for total plant cost–capacity correlation.

Table 15 summarizes capacity exponents and estimated plant costs [32]. It is important to note that the basis is essentially a "battery-limits" plant (no off-sites) using a previously developed site. This means that all off-site utilities, including general service buildings, roads, drainage, etc., are already available, and capital costs are devoted to specific requirements of the particular plant. Again, keep in mind that a lot of broad general plant requirements have to be encompassed in a correlating coefficient developed from published data not subject to close segregation for detail. It is difficult to locate discrepancies in tables such as 15; however, one based on published records indicates that the 200,000 tons/yr high-pressure polyethylene plant would cost in the range of $50 to 75,000,000 rather than the $14,000,000 indicated.

Additional references [5, 16] provide capacity exponents and, as expected, there is not complete agreement within the sources. Table 16 presents some of the more significant differences in exponent values. Similar cost charts are presented by Berk and Haselbarth [33].

Figures 13a–i present a selected group of cost–capacity charts from Guthrie [16] which represent the capital investment of a battery limits process plant for equipment, basic storage, field materials and erection labor, engineering, indirect costs including contractor's overhead, and profit. In addition, 15 to 20% has been added to represent cost of start-up, working capital, and other capitalized costs. The total is intended to reflect a Gulf Coast mid-1968 installation.

Additional cost charts and exponents are given by Drayer [17] using some foreign construction data.

Auxiliary costs can be conveniently estimated using capacity costs as given in Table 17. Comparative 1962 costs for various utility services are given by Page in Table 18.

Building Costs

Any short-cut estimating technique normally does not include buildings or any special services outside the immediate process area.

Typical building costs vary widely and are summarized in Table 19.

More accurate building cost estimates are usually obtained by describing the building type and construction to a contractor, who can often quote a rather reliable estimating figure. If drawings of similar construction are available, an estimator can provide an estimate specific to that detail. Other valuable estimating references include many building services, such as Boeckh [34]. In general, direct plant-related operational buildings as listed in Table 19 account for 2.5 to 12% of the total capital cost of a project.

Plant process structures to support equipment may be in many shapes and with several floor levels. They may be enclosed with corrugated plastic or asbestos siding on structural steel or left open. The costs of such structures can be estimated by a good estimator with a sketch from a structural designer. Approximate costs may be developed by using Ref. 22.

Indirect Costs

Indirect costs include engineering for the project, project management, purchasing, construction field supervision, and general overhead costs. Various

TABLE 15 1967 Capital-Cost Data for Processing Plants[a]

Compound	Source or Route	Typical Plant Size (tons/yr)	Investment Cost ($)	Investment, ($/annual ton)	Size Factor f^b	Remarks
Acetaldehyde	Ethylene	50,000	3,500,000	70	0.70	Metallic catalyst required
Acetylene	Natural gas	75,000	9,500,000	127	0.70	High purity
Alumina	Bauxite	100,000	9,000,000	90		
Aluminum sulfate		75,000	2,000,000	27		
Ammonia		500,000	16,000,000	32		
Ammonium phosphate		250,000	2,500,000	10	0.70	Fertilizer grade
Ammonium sulfate		140,000	1,200,000	9	0.68	
Carbon black		30,000	3,000,000	100	0.68	
Carbon dioxide		200,000	2,400,000	12		
Carbon tetrachloride		30,000	2,500,000	85		
Butadiene	Butane	100,000	50,000,000	500	0.70	
Butadiene	Butylenes	200,000	70,000,000	350	0.70	
Chlorine/caustic	Cl₂	70,000	13,000,000		0.69	
	NaOH	78,000				
Cyclohexane		100,000	750,000	8	0.70	Does not include hydrogen plant
Diphenylamine		10,000	2,400,000	240		
Ethanolamine		25,000	1,750,000	70		
Ethyl alcohol	From ethylene by direct hydration or via ethyl sulfuric acid	75,000	3,750,000	50	0.72	Manufacturing costs are lower in the direct hydration process
Ethylbenzene } p-Xylene }		20,000 / 8,500	1,800,000 / 1,100,000	200		These chemicals are produced simultaneously
Ethyl chloride		15,000	3,000,000	200		
Ethyl ether		35,000	1,200,000	35		
Ethylene	Refinery gases or hydrocarbons	300,000	15,000,000	50	0.71	
Ethylene dichloride		25,000	3,200,000	127	0.71	
Ethylene oxide	Direct oxidation of ethylene	100,000	9,000,000	90	0.67	Cost also includes conversion to ethylene glycol as needed P_2O_5
37% formaldehyde	Hydrocarbons	100,000	13,000,000	130		
Glycerin (synthetic)		35,000	5,500,000	157	0.67	
Hydrofluoric acid		15,000	2,600,000	175		
Hydrogen		60,000	6,500,000	108		
Isopropyl alcohol		150,000	7,500,000	50	0.80	
Maleic anhydride		50,000	18,000,000	360		

Product	Basis	Capacity	Investment ($)		Size factor	Remarks
Melamine		70,000	11,500,000	164		
Methanol	Natural gases	210,000	9,000,000	43	0.71	
Methyl chloride	Methanol	10,000	500,000	50	0.72	
Methyl ethyl ketone		35,000	3,750,000	107		
Methyl isobutyl ketone		25,000	1,250,000	50		
Methyl isobutyl carbonal		10,000	750,000	75		
Nitric acid		50,000	5,000,000	100		
Oxygen plants		150,000	2,250,000	15	0.71	
Phenol		45,000	9,000,000	200		
Phosphoric acid (as P_2O_5)		100,000	2,400,000	24	0.66	Wet process—contains 30%
cis-Polybutadiene		50,000	12,000,000	240	0.67	
Polyethylene (high-pressure)		200,000	14,000,000	70	0.70	
Polyethylene (low-pressure)		50,000	22,000,000	440	0.70	High-purity ethylene required
Polyisoprene (includes manufacture of the monomer)		30,000	5,000,000	320	0.74	
Soda ash	Natural brine	400,000	34,000,000	85		No synthetics plants built since 1934
Sodium metal		20,000	7,000,000	350		
Styrene		20,000	8,500,000	425		
Sulfuric acid	Contact process	280,000	2,100,000	8	0.67	
Sulfur recovery	Refinery gases	15,000	1,500,000	100		
Toluene diisocyanate		12,500	7,500,000	600		
Urea		140,000	4,300,000	31		
Vinyl acetate		40,000	7,000,000	175		
Vinyl chloride monomer		100,000	2,000,000	20		
Refinery Products		(bbl/d)				
Alkylation units (H_2SO_4 or HF)	From reformer streams, e.g., Udex	10,000	7,750,000	775		
BTX extraction		10,000	3,400,000	340	0.70	
Catalytic cracker (fluid)	Cost based on fresh feed	35,000	14,000,000	400		Includes vapor recovery and CO boiler
Catalytic reformer		23,000	7,500,000	375		
Crude distillation units		100,000	4,700,000	47		
Delayed coker		14,000	5,000,000	357		
Hydrocracker		28,000	21,000,000	750		
Wax plants		7,500	900,000	120		
Gas absorption and dehydration plants		50 MM ft³/d	2,000,000			

a Source: J. E. Haselbarth, *Chemical Engineering* (December 4, 1967). Reprinted by special permission. Copyright © by McGraw-Hill, Inc., New York.
b Where no size factor appears, assume a value of 0.70. Use: To obtain investment for a capacity other than the one shown, multiply the stated investment cost by the ratio of the desired capacity to the stated capacity, raised to the power e.

TABLE 16   Selected Capital-Cost Plant Capacity Exponents[a]

Plant Product	Process Route	Size Exponent, f
ABS[b]	Batch reaction, emulsion, resin/graft	—
Acetylene	Hydrocarbon pyrolysis	0.65
Acrylonitrile	Acetylene and HCN	0.60
Ammonia	Reform natural gas	0.58
Butanol	Propylene, CO, water	0.40
Chlorine	Electrolysis NaC	0.45
Ethylene	Steam pyrolysis refinery gases, naptha, gas oil	0.83
Ethylene oxide	Catalytic oxidation ethylene	0.78
Ethylene glycol	Ethylene, chlorine, including ethylene oxide	0.75
Methanol	CO_2, natural gas reform	0.60
Phenol	Toluene oxidation	0.75
Phthalic anhydride	Naphthalene, air	0.70
Polyethylene (low pressure)	Ethylene, catalyst	0.65
Polyethylene (high pressure)[b]	Ethylene, catalyst	—
Polypropylene[b]	Propylene, catalyst	0.70
Polyvinyl chloride	Vinyl chloride polymerization	0.60
Styrene	Ethylene, benzene	0.60
Urea	Ammonia, carbon dioxide	0.70
Vinyl acetate	Ethylene, oxygen, catalyst	0.65
Vinyl chloride	Ethylene, chlorine, air	0.80

[a]Source: K. M. Guthrie, *Chemical Engineering* (March 23, 1970). Reprinted by special permission. Copyright © by McGraw-Hill, Inc., New York. Note: The capital costs include all major equipment, basic storage, field materials and erection labor, engineering, indirect costs including taxes, freight, contractor's overhead and profit. Site development, off-site facilities (warehouses, utility buildings, etc.) and off-site utilities not included.
[b]Suggestion: PE = 0.68–0.71; ABS = 0.65–0.75; PP = 0.75–0.85. Use 0.80 = 0.95 for separate extrusion and large volume storage finished product, then add back to balance of plant costs.

techniques are used to develop these depending upon the available records and history of various projects from the simple to the complex. Without back-up records it is necessary to resort to averages developed from "typical" process plants.

Engineering

Engineering costs are expressed using a number of different reference points in the published literature and therefore should be examined before using in general form.

Estimating is easier and often more consistent when a constant reference cost category is used. The delivered cost of all equipment, or all equipment and job materials, are often the most common selected.

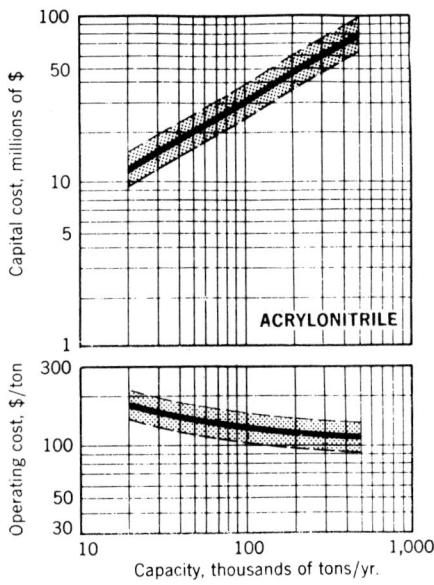

Acrylonitrile: Via catalytic reaction involving acetylene
and hydrogen cyanide feedstock.
Size Exponent: 0.60   **Data:** A = 0; E = 1; P = 2
Included: Process unit and storage facilities

FIG. 13a. Capital and operating costs for acrylonitrile. [Reprinted by special permission from
K. M. Guthrie, *Chem. Eng.*, 77, 140 (June 15, 1970), copyright © by McGraw-Hill,
Inc., New York.]

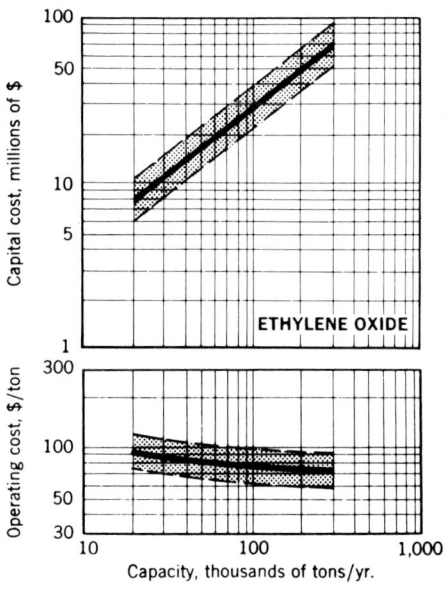

Ethylene oxide (high-purity): From commercial ethylene, air
or oxygen, via catalytic oxidation and isothermal process.
Byproducts: High and low grades of ethylene glycols.
Size Exponent: 0.78   **Data:** A = 0; E = 2; P = 2
Included: Process unit and storage facilities

FIG. 13b. Capital and operating costs for ethylene oxide. [Reprinted by special permission from
K. M. Guthrie, *Chem. Eng.*, 77, 140 (June 15, 1970), copyright © by McGraw-Hill,
Inc., New York.]

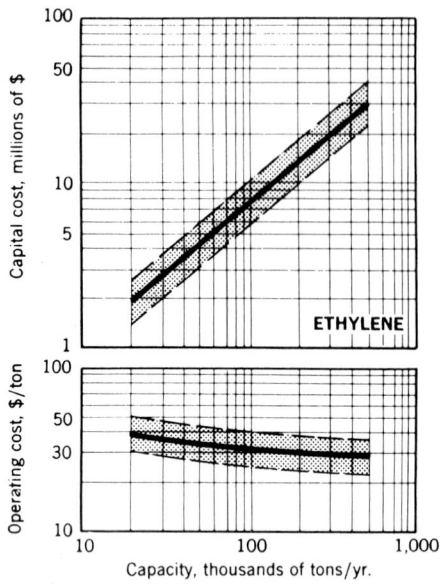

Ethylene (high-purity): From refinery gases, naphthas and gas oils via steam pyrolysis and low-temperature distillation. Byproducts: Propylene, methane, H_2.

Size Exponent: 0.83   **Data:** A = 1; E = 3; P = 4

Included: Process unit and storage facilities

FIG. 13c. Capital and operating costs for ethylene. [Reprinted by special permission from K. M. Guthrie, *Chem. Eng.*, 77, 140 (June 15, 1970), copyright © by McGraw-Hill, Inc., New York.]

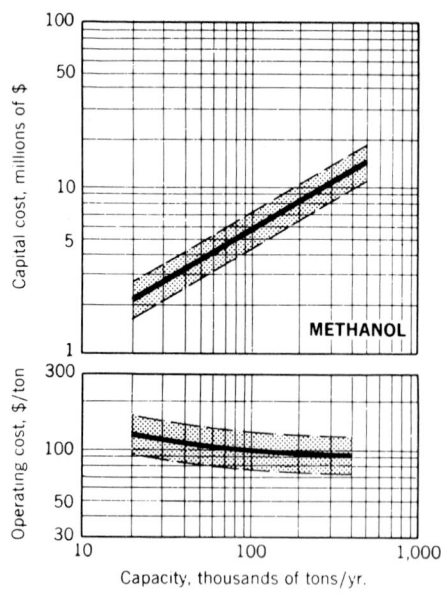

Methanol: From carbon dioxide, natural gas and steam, via a reforming and synthesis process. Byproduct: Fuel gas

Size Exponent: 0.60   **Data:** A = 2; E = 2; P = 4

Included: Process unit and storage facilities

FIG. 13d. Capital and operating costs for methanol. [Reprinted by special permission from K. M. Guthrie, *Chem. Eng.*, 77, 140 (June 15, 1970), copyright © by McGraw-Hill, Inc., New York.]

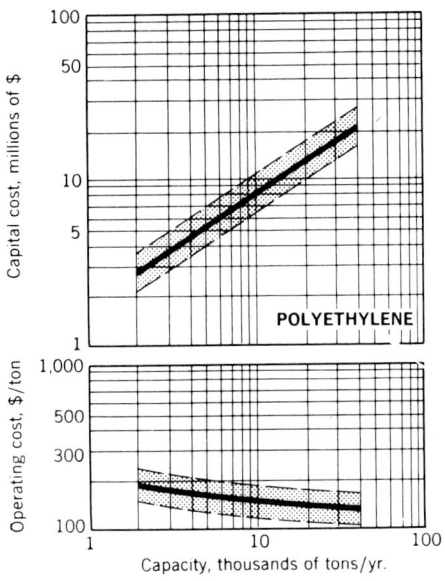

Polyethylene (high-density): From pure ethylene,
via a catalytic reaction and high-pressure separation.
Size Exponent: 0.65   **Data:** A = 2; E = 2; P = 4
Included: Process unit and storage facilities

FIG. 13e. Capital and operating costs for polyethylene. [Reprinted by special permission from
K. M. Guthrie, *Chem. Eng.*, 77, 140 (June 15, 1970), copyright © by McGraw-Hill,
Inc., New York.]

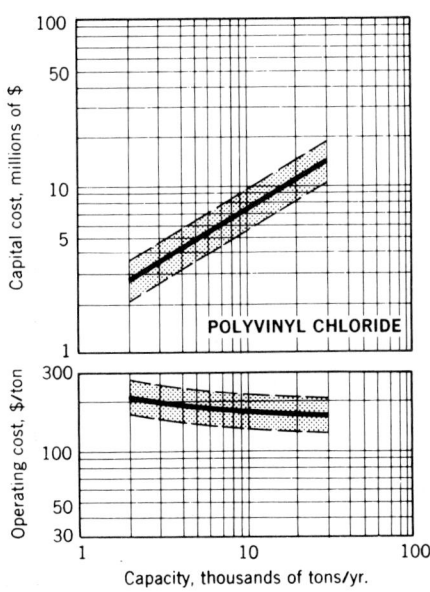

Polyvinyl chloride: From vinyl-chloride-rich monomer
streams, via polymerization and filtration.
Size Exponent: 0.60   **Data:** A = 0; E = 2; P = 4
Included: Process unit only

FIG. 13f. Capital and operating costs for polyvinyl chloride. [Reprinted by special permission
from K. M. Guthrie, *Chem. Eng.*, 77, 140 (June 15, 1970), copyright © by McGraw-
Hill, Inc., New York.]

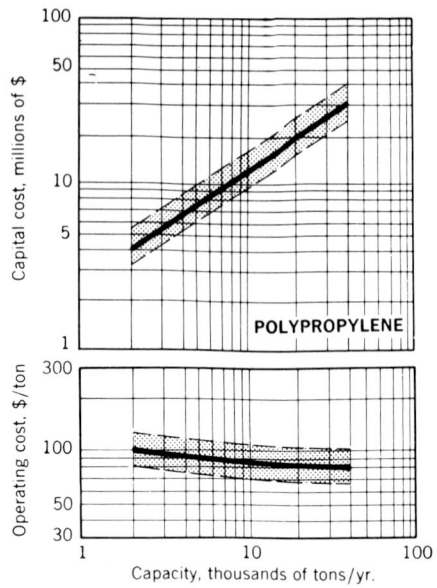

Polypropylene: From propylene-rich streams, via a
low-pressure catalytic reaction and polymerization.
Size Exponent: 0.70   **Data:** A = 0; E = 2; P = 3
Included: Process unit only

FIG. 13g.   Capital and operating costs for polypropylene. [Reprinted by special permission from
K. M. Guthrie, *Chem. Eng.*, *77*, 140 (June 15, 1970), copyright © by McGraw-Hill,
Inc., New York.]

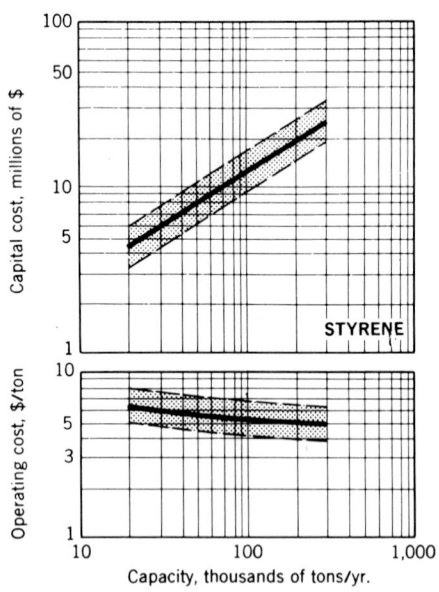

Styrene: From benzene, ethylene, steam, via a catalytic
reaction, with alkylation and distillation to recover
styrene monomer.
Size Exponent: 0.60   **Data:** A = 0; E = 2; P = 4
Included: Process unit only

FIG. 13h.   Capital and operating costs for styrene. [Reprinted by special permission from K. M.
Guthrie, *Chem. Eng.*, *77*, 140 (June 15, 1970), copyright © by McGraw-Hill, Inc.,
New York.]

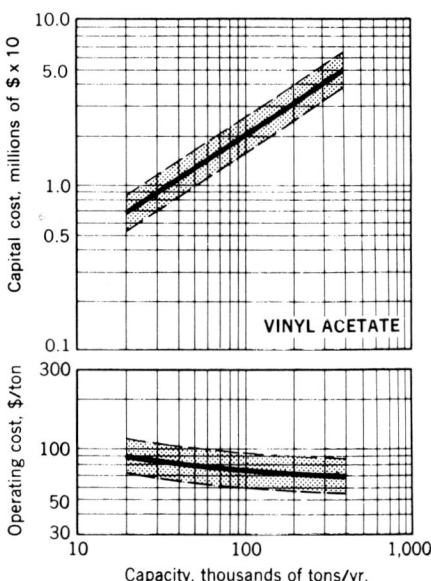

Vinyl acetate: From ethylene and oxygen, via a catalytic reaction and distillation.
Size Exponent: 0.65   **Data:** A = 2; E = 3; P = 2
Included: Process unit only

FIG. 13i.   Capital and operating costs for vinyl acetate. [Reprinted by special permission from K. M. Guthrie, *Chem. Eng.*, 77, 140 (June 15, 1970), copyright © by McGraw-Hill, Inc., New York.]

TABLE 17   Estimating Costs for Auxilliaries[a]

Service	Size Range	Cost Basis ($)
Package steam generators	75,000–10,000 lb/h	2.50–5.00/lb
Large steam generators	300,000–80,000 lb/h	4.00–14.00/lb
Refrigeration systems	1,000–100 tons	250–500/ton
Air conditioning systems	1,000–50 tons	900–1,350/ton
Cooling towers	25,000–1,000 gal/min	15–35/gal
Electrical distribution systems	10,000–1,000 kW	60–120/kW
CS storage spheres		
50 (lb/in.^{2}gauge)	5,000–50 tons	20–80/ton (water)
CS storage tanks (atm)	5,000–50 tons	10–70/ton (water)
SS storage tanks (atm)	500–50 tons	90–240/ton (water)

[a]Source: J. M. Berk and J. E. Haselbarth [33]. Reprinted by special permission. Copyright © by McGraw-Hill, Inc., New York. Note: Costs are approximately 1960 and must be adjusted for later years by cost indexes.

TABLE 18   Installed Cost of Utilities[a]

| | Total Installed |
Type of Service	Cost ($)
Steam	6.00 per lb/h
Electricity	240.00 per kVA
Refrigeration	600 per ton
Compressed air	50 per ft^3/min
Manufactured gas	5.00 per ft^3/h
Sewage disposal	3.00 per gal/h
Cooling towers	0.75 per gal/h
Well water	0.60 per gal/h
Filtered water	1.50 per gal/h
Softened water	3.00 per gal/h

[a]Source: J. S. Page [27]. Data used by permission. Reference costs to June 1962; Marshall and Stevens. All Industry Index = 238.7; Process Industry Index = 236.2.

Guthrie [22] uses the direct equipment and all other materials costs as a reference to build an estimate. For typical petrochemical plant estimates, see Table 20.

To correct for other labor/materials relationships, use Fig. 14 and the various categories of plant types shown in Table 21. Then,

$$\text{Total engineering costs (dollars)} = (M + L)(0.10)\, F_{ce} F_{me} F_{pt}\,(\text{Index})$$

where M = direct equipment plus all other direct materials costs
L = direct labor costs
F_{ce} = labor/materials ratio factor from Fig. 14
F_{me} = materials + labor magnitude factor from Fig. 14
F_{pt} = project type factor from Table 21
Index = Mid-1968 Gulf Coast costs, referenced to a compatible index such as *Chemical Engineering* Index

Engineering costs, including the necessary supervision and office costs noted in the listing above, will run 32 to 33% of delivered equipment costs for most process industry facilities [35].

When engineering costs are expressed as a percent of total project costs, the following ranges are based on historical data [36]: 6 to 28%, with a median of about 12%. This then fits a typical refinery or petrochemical plant giving an order-of-magnitude distribution of costs as 12–63–25% for engineering, materials, and labor. By making a relative comparison of the complexity of engineering work in a project, the extent of engineering services can be forecast and then referenced to the breakdown suggested in Table 22.

TABLE 19 Estimating Costs for Buildings, Single Story[a]

Type Building	Cost Range Dollars ($/ft^2 floor area)
Warehouse, single story: steel frame, corrugated asbestos walls, reinforced concrete floors and roof, toilet facilities, lighting, heating, sprinklers	9–15; median 11
Shops and maintenance buildings (less tools and equipment): steel frame, corregated asbestos walls (or concrete block), concrete roof and floors, lighting, heating, toilet facilities, sprinklers	9–18; median 12
Add for average shop equipment	7–14
Change house (less equipment, lockers): steel frame, concrete block walls, concrete roof and floors, lighting, heating, toilet and showers, sprinklers	9–18; median 18
Laboratory buildings (less equipment): steel frame, concrete block walls, concrete roof and floors, toilet facilities, special plumbing, heating, lighting, sprinklers, air conditioned	18–30; median 21
Add for laboratory furniture and equipment	6–20
Cafeterias (less equipment): steel frame, concrete block walls, concrete floor and roof, toilet facilites, special plumbing, heating, lighting, sprinklers, air conditioned	18–45; median 35
Add for cafeteria equipment	5–10
Office buildings: steel frame, concrete block (or concrete and brick) walls, concrete floor and roof, toilet facilities, heating, lighting, sprinklers, air conditioned	20–35; median 28
Add for normal office furniture and equipment	7–11
General notes: add for	
Floor tile	1.50
Special painting	1.25
Special interior finishes	4–7
Communicating systems, phones	0.75
Ventilating only (fans), for nonair conditioned building	0.75–1.25

[a]Note: Costs suggested above give average building finish; for special architectural effects, costs may be double those indicated. Reference years: 1970–1972.

TABLE 20   Breakdown of Engineering Related Costs[a]

Costs	Percentage of Equipment and Other Materials[b]
Total engineering	13.6
Project engineering	1.9
Process engineering	0.6
Design and drafting	3.6
Procurement	0.4
Home office construction	0.2
Office indirects and overhead	6.9

[a]Source: K. M. Guthrie, *Chemical Engineering* (March 24, 1969). Reprinted by special permission. Copyright © by McGraw-Hill, Inc., New York.
[b]For project with $3.3 million direct materials and labor/materials of 0.36; and total engineering costs ratioed to total equipment plus materials plus direct labor of 0.10 [that is (direct equipment plus other materials plus direct labor) times 0.10 gives engineering costs] other materials run 0.62 times direct equipment costs.

Construction Related Costs

Other than direct labor, the construction costs are included in the indirect cost category. These "indirects" must collect all the extraneous costs required to execute the erection phase. The items in this category normally include those listed below and are often peculiar to each job, its geographical location, labor conditions, and other factors.

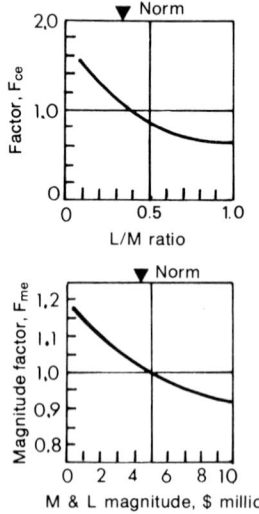

FIG. 14.   Labor/materials adjustment factors for developing engineering costs. [Reprinted by special permission from K. M. Guthrie, *Chem. Eng.*, *76*, 120 (March 24, 1969), copyright © by McGraw-Hill, Inc., New York.]

TABLE 21   Project-Type Classification[a]

	Classification Factor
Chemical complex	1.4
Chemical processing plant	1.0
Solids/fluid processing	0.8
Solids handling	0.6
Buildings only	0.4

[a]Source: K. M. Guthrie, *Chemical Engineering* (March 24, 1969). Reprinted by special permission. Copyright © by McGraw-Hill, Inc., New York.

Fringe benefits include the employer's payments to employee funds for all craftsmen and laborers (nonsupervisory) such as union or other health, hospital, etc., plus vacations, holidays, sick leave, and retirement. If the job requires travel pay or subsistence pay for craftsmen, these are added here.

Labor burden includes the employer's required contribution to Federal Social Security, Federal Unemployment Insurance, State Unemployment Insurance, and Workmen's Compensation.

Field supervision includes the salaries, travel, and subsistence when applicable, fringe benefits, and payroll burdens for the supervisory, inspection, and support personnel.

Temporary facilities includes all support facilities needed to construct, office, and fabricate the permanent plant—buildings; rest rooms; first-aid; roads; parking areas; storage areas; fences; scaffolding; water, gas, oil and services.

Construction equipment includes all costs and rental or lease charges for construction machinery required at various stages of a project—cranes, bulldozers, road graders, welding machines, etc.

Small tools includes the hundreds of miscellaneous tools used, broken, and lost during the course of a job and usually considered an expendable commodity for the project. Individual values are set at a maximum of $150 to 250 to cover pipe wrenches, crescent wrenches, welders gloves and helmets, safety hard hats, safety glasses, electric drills, etc.

TABLE 22   Percentage of Engineering Costs for Total Project

Size Project ($)	Most Inorganic, Not Complicated	Organic with Average Piping, Complex Inorganic	High Detail for Piping and Other
< $500,000	12–18; 12%[a]	20–25; 21%[a]	20–35; 25%[a]
500 M–2 MM	8–15; 11%	10–20; 15%	12–25; 18%
3 MM–10 MM	5–12; 10%	10–15; 13%	12–18; 15%
11 MM–20 MM	3–10; 6%	8–13; 10%	10–15; 12%

[a]Suggested median

TABLE 23   Suggested Indirect Construction Costs[a]

	%	Percentage Direct Labor Cost	Percentage Direct Labor Plus All Material[b]
Fringe benefits	14.8	10.0	2.6
Labor burden	22.4	15.0	4.0
Field supervision	17.8	12.0	3.2
Temporary facilities	8.9	6.0	1.6
Construction equipment	14.8	10.0	2.6
Small tools	3.6	2.4	0.6
Miscellaneous	17.7	12.0	3.2
Total construction overhead	100.0	67.4	17.8

[a]Source: K. M. Guthrie, *Chemical Engineering* (March 24, 1969). Reprinted by special permission. Copyright © by McGraw-Hill, Inc., New York.
[b]Equipment and field materials (average) = 1.62 (equipment only).

Miscellaneous includes all the other hundreds of costs required for medical services, trash pick-up, job area cleaning, consumable supplies such as welding rod and paper cups, job insurance for public liability, and automobile damage and liability.

Guthrie gives an average project indirect cost breakdown in the $2 million dollar range in Table 23.

Total construction costs may be estimated using Fig. 15.

$$\text{Total construction costs} = (M + L)(0.178) \, F_{co} F_{mo} \, (\text{Index})$$

where M = direct equipment plus all other direct materials costs
    L = direct labor costs
    F_{co} = labor/materials ratio factor from Fig. 15
    F_{mo} = materials plus labor magnitude factor from Fig. 15

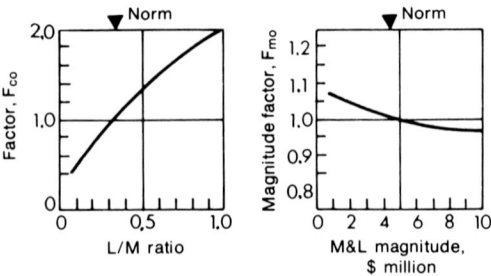

FIG. 15.   Construction overhead correction factors. [Reprinted by special permission from K. M. Guthrie, *Chem. Eng.*, 76, 119 (March 24, 1969), copyright © by McGraw-Hill, Inc., New York.]

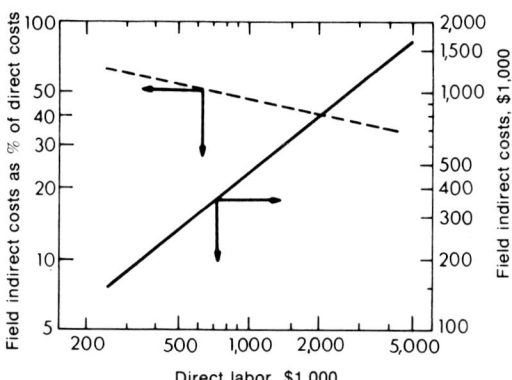

FIG. 16. Relationship between field indirect costs and direct labor costs for construction projects in a range of sizes. [Reprinted by special permission from K. M. Guthrie, *Chem. Eng.*, *75*, 190 (May 20, 1968), copyright © by McGraw-Hill, Inc., New York.]

The average L/M ratio is 0.36 for direct labor to all direct materials (not just equipment), and (delivered equipment)(1.62) = M.

The difficulty in using some estimating techniques is the lack of good translation from equipment costs to the direct labor for installation. Guthrie [16, 22, 37] offers a wide cross-section of such needed data. Once a direct labor figure is established, the chart (Fig. 16) of Gallagher makes an overall estimate of indirect field or construction costs for general use and gives guidance as to the order of magnitude.

Table 24 gives a reasonably detailed breakdown of indirect field costs and is necessary to properly collect field costs for construction control. Figures 17, 18, and 19 show the effects of project size on indirect costs associated with the schedule of Table 24. The costs are based on $5.00/h and 1968 material prices, and need to be adjusted when used for other conditions.

Example 4. Determining Total Field Staff Costs. Determine the total field staff costs and total construction support costs for a project estimated to have 150,000 direct labor man-hours. The plant construction is expected to be about at its midpoint by January 1972. The average hourly wage rate is projected from wage rates for the area to be $6.10/h.

$$\text{Adjusted direct field labor at } 5.00/h = (5.00)(150,000) = \$750,000$$

From Fig. 17, percent total field staff is 14.5%:

$$\text{Field staff cost} = (0.145)(750,000)$$

$$= \$108,750$$

Correct for actual wage rate:

$$\text{Actual field staff cost} = (6.10/5.00)(108,750) = \$132,800$$

TABLE 24 Indirect Field Cost Account Schedule[a]

210000	*Field staff*
210100	Supervision
210200	Accounting
210300	Field engineering
210400	Staff engineering
210500	Warehousing
210600	Service personnel
220000	*Construction support*
220100	Temporary buildings
220200	Temporary roads
220300	Construction utilities
220310	Utility installation
220320	Utility operation
220330	Field communications
220400	Construction supplies
220410	Consumable supplies
220420	Welding supplies
220430	Safety supplies
220440	Office supplies
220450	Scaffolding
220500	Cleanup
230000	*Labor benefits*
230100	Craft benefits
230110	Initial and terminal
230120	Daily transportation
230130	Fringe benefits
230140	Subsistence
230150	Showup time
230200	Payroll taxes and insurance
230300	Construction camp
230310	Camp setup
230320	Camp utilities
230330	Camp operation
230340	Camp buildings
240000	*Equipment and tools*
240100	Construction equipment
240110	Earthmoving equipment
240120	Batch plant equipment
240130	Building and steel erection
240140	Process equipment setting
240150	Pipe erection
240160	Cars and pickup trucks
240170	Other equipment
240200	Small tools
240300	Equipment servicing

[a]Reprinted, by permission, from W. R. Weinheimer, *Trans. Am. Assoc. Cost Eng.*, p. 59 (1969).

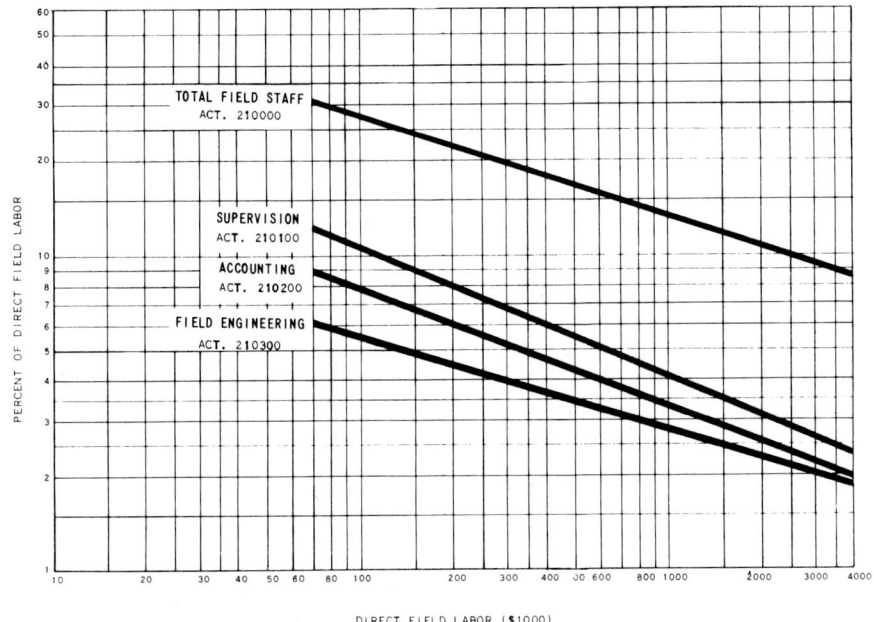

FIG. 17. Field staffing costs shown as a percentage of direct field labor. Craft labor adjusted on the basis of a $5.00/man-h average craft wage rate. Staff labor adjusted to 1968 rates. [Reprinted by permission from W. R. Weinheimer, *Trans. Am. Assoc. Cost Eng.*, p. 60 (1969).]

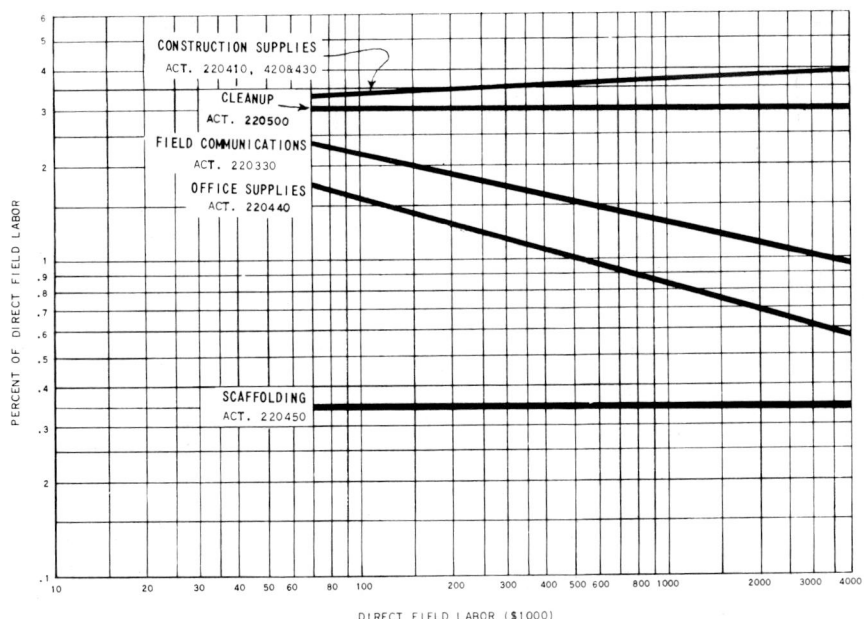

FIG. 18. Construction support costs shown as a percentage of direct field labor. All data adjusted on the basis of a $5.00/man-h average wage rate and 1968 material prices. [Reprinted by permission from W. R. Weinheimer, *Trans. Am. Assoc. Cost Eng.*, p. 60 (1969).]

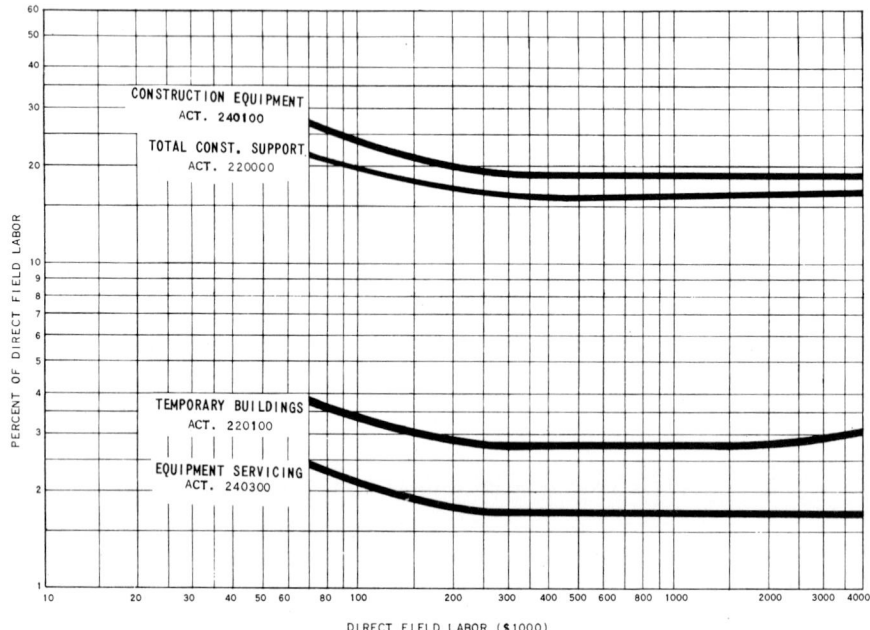

FIG. 19. Construction support and construction equipment costs shown as a percentage of direct field labor. All data adjusted on the basis of a $5.00/man-h average wage rate and 1968 material prices. [Reprinted by permission from W. R. Weinheimer, *Trans. Am. Assoc. Cost Eng.*, p. 61 (1969).]

From Fig. 19:

$$\text{Total construction support} = 16\%$$
$$\text{Costs} = (0.16)(750,000) = \$120,000$$
$$\text{Corrected for wage rate} = (6.10/5.00)(120,000) = \$146,200*$$

Table 12b also lists suggested percentages for indirect cost factors as related to other costs and can be a useful way of developing these costs when other information is not available.

Detailed or Definitive Capital Estimates

Detailed or definitive estimates require more time to prepare, but they provide a more complete breakdown of the individual components of the project. The analysis is more suited to the specific scope and conditions of the project than any made by ratios or percentage relationships. Detailed estimates yield a higher accuracy, usually ± 3 to 8%.

*This account is a mixture of labor and pure material costs and may best be "corrected" by using 50% of (6.10/5.00) ratio *increase*.

Such estimates cannot be prepared at the initial concept planning of a project, since they require not only knowledge of specific items of the project but comprehensive understanding of the scope decisions that make a project firm. If sufficient engineering, estimating, and other needed talents are invested into detailed planning and design, including drawings, a reasonably detailed estimate can be prepared before a project is authorized for engineering and construction. Of course, this might require several months to complete since all items would require listing and pricing. Often *progressive* estimates made as a project evolves from initial planning allow the definitive estimate to be prepared after firm engineering details, specifications, and purchases are about 50 to 75% complete. All the decisions on layouts, equipment, types of construction, and materials should be firm by this time. Some projects move at such a pace that field work is commenced before the definitive estimate is prepared.

Cost of Preparing Cost Estimates

Simple projects that can be estimated by percentages or ratios cost the process company engineering department or the engineering contractor only a few hundred dollars. However, as the project complexity increases and the degree of detail needed to improve the accuracy of the assembled estimate increases, the costs climb to thousands of dollars.

Reported costs for contractor preparation of proposals for construction of various plants are given by Loring [38] as varying from $5,000 for ethylene projects costing $15 to 70 million; $20,000 to $60,000 for a petrochemical project of $100 to 200 million; and $7000 to $90,000 for $70 to 100 million refinery projects. So much is dependent on the scope of work, location, etc. that the costs listed merely serve to demonstrate that developing the cost estimate of a project also costs money. Depending on the type of arrangement with a prospective client, most contractors must absorb the costs of preparing proposals. It is for this reason, in part, that a contractor will often decline to bid, particularly if the field of bidders is more than three or if they recognize that the practices of some competitors may be questionable. For example, some contractors will bid low on certain types of contracts and will have reopening clauses whereby they can add on high extras after the bid is issued. When a contractor is supplying an estimate, proposal costs will usually be greater if the owner's standards are used rather than the contractor's.

Table 25 indicates a typical spread of man-hours used by various contractor departments in preparing a proposal for a plant.

Sources of Cost Data

For estimates expected to be 3 to 8% accurate it is essential to assemble the details using:

1. Current quotations from manufacturers for the specific equipment and materials involved
2. Updated material costs for commodity type items as well as purchased tools, etc.
3. Updated rental/lease costs for construction machinery, tools, etc.
4. Updated labor requirements such as man-hours to perform various work functions based on historical records and accurate percentage analysis of similar projects or segments of projects
5. Analysis of construction assembly/erection sequences by experienced construction field personnel
6. Updated field overhead and services costs based on historical records
7. Corrections of available historical data/costs for differences in geographical, climate (weather), and labor conditions
8. Known peculiarities in working with a particular client's organization (if a contractor) or with a particular contractor (if a process company estimate)

TABLE 25 Contractor Man-hours Allocated to Proposal Preparation by Various Departments[a]

| | Departmental Man-Hour Allocations | | | | |
Type of Plant	Process	Engineering	Proposal	Estimating	Others[b]
Ammonia	700	1200	1200	500	300
Ammonia + urea	150	200	500	300	100
Ammonia + urea	600	340	200	160	200
Chemical	300	1000	800	350	150
Chemical	350	200	300	250	100
Chemical	400	400	400	500	200
Ethylene	1000	100	200	100	100
Ethylene	700	350	1000	500	100
Nitric acid	250	200	250	350	150
Nitric acid	280	180	230	460	200
Petrochemical	—	100	150	250	50
Petrochemical	800	600	850	300	250
Petrochemical	800	3000	800	1200	1000
Refinery (oil)	820	2100	2000	1100	400
Refinery (oil)	1100	300	800	300	200
Sulfuric acid	200	50	100	100	100
Sulfuric acid	200	500	500	600	375
Sulfuric acid	300	450	450	900	350
Sulfuric acid	200	150	300	200	250
Urea	250	400	400	150	100
Urea	1100	6800	2000	2400	2000
Urea	1000	3600	1000	3500	1200
Waste treatment	120	175	200	200	100

[a]Source: R. E. Loring, *Chemical Engineering* (November 16, 1970). Reprinted by special permission. Copyright © by McGraw-Hill, Inc., New York.
[b]Includes departments such as Purchasing and Construction.

Equipment and Material Quotations

Firm quotations based on written job specifications give the most accurate data to include in the estimate. Since these costs amount to 35 to 65% of many projects, they form a solid backup for the reliability of the assembled cost estimate. Sometimes quotations take longer to obtain; however, most manufacturers can furnish quite reliable estimating or preliminary costs ± 2 to 5% on a few days notice. They are accustomed to providing this information from their pricing books and often issue such books or price lists to reputable contractors and process company estimating departments.

The records of recent purchases are also a reliable and quick source of good estimating information for new projects. It is quite evident that to perform rapid, reliable estimates, a regular program of record keeping, updating and integration, and evaluation of information is essential.

Well-documented descriptive cost charts can be sufficiently accurate for many estimates when quotations are not available. Care must be given to the age of the data in order to update for a current estimate using an appropriate cost index. For accurate estimates it is not advisable to update over 5 yr.

Since the definitive estimate is intended to represent as much firm cost information for all categories as can be accumulated, the use of general charted cost data destroys some of the confidence in the results and, in effect, suggests that the estimate loses the expected 3 to 8% accuracy classification.

Good cost information is found in Refs. 5, 19, 22, 27–31, 37, 39, and 40; however, these are not equivalent to firm or preliminary quotations. Charts or graphs of costs are not included in this section, since they usually are not suitable for the complete preparation of a definitive estimate.

Instruments are an important segment of nearly all process industry estimates. The proper estimation of these costs requires special understanding of the many details involved. Liptak [20] has a fine presentation of the development of instrumentation costs for a process plant.

Cost Estimate Presentations

The forms used for both the accumulation of details as well as the summary of the cost categories vary to suit the individual companies. The first phase of developing the definitive or detailed estimate is to list and price or cost out the individual items and segments of work. Forms similar to Figs. 20, 21, 22, and 23 are useful for detailing the items and work to be performed. When estimates are prepared by computers, the printout can be quite detailed as will be illustrated later.

Estimate summary forms are essential for accumulating the detailed information (Figs. 24–28). Once data are accumulated, major component classifications can be reviewed and examined with reference to other similar estimates. When data is properly grouped by general size and type/complexity of project, the calculation of percentages for the categories can provide a quick order-of-magnitude check on the validity of the estimate. If some category is significantly high or low, it should be reexamined.

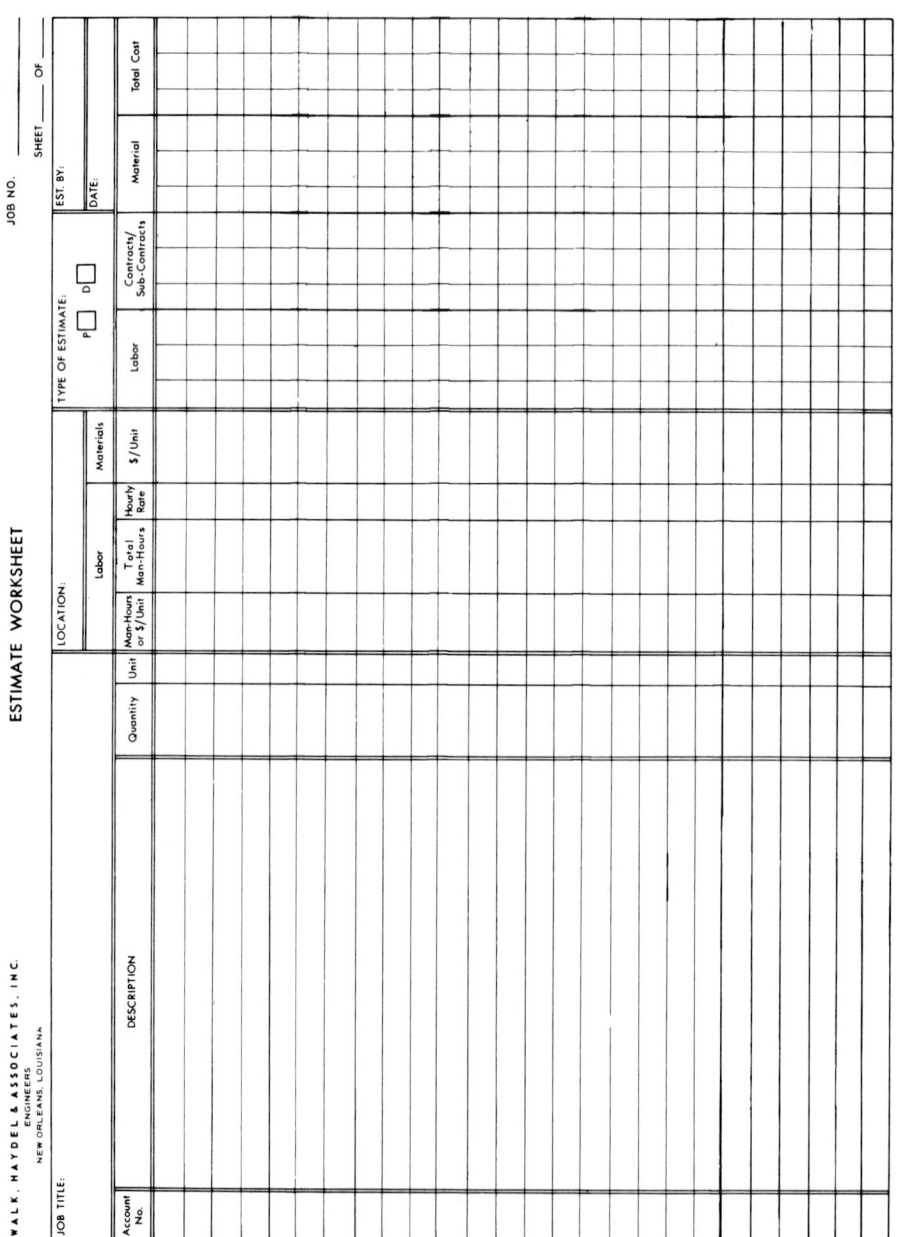

FIG. 20. Estimate work sheet. (Reprinted by permission from Walk, Haydel and Associates, New Orleans.)

COST ESTIMATE
WORK SHEET

ENG. DEPT. JOB NO.______
______OF______WORK SHEETS

PLANT______ TITLE______ SCHD. NO.______ SUB-DIV. NO.______

DESCRIPTION	QUAN	UNIT	UNIT MATL. COST	UNIT LABOR COST	MATL. COST	LABOR COST	TOTAL

ESTIMATED BY______ ______19____
CHECKED BY______ ______19____
ENG. APPROVAL______ ______19____

FIG. 21. Cost estimate work sheet.

THE DOW CHEMICAL COMPANY
PITTSBURG, CALIFORNIA

ENGINEERING DEPARTMENT
COST ESTIMATE - BREAKDOWN

JOB NO. 5916
CHARGE
CONSTRUCTION BY DOW ☐ CONTRACTOR ☒
CLASSIFICATION B & A ☒ EXPENSE ☐

DATE Oct. 8 - 1958
PAGE 6 OF 16
PREPARED BY W. G. Clark
CHECKED BY

ACCOUNT NO.	DESCRIPTION	UNIT	QUAN.	UNIT LABOR	UNIT MATERIAL	LABOR	MATERIAL	OTHER	TOTAL
4000	EQUIPMENT								
4060	COLUMNS								
	B-1 Batch Still								
	1956 Cost $4800.00	ea	1		5750.-	—	5750.-	—	5750.-
	B-1 Dip Pipes								
	1956 Cost $300.00	ea	1		360.-	—	360.-	—	360.-
4140	HEAT EXCHANGERS								
	H-1 Shell & Tube								
	1956 Cost $6200	ea	1		7450.-	—	7450.-	—	7450.-
4230	SEPARATOR								
	T-32 Phase Separator								
	1956 Cost $700.00	ea	1		840.-	—	840.-	—	840.-
4240	TANKS								
	X-11 De-Super heater	ea	1		300.-	—	300.-	—	300.-
4270	PLATFORMS								
	B-1 Operating Platform	ton	0.5	150.-	400.-	75	200.-		275.-
	Floor Plate	#	40	0.50	1.50	20.-	60.-		80.-
	Hand Railing	lf	20	2.-	1.50	40.-	30.-		70.-
	Ladder	lf	20	3.-	10	60.-	200.-		260.-
4300	INSTALLATION								
	B-1	ea	1	400.-	50.-	400.-	50.-		450.-
	H-1	ea	1	150	10.-	150.-	10.-		160.-
					SUB TOTAL	785.-	15250		16035.-

FIG. 22. Cost estimate breakdown form. [Reprinted by permission from W. G. Clark, *Pet. Refiner*, *38*(9), 268 (1959).]

THE DOW CHEMICAL COMPANY
PITTSBURG, CALIFORNIA

ENGINEERING DEPARTMENT
COST ESTIMATE - BREAKDOWN

JOB NO. _____5916_____   DATE __Oct. 8 - 1958__
CHARGE _____   PAGE __7__ of __16__
CONSTRUCTION BY DOW [ ] CONTRACTOR [X]   PREPARED BY __W. G. Clark__
CLASSIFICATION B & A [X] EXPENSE [ ]   CHECKED BY __WGC__

ACCOUNT NO.	DESCRIPTION	UNIT	QUAN.	UNIT LABOR	UNIT MATERIAL	LABOR	MATERIAL	OTHER	TOTAL
4000	EQUIPMENT								
	SUBTOTAL PG #6					785	15250		16035
4300	INSTALLATION (cont'd)								
	T-32	ea	1	75	10	75	10	?	85
	X-11	ea	1	30	5	30	5	?	35
4260	INSULATION								
	B-1 85% MAG	#	650		3	~	~	1950	1950
4320	PAINTING								
	ALL EQUIPMENT	#	1000		0.25	~	~	250	250
4330	TESTING								
	Hydrostatic	allow				200	50		250
	NO 4000 SUB TOTAL					1090	15315	2700	18605
	SALES TAX	%	4			?	610	?	610
	FREIGHT	%	5			?	765	?	765
	CONTRACTOR OVERHEAD	%	65	*6		700		130	830
	CONTRACTOR PROFIT	%	10	*5		180	830	~	1010
	CONTINGENCY	%	15	*8 *10		300	1400	230	1930
	NO 4000 TOTAL					2270	18920	2560	23750

SUB TOTAL

FIG. 23. Cost estimate breakdown form showing estimate total. [Reprinted by permission from W. G. Clark, *Pet. Refiner*, *38*(9), 268 (1959).]

As estimates are updated during the progress of either design engineering or construction, it is helpful to compare the cost categories as changes occur. The changes may be the result of unforeseen conditions or specific decisions to add to or otherwise change the project in some manner. Figures 29, 30, and 31 suggest forms that provide for identifying the areas of change, as well as the anticipated revised new total cost.

Computers in Cost Estimating

Sophisticated cost estimating systems are now programmed for the electronic computer. These programs require large internal storage banks for pricing data; that is, tables of individual items of material and labor components for thousands of specific materials are entered and stored. In addition, there are assembly subprograms for developing the estimates of pressure vessels and other equipment based on size, design pressure, and materials of construction. It is obvious that this data must be kept up to date by specially assigned personnel.

When an estimate is required, the basic conditions are entered into the computer from an estimate take-off assembled in a manner consistent with the capability of the estimating program in the computer. The material take-off steps are required even if the estimate is prepared without the use of the computer. The advantage of the computer is that it can take a straightforward take-off and simulate for estimating purposes all the cost-related steps, price them, and print out all the details for review. This is extremely time saving for large or complex projects involving a wide variety of pipe sizes and construction materials. Several examples are given in Fig. 32 (civil or structural) and 33 (piping).

COST ESTIMATE SUMMARY

CUSTOMER						PROP. NO.
LOCATION						JOB NO.
PROJECT						DATE
						BY
REV. NO.	REV. DATE					BY

ACT	DESCRIPTION	CRAFT HOURS	LABOR	MATERIAL	OTHER	TOTAL
11	Earthwork & Concrete					
12	Buildings & Structural					
13	Process Equipment					
14	Piping					
15	Electrical					
16	Instrumentation					
17	Insulation					
18	Painting					
19	Plant Facilities					
	TOTAL DIRECT FIELD COST					
21	Field Staff					
22	Construction Support					
23	Craft Benefits					
24	Construction Equipment					
	TOTAL INDIRECT FIELD COST					
	TOTAL FIELD COST					
30	Engineering					
	TOTAL FIELD & ENGINEERING COST					
41	Premium Pay					
42	Legalities					
43	Escalation					
44	Contingency					
	SUB TOTAL					
50	Fee					
	TOTAL					

FIG. 24. Cost estimate summary. [Reprinted by permission from W. R. Weinheimer, *Trans. Am. Assoc. Cost Eng.*, p. 62 (1969).]

Cost Estimate Summary

PLANT: _____ JOB NO. _____ ACCT. NO. _____

TITLE: _____ EST. BY _____ DATE _____

                                 CHK'D BY _____ DATE _____

	*MATERIAL	*LABOR	OVERHEAD	TOTAL
BUILDINGS _____				
EARTHWORK – WAYS – YARDS_____				
FOUNDATIONS _____				
SUPERSTRUCTURES_____				
CHEMICAL PROCESS EQUIPMENT_____				
PIPELINES _____				
GENERAL MECHANICAL _____				
INSTRUMENTATION _____				
_____				
ELECTRICAL – APPARATUS – DIST. _____				
MOBILE EQUIPMENT _____				
PAINTING _____				
TOTALS				

*Costs include Contractor's Job Costs and General Contractor's Fee on Sub-Contract.

OVERHEAD COSTS:

FIELD INSPECTION   ____ %   _____

ENGINEERING   ____ %   _____

CONSTRUCTION OVERHEAD   ____ %   _____

CONTINGENCIES   ____ %   _____

Total   _____

FIG. 25.   Cost estimate summary form.

COST ESTIMATE

CUSTOMER _____   DESCRIPTION _____   NO. _____
                                 _____   W.O. NO. _____
LOCATION _____   _____   CONT. NO. _____
                                 _____   MADE BY _____
PROJECT _____   _____   APPROVED _____

A/C NO.	ITEM & DESCRIPTION	MANHOURS	ESTIMATED COST			
			LABOR	SUB-CONTRACTS	MATERIALS	TOTAL
	Excavation					
	Concrete					
	Structural Steel					
	Buildings					
	Machinery & Equipment					
	Piping					
	Electrical					
	Instruments					
	Painting & Scaffolding					
	Insulation					
	DIRECT FIELD COSTS					
	International Expense					
	Temporary Construction Facilities					
	Constr. Services, Supplies & Expense					
	Field Staff, Subsistence & Expense					
	Craft Benefits, Payroll Burdens & Insur.					
	Equipment Rental					
	Small Tools					
	INDIRECT FIELD COSTS					
	TOTAL FIELD COSTS					
	Home Office Construction					
	Project Engineering					
	Process Engineering					
	Design					
	Purchasing					
	Business Services					
	Office Expense					
	Office Payroll Burdens					
	Indirect Office Costs					
	TOTAL OFFICE COSTS					
	TOTAL FIELD & OFFICE COSTS					
	Fee					
	Sales Tax					
	Escalation					
	Contingency					
	TOTAL					

DATE _____ REVISION NO. _____ REVISION DATE _____ PAGE NO. _____

FIG. 26.   Construction cost estimate form.

WORK SHEET SUMMARY

PLANT _____ JOB NO. _____ ACCT. NO. _____
TITLE _____ EST. BY _____ DATE _____
                                                       CHK'D. BY _____ DATE _____

LINE OR SUB-DIV NO.	DESCRIPTION	MATERIAL	LABOR	TOTAL
1	SITE DEVELOPMENT:			
2	FILL, GRADING, SITE PREPARATION			
3	ROADS AND PAVING			
4	PIPELINES TO BLDG. LINE			
5	SEWERS TO BLDG. LINE			
6	RAILROAD			
7	ELECTRICAL DISTRIBUTION TO BLDG. LINE _____			
8	SUB-TOTAL (LINES 2-7)			
9	CONTRACTOR'S JOB COST (%M %L LINE 8)			
10	CONTRACTOR'S FEE ON SUB-CONTRACTS (%)			
11	SUB-TOTAL (LINES 8-10)			
12				
13	BUILDINGS:			
14	PILING			
15	CONCRETE			
16	STRUCTURAL STEEL AND STEEL SASH			
17	MASONRY			
18	CARPENTRY, MILLWORK AND MISC. COSTS			
19	ROOFING AND SIDING			
20	PAINTING			
21	ELECTRICAL			
22	PLUMBING, HEATING AND VENTILATING			
23	FIRE PROTECTION			
24	SUB-TOTAL (LINES 14-23)			
25	CONTRACTOR'S JOB COST (%M %L LINE 24)			
26	CONTRACTOR'S FEE ON SUB-CONTRACTS (%)			
27	SUB-TOTAL (LINES 24-26)			
28				
29	EQUIPMENT:			
30	FOUNDATIONS			
31	SUPERSTRUCTURES			
32	PROCESS EQUIPMENT			
33	INSTRUMENTATION			
34	MECHANICAL EQUIPMENT (A) COMPRESSORS, PUMPS, ETC.			
35	MECHANICAL EQUIPMENT (B) OTHER			
36	PIPING			
37	ELECTRICAL			
38	PAINTING			
39	INSULATION			
40	FIRE PROTECTION			
41	SUB-TOTAL (LINES 30-40)			
42	CONTRACTOR'S JOB COST (%M %L LINE 41)			
43	CONTRACTOR'S FEE ON SUB-CONTRACTS (%)			
44	SUB-TOTAL (LINES 41-43)			
45	SUB-TOTAL (LINES 11, 27 AND 44)			
46				
47	OVERHEAD:			
48	FIELD INSPECTION %LINE 45			
49	ENGINEERING %LINE 45			
50	CONSTRUCTION OVERHEAD %LINE 45			
51	SUB-TOTAL (LINES 48-50)			
52	SUB-TOTAL (LINES 45 AND 51)			
53				
54	CONTINGENCIES: %LINE 45			
55	TOTAL: (LINES 52 AND 54)			

FIG. 27.   Cost estimate work sheet summary.

Figure 34 illustrates a block flow diagram of the use of the computer to develop the cost estimate along with tying into inventory of materials required and requisitioning materials not in stock. An additional feature is the tie-in to the automated drafting program for drawing preparation and simultaneous material take-off.

The American Association of Cost Engineers is developing a Cost of Major Equipment System Data Bank program. This project provides for a definition of the cost elements, collection and classification of cost records, and analysis and utilization of the data [41].

Specialized Costs

The costs associated with many of the details in a process plant are quite specialized. They require good technical understanding and the ability to include judgment into the cost development operation. Specifically developed cost data are needed for such important cost steps as:

WALK, HAYDEL & ASSOCIATES, INC.
ENGINEERS
NEW ORLEANS, LOUISIANA

ESTIMATE SUMMARY

PROJECT	TYPE OF ESTIMATE	Sheet 1 of
		REVISION #
LOCATION	P ☐	JOB NO.
		DATE
STRUCTURE	D ☐	ESTIMATOR
		CHECKED BY

ACCT. NO.	DESCRIPTION	LABOR	CONTRACT/ SUBCONTRACT	MATERIAL	TOTAL
	TOTAL DIRECT COSTS				
	GRAND TOTAL ESTIMATED COST				

FIG. 28. Estimate summary form. (Reprinted by permission from Walk, Haydel and Associates, New Orleans.)

REVISED ESTIMATE AND JOB COST REPORT

SAMPLE COPY

ACCOUNT NUMBER	DESCRIPTION	COMMITMENTS		ESTIMATE TO COMPLETE		REVISED
		MATERIAL	LABOR	MATERIAL	LABOR	MATERIAL
	PROCESS EQUIPMENT					
A	TOWERS	895,000	90,000	15,000	25,000	910,000
B	BOILERS, STEAM SUPERHEATERS	-	-	-	-	-
F	PROCESS FURNACES	1,050,000	53,000	160,000	62,000	1,210,000
G	GENERAL EQUIPMENT	170,500	31,000	39,500	33,000	210,000
L	REACTORS	75,000	4,000	5,000	1,000	80,000
M	DRUMS	306,000	12,000	12,000	6,000	318,000
Q	STORAGE TANKS	204,000	8,000	8,000	4,000	212,000
P	PUMPS AND DRIVERS	525,000	11,500	22,000	5,000	547,000
R	COMPRESSORS AND DRIVERS	1,225,000	26,500	53,000	17,000	1,278,000
S	STACKS	10,000	1,000	-	-	10,000
T	HEAT EXCHANGERS	1,675,000	22,000	40,000	8,000	1,715,000
	TOTAL — PROCESS EQUIPMENT	6,135,500	259,000	354,500	161,000	6,490,000
	PROCESS MATERIALS					
C	PIPING	1,718,500	806,000	1,081,500	1,094,000	2,800,000
D	STRUCTURES	255,500	100,500	14,500	39,500	270,000
E	ELECTRICAL	287,500	327,500	102,500	272,500	390,000
H	BUILDINGS	112,300	90,500	27,700	24,500	140,000
J	CIVIL	438,500	411,500	66,500	73,500	505,000
K	INSTRUMENTS	377,500	25,500	172,500	64,500	550,000
N	INSULATION AND PAINTING	485,000	636,000	95,000	94,000	580,000
	TOTAL — PROCESS MATERIALS	3,674,800	2,397,500	1,560,200	1,662,500	5,235,000
	DISTRIBUTABLE ACCOUNTS					
V	INSURANCE AND TAXES	190,000	205,000	130,000	235,000	320,000
O	OTHER DISTRIBUTABLE ITEMS	57,000	35,000	163,000	245,000	220,000
X	TEMP. CONSTRUCTION FACILITIES	55,000	60,000	15,000	20,000	70,000
Y	FIELD OFFICE	30,500	190,500	29,500	189,500	60,000
Z	CONSTRUCTION TOOLS & EQUIPMENT	450,000	150,000	130,000	40,000	580,000
	TOTAL — DISTRIBUTABLE ACCOUNTS	782,500	640,500	467,500	729,500	1,250,000
	INDIRECT ACCOUNTS					
U	HEADQUARTERS OFFICE	70,500	709,700	34,500	135,300	105,000
	TOTAL — PROJECT COSTS	10,663,300	4,006,700	2,416,700	2,688,300	13,080,000

FIG. 29.   Revised estimate and job cost report. [Reprinted by permission

1. Painting (type, number of coats, type cleaning, areas)
2. Insulation (type, thickness, weather covering, cements, areas)
3. Furnace linings (bricks, hangers, etc.)
4. Brick linings and floors (acidproof, stacks, cements, etc.)
5. Instrumentation installation details
6. Cathodic protection
7. Cost effects when building scale models
8. Others unique to the process and facilities

| JOB TITLE | | ABC CHEMICAL COMPANY | | | J.O. NO. 0000 | | |
| | | PROCESS UNIT | | | REPORT NO. | DATE | |

ESTIMATE		ESTIMATE AND AUTHORIZED CHANGES			DIFFERENCE		
LABOR	TOTAL	MATERIAL	LABOR	TOTAL	MATERIAL	LABOR	TOTAL
115,000	1,025,000	930,000	125,000	1,055,000	- 20,000	- 10,000	- 30,000
-	-	-	-	-	-	-	-
115,000	1,325,000	1,190,000	110,000	1,300,000	+ 20,000	+ 5,000	+25,000
64,000	274,000	195,000	57,500	252,500	+ 15,000	+ 6,500	+21,500
5,000	85,000	85,000	6,000	91,000	- 5,000	- 1,000	- 6,000
18,000	336,000	330,000	20,000	350,000	-12,000	- 2,000	-14,000
12,000	224,000	220,000	15,000	235,000	- 8,000	- 3,000	-11,000
16,500	563,500	550,000	20,000	570,000	- 3,000	- 3,500	- 6,500
43,500	1,321,500	1,270,000	50,000	1,320,000	+ 8,000	- 6,500	+ 1,500
1,000	11,000	10,000	1,500	11,500		- 500	- 500
30,000	1,745,000	1,710,000	30,000	1,740,000	+ 5,000		+ 5,000
420,000	6,910,000	6,490,000	435,000	6,925,000		-15,000	-15,000
1,900,000	4,700,000	2,895,000	1,980,000	4,875,000	-95,000	-80,000	-175,000
140,000	410,000	290,000	150,000	440,000	-20,000	-10,000	- 30,000
600,000	990,000	410,000	620,000	1,030,000	-20,000	-20,000	- 40,000
115,000	255,000	135,000	125,000	260,000	+ 5,000	-10,000	- 5,000
485,000	990,000	500,000	490,000	990,000	+ 5,000	- 5,000	
90,000	640,000	575,000	110,000	685,000	-25,000	-20,000	-45,000
730,000	1,310,000	585,000	730,000	1,315,000	- 5,000		- 5,000
4,060,000	9,295,000	5,390,000	4,205,000	9,595,000	-155,000	-145,000	-300,000
440,000	760,000	315,000	450,000	765,000	+ 5,000	-10,000	-5,000
280,000	500,000	190,000	285,000	475,000	+30,000	- 5,000	+25,000
80,000	150,000	80,000	90,000	170,000	-10,000	-10,000	-20,000
380,000	440,000	60,000	390,000	450,000		-10,000	-10,000
190,000	770,000	590,000	210,000	800,000	-10,000	-20,000	-30,000
1,370,000	2,620,000	1,235,000	1,425,000	2,660,000	+15,000	-55,000	-40,000
845,000	950,000	100,000	870,000	970,000	+ 5,000	-25,000	-20,000
6,695,000	19,775,000	13,215,000	6,935,000	20,150,000	-135,000	-240,000	-375,000

from J. Alcabes, *Trans. Am. Assoc. Cost Eng.*, p. 98 (1969).]

Plant Models Related to Costs

Small projects costing less than approximately $500,000 cannot usually justify the costs of a scale plant model complete with detail. Of course, simple layout blocks for general arrangement studies are valuable to good design.

Larger projects often benefit by detailed model studies, since the complexity is more easily visualized with the scale arrangements of pipe and equipment. A

ENGINEERING DEPARTMENT · ESTIMATE OF COST

THE DOW CHEMICAL COMPANY
PITTSBURG, CALIFORNIA

DATE:
CHARGE:
PROJECT ENGINEER:

Job No. 5916 – Plant Expansion
Summary Sheet

Amount Paid
To:

ITEM	DESCRIPTION	AUTHORI-ZATION	TRANSFERS	REVISED AUTHORI-ZATION	PREVIOUS ESTIMATE	CURRENT BEST ESTIMATE	COMMIT-MENTS	BALANCE TO BE COMMITTED		NET (OVER) OR UNDER AUTH.
4060	B-1	6 280		6 280		5 603	4 550	1053	4 620	1 660
4061	B-1 Dip Pipes	395		395		596	585	11	596	(201)
4140	H-1	8 140		8 140		8 428	8 356	72	8 428	(288)
4230	T-32	920		920		743	743	~~	743	177
4240	X-11	330		330		252	252	~~	252	78
4260	B-1 Insulation	1 950	(1 950)			~~	~~	~~	~~	~~
	Total Dow Purchases	18 015		16 065		15 622	14 486	1136	14 639	1 426
9600	Contract	3 805	1950	5 755		4450	4450		4450	1 305
	Change Orders	---	1 140	1 140		1 140	560	580	560	580
	Total Contract	3 805		6 895		5 590	5 010	580	5 010	1 885
	Contingency	1 930	(1 140)	790		300	300	~~	~~	790
	Total Job Cost	23 750		23 750		21 512	19 796	1 716	19 649	4 101
TOTALS										

FIG. 30. Job status cost estimate report. [Reprinted by permission from W. G. Clark, *Pet. Refiner*, *38*(9), 270 (1959).]

study by Lindeman [42], Table 26, indicates that the modeling technique has its greatest benefit for "high-density" piping systems in projects costing over 1 million dollars.

Contingency

The inclusion of a contingency amount in any estimate reflects the degree of uncertainty regarding the accumulated costs. Factored and preliminary estimates do not justify the identification of a contingency due to the broad range of possible accuracy in the basic technique of developing the estimate.

As the estimate becomes more detailed, the contingency is intended to cover the overlooked details that cannot possibly be covered 100% in the breakdown. The contingency can be considered to cover minor design details needed to complete a system but not the omission of significant equipment or an entire process or utility system. Quite often contingency costs are needed to cover extra construction labor due to weather delays, soil conditions, or overtime to catch up a section of the schedule.

Practices in establishing contingency values or percentages vary; however, they should be based on a considered evaluation of the scope of the project and its size and complexity. It is best to separate the material and labor contingencies, since they do not need to be the same. If subcontracts are a significant part of the job cost, contingency should also be allowed for these. Table 27 suggests values of contingencies to be considered for different sizes of projects. These must be weighed against the known and unknown features of the project at the time the estimate is prepared.

As a general rule, contingency figures for detailed estimates range from 8 to

	UNION CARBIDE CANADA LIMITED. CHEMICALS AND RESINS, DESIGN AND CONSTRUCTION	PROJECT COST CONTROL STATEMENT					
		PLANT: Montreal PROJECT: 69-10 WORK ORDER: 930-4782	PERIOD ENDING: March 1971 PREPARED BY: J. Doe DATE ISSUED: April 3, 1971			ISSUE NO. 8	
Area	DESCRIPTION	DEFINITIVE ESTIMATE	EXPEN- DITURES	COMMIT- MENTS	BALANCE	EXPECTED FINAL COST THIS MO.	LAST MO.
0100	Power Oil Room	575.7	42.3	210.0	343.1	595.4	625.0
0200	High Pressure Room	630.5	38.0	460.5	112.1	610.6	610.6
0300	Reactor Cell No.A	375.0	15.2	135.7	229.8	380.7	375.0
0400	Reactor Cell No.B	260.5	11.5	87.2	156.3	255.0	255.0
0500	Product Room	275.8	15.6	102.1	172.5	290.2	286.5
0600	Bulk Bins & Conveyors	273.2	83.5	165.3	24.4	260.1	260.1
0700	Building Extension	125.7	86.0	38.0	2.0	125.0	125.0
0800	Control Room	165.4	32.3	65.4	72.3	170.0	173.0
0900	Cooling Section	135.6	12.6	95.1	12.3	120.0	120.0
8060	Engineering	462.4	210.7	–	274.3	485.0	492.0
8070	Purchasing, Acct'g, Stores	35.7	18.6	–	13.4	32.0	32.0
7500	Field Inspection	35.0	12.0	–	23.0	35.0	36.0
7000	Contingency	350.0	–	–	350.0	350.0	350.0
		($'000)					
	NOTE: Cost Control type statement that is issued to Management for review. Figures are only for purposes of illustration.						
	GRAND TOTALS	3700.5	578.3	1359.3	1785.5	3710.0	3640.2
	AUTHORIZATION	3800.0					

FIG. 31. Project cost control statement. [Reprinted by permission from A. Lukaweske, *Trans. Am. Assoc. Cost Eng.*, p. 297 (1971).]

LOCATION JAPAN   FACILITY C8-C9 SPLITT ESTIMATE 700   DATE 3-16-70 REVISION   DATE 03/25/70   CIVIL - CODES 307, 308, & 309

CIVIL - CODES 307, 308, & 309

DESCRIPTION	UNITS	AMOUNT	CONCRETE MATL $	308 LABOR MHRS	LABOR $	AMOUNT	EXCAVATION MATL $	307 LABOR MHRS	LABOR $	PILING - 309 NUMBER PILES	SUBCONT $
HEAT EXCHANGERS											
E-701 FOUNDATION	CY	6.3	167	47	136	41.2		39	114	2	600
E-701 SUPPORT SADDLES	CY	1.4	91	39	113						
E-703 FOUNDATION	CY	5.4	142	40	115	35.0		33	96	2	600
E-703 SUPPORT SADDLES	CY	1.0	64	27	79						
TOWERS AND VERTICAL DRUMS											
T-701 FOUNDATION	CY	92.2	1,918	774	2,261	138.3		131	382	37	11,115
T-701 SKIRT FIREPROOF, 2" GUNITE	SF	469.1	86	139	403						
HORIZONTAL DRUMS											
D-702 FOUNDATION	CY	2.8	75	21	61	27.1		26	75		
D-702 SUPPORT SADDLES	CY	2.6	166	71	206						
PUMPS											
P-701AR FOUNDATION	CY	10.7	231	169	493	21.4		20	59	2	600
P-702AR FOUNDATION	CY	14.8	320	233	681	29.6		28	82	6	1,800
PAVING											
4" CONCRETE PAVING, NO REBAR	SY	284.2	443	254	741	31.6		30	87		
CONCRETE TOEWALLS	LF	329.4	237	242	707						
2" THICK CRUSHED STONE	SY	284.2	61	45	131	284.2		3	8		
PIPE SUPPORTS											
MAIN PIPE RACK FOUNDATIONS	CY	10.5	335	362	1,053	62.7		59	173		
SEWERS											
SEWERS TRENCHING	CY					43.1		56	163		
SEWERS SHORING, FEET OF TRENCH	LF					29.7		19	54		
SEWERS MANHOLES	CY	0.67	19	19	55	5.3	9	5	14		
SEWERS MANHOLE FRAMES AND COVERS	FA	0.13	3								
SEWERS CATCH BASINS	CY	2.0	71	64	186	16.2		15	45		
SEWERS CATCH BASIN FRAMES	EA	1.3	124								
LIGHTING, WELDING, AND GROUNDING											
CABLE TRENCHING AND CONCRETE COVER	CY	2.3	41	17	50	38.0		60	175		
ELECTRICAL											
POWER CABLE TRENCH AND CONCRETE COVER	CY	8.0	141	59	172	48.0		71	206		
DRUM EXCHANGER COMMON STRUCTURE											
COM STRC SUPPORT STEEL FOUNDATION	CY	32.6	886	410	1,198	81.4		77	225		
COM STRC SUPPORT STEEL FIREPROOF	CY	27.0	1,251	1,982	5,786						
TOTAL			6,872	5,011	14,632		9	671	1,958	49	14,715
ROUND UP			6,900	5,020	14,700		100	680	2,000		14,800

FIG. 32. Civil construction costs estimated by computer. [Reprinted by permission from A. S. Correll, *Trans. Am. Assoc. Cost Eng.*, p. 118 (1970).]

GASOLINE REFORMER
MARCUS HOOK
ENG. JOB NO. 90-145
PRINT OUT - BY LINE SIZE AND SCHEDULE
SUMMARY - ALL SIZES
ABOVE GROUND + BELOW GROUND
DATE - 01/26/70

CRAFT HOURLY RATE = 6.70 PER HOUR
CRAFT SUPERVISION HOURLY RATE = 7.20 PER HOUR
PERCENTAGE ADJUSTMENTS

WELDING	INSTALL.	TESTING	MISC. SUPPORTS	EXCAV.	CRAFT SUPV.
1.00	0.40	1.00	1.10	1.00	1.00

SHEET 1A

LINE SIZE	SCH.	STRAIGHT PIPE FIELD LIN.FT.	FIELD TONS	STRAIGHT PIPE SHOP LIN.FT.	SHOP TONS	VALVES	FTGS,FLGS P,N,G	STRAIGHT PIPE (DIRECT FIELD MATERIAL - NET COST)	TESTING	MISC. SUPPORTS	RADIOGRAPH S/C	EXCAV, BACKFILL	TOTAL	SEQ NO
1/2	80	1	0.00	0	0.0	40	4	0	1	1	0	0	46	1
3/4	80	17	0.01	0	0.0	243	11	7	5	9	0	0	275	2
1	40	1	0.00	0	0.0	0	0	0	0	1	0	0	1	3
1	80	12	0.01	0	0.0	41	18	7	1	7	0	0	74	4
1-1/2	40	0	0.0	0	0.0	23	0	0	0	0	0	0	23	5
2	40	254	0.46	0	0.0	338	87	130	11	90	0	45	701	6
2	80	0	0.0	0	0.0	0	9	0	0	0	0	0	9	7
3	40	0	0.0	53	0.20	843	10	0	20	29	2	0	904	8
4	40	0	0.0	89	0.48	1177	37	0	29	49	4	0	1296	9
4	STD	0	0.0	0	0.0	150	0	0	3	0	0	0	153	10
6	40	0	0.0	287	2.60	1807	84	0	58	123	18	30	2120	11
6	80	0	0.0	3	0.04	0	0	0	0	2	0	0	2	12
6	STD	0	0.0	0	0.0	0	6	0	1	0	0	0	7	13
8	40	0	0.0	51	0.73	371	19	0	16	28	4	0	438	14
10	40	0	0.0	0	0.0	0	3	0	1	0	0	0	4	15
10	XS	0	0.0	41	1.12	1892	20	0	48	23	6	0	1979	16
12	XS	0	0.0	0	0.0	0	0	0	2	0	0	0	2	17
14	STD	0	0.0	15	0.41	2298	61	0	77	8	2	0	2446	18
20	STD	150	5.92	0	0.0	0	660	1276	39	0	97	75	2147	19
TOTALS		435	6.40	535	5.66	9213	1029	1420	312	370	133	150	12627	

AVERAGE PIPE SIZE, FIELD + SHOP = 7.00

TOTAL FIELD LABOR MANHOURS PER TON OF STRAIGHT PIPE, FIELD + SHOP = 186

FIG. 33. Piping estimate by computer. [Reprinted by permission from H. J. Baltzell and R. G. Jones, *Trans. Am. Assoc. Cost Eng.*, p. 126 (1970).]

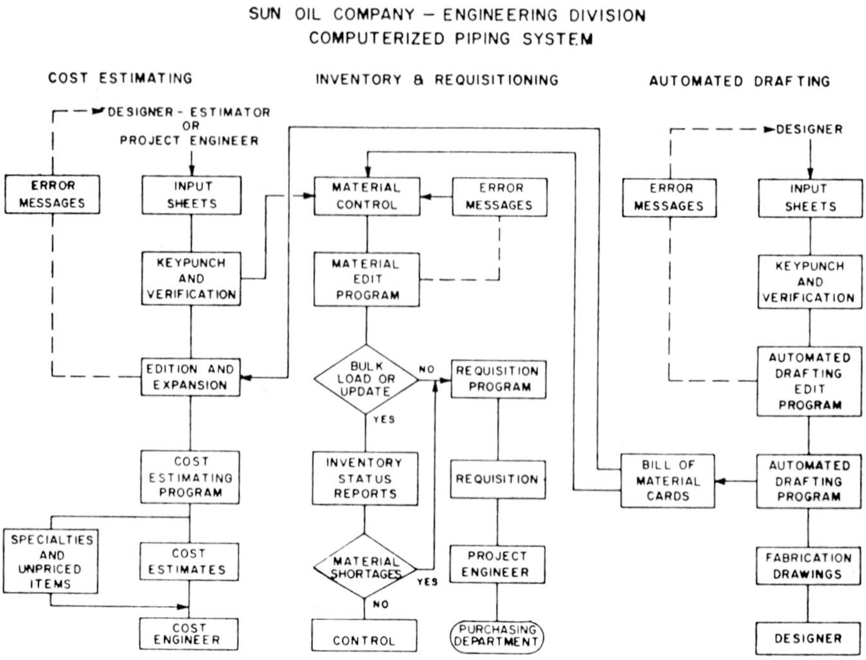

SUN OIL COMPANY — ENGINEERING DIVISION
COMPUTERIZED PIPING SYSTEM

FIG. 34. Computerized piping system for estimating, inventory, and requisition of materials. [Reprinted by permission from H. J. Baltzell and R. G. Jones, *Trans. Am. Assoc. Cost Eng.*, p. 122 (1970).]

15% of all direct costs or 5 to 10% of direct plus indirect costs (excluding engineering costs and contractor profit). As a percentage of total project cost, the contingency ranges from 2 to 6% for projects in the $3 to 15 million range and possibly up to 10 or 20% for smaller projects.

Procedure for Developing Detailed or Definitive Cost Estimate

A general guide for developing the detailed cost estimate is summarized:

1. List and price all equipment on the detailed mechanical, piping and instrument flow sheet.
2. Price labor for installation of individual items or group as a total labor installation cost, using historical records, labor factors, and experience.
3. List and price material commodity take-offs for such items as pipe, insulation, structural steel, concrete, electrical, and instrument installation materials in their proper categories.

TABLE 26 Cost Analysis of Effects of Plant Modeling[a]

Installed Cost[b]	Low-Density Piping System — Conventional		Low-Density Piping System — Modeling		High-Density Piping System — Conventional		High-Density Piping System — Modeling	
	%	$	%	$	%	$	%	$
$100,000								
Design engineering[c]	10%	$10,000	10%	$10,000	10%	$10,000	10%	$10,000
Piping design	3	3,000	5.5	5,500	5	5,000	6.5	6,500
Installed cost	100	100,000	100	100,000	100	100,000	99.8	99,800
Total[d]		$113,000		$115,500		$115,000		$116,300
$1,000,000								
Design engineering[c]	6.6%	$66,000	6.6%	$66,000	6.6%	$66,000	6.6%	$66,000
Piping design	2.7	27,000	3.2	32,000	4.1	41,000	3.7	37,000
Installed cost	100	1,000,000	99.6	996,000	100	1,000,000	96.7	967,000
Total		$1,093,000		$1,094,000		$1,107,000		$1,070,000
$10,000,000								
Design engineering[c]	5%	$500,000	5%	$500,000	5%	$500,000	5%	$500,000
Piping design	2.5	250,000	2.4	240,000	3.4	340,000	2.6	260,000
Installed cost	100	10,000,000	99.5	9,950,000	100	10,000,000	95.6	9,560,000
Total		$10,750,000		$10,690,000		$10,840,000		$10,320,000

[a] Reprinted, by permission, from C. G. Lindeman, *Trans. Am. Assoc. Cost Eng.*, p. 298 (1969).
[b] Installed cost referred to estimated cost (of portion of project modeled) based on the conventional (drawing) approach not including engineering, buildings, site, site preparation, roads, procurement, expediting, inspection, field supervision, and startup.
[c] All percentages based on conventional installed cost.
[d] Design engineering is total engineering less piping design. Does not include procurement or field supervision.

TABLE 27 Suggested Contingency Percentage Ranges for Detailed Estimates

Project Size	Contingency (%)[a]			
	Equipment	Commodities[b]	Subcontracts	Field Labor
To $500,000	1.5–5	10–15	3–15	5–20
$500,000 to $5,000,000	1.5–5	7–14	2–10	5–18
$6,000,000 to $15,000,000	1.5–4	4–8	2–8	3–16

[a]Percentages refer to the total of *direct costs* only for the particular category, not to the project size. Engineering costs not included.
[b]Refers primarily to bulk items such as pipe, pipe supports, electrical, and instrument materials.

4. List subcontracts for labor only or labor and materials for any categories to be installed in this manner. Some examples include refrigeration systems, pipe and equipment insulation, painting materials, and labor.

5. Determine indirect costs for field construction using historical records or approximate data presented in previous sections of this article. For labor and supervision benefits and other indirect costs, use an estimated manpower requirement schedule developed by the construction supervision. If this is not available, the percentage approach will be reasonably accurate.

6. Establish contingencies.

7. Establish cost–time escalation of materials and/or labor anticipated for the period of purchase and construction.

8. Establish the *premium* portion of any field overtime worked or to be worked to achieve the schedule.

9. If sales tax has not been included in the indirect costs, it should be determined separately.

10. Engineering costs and other office expenses including purchasing and expediting are determined from the estimates and records of man-hours required to perform the work.

11. Royalty or know-how charges for licensed information or process data are usually included as a part of the project's capital cost.

12. Contractor construction fees or profit can be developed with the help of the contractors bidding or by a percentage ranging from 5 to 15% of the total project cost, excluding the fee, contingency, royalty, and land.

Summary Assembly of Estimate

Referring to the estimate summary sheets such as Fig. 24 or 26, the costs may be assembled:

TABLE 28 Relative Costs of Foreign Plant Construction[a]

	United States (Gulf Coast)	England	Holland	Belgium	Germany	France	Italy	Japan
Relative Costs (1968) of Complete Process Plants								
Materials	1.00	0.96	0.97	0.96	0.91	0.92	0.89	0.91
Field	1.00	0.95	0.95	0.95	0.90	0.90	0.85	0.80
Engineering	1.00	0.75	0.80	0.90	0.80	0.90	0.80	0.70
Weighted average	1.00	0.91	0.92	0.94	0.88	0.91	0.86	0.83
Relative Costs for Material and Process Equipment								
Furnaces	1.00	0.95	0.90	0.90	0.90	1.00	1.00	0.90
Vessels	1.00	1.00	1.10	1.00	0.98	0.95	0.85	0.80
Exchangers	1.00	1.10	1.00	1.00	0.90	1.10	0.85	1.15
Pumps	1.00	0.90	0.90	0.85	0.85	0.82	0.80	0.80
Compressors	1.00	0.85	0.85	0.85	0.75	0.80	0.80	1.00
Piping	1.00	0.91	1.00	0.95	0.91	0.88	0.87	0.90
Structural	1.00	0.90	0.95	0.95	0.90	0.90	0.85	0.85
Instruments	1.00	1.05	1.05	1.05	1.10	1.05	1.00	1.00
Insulation	1.00	0.95	1.00	1.00	0.90	0.90	0.90	0.95
Electrical	1.00	0.95	0.95	0.93	0.93	0.90	0.88	0.85
Weighted average	1.00	0.96	0.97	0.96	0.91	0.92	0.89	0.91
Relative Costs of Engineering Work								
Base average man-hour rate, $	5.50	2.50	2.25			3.25		—
Payroll burden, %	25	20	55			50		—
Average man-hour rate with benefits, $	7.00	3.00	3.50			5.05		—
Nonpayroll items, $	0.90	0.70	0.60			0.70		—
Relative efficiencies	1.00	0.85	0.85			0.90		0.85
Relative overall cost + overhead + fee	1.00	0.75	0.80			0.90		0.70

[a]Source: R. J. Johnson, *Chemical Engineering* (March 10, 1969). Reprinted by special permission. Copyright © by McGraw-Hill, Inc., New York.

A.   Direct costs
     Collect items listed in Paragraphs 1, 2, 3, and 4; subtotal.
B.   Indirect costs: field
     Collect items listed in Paragraph 5; subtotal.
C.   Total field costs
     A + B above; subtotal.
D.   Total office costs
     Collect items included in Paragraph 10 above.
E.   Collect special items 6, 7, 8, 9, 11, and 12; subtotal.
F.   Total project estimate is A + B + C + D + E.

Estimating Worldwide Plant Costs

The preparation of any cost estimate intended for foreign situations requires specialty knowledge of the country and the location. Design and construction standards are different than in the United States, as are labor costs, equipment costs, labor efficiency, and general methods of construction. A few general guidelines can be cited from Johnson [43] and from Gallagher [44]; however, the development of decision-making costs for management should be established using experienced personnel with good knowledge of the locations. Table 28 lists some relative costs for engineering design, equipment purchase, and construction in several foreign countries.

This material appeared in Ernest E. Ludwig, *Applied Project Management for the Process Industries*, Gulf Publishing Co., Houston, Texas, 1974.

Selected Bibliography

Blecker, H. G., "Cost Models: A Computerized Approach to the Simulation of Equipment Costs," *Trans. Am. Assoc. Cost Eng.* p. 177 (1972).

Bresler, S. A., and Kuo, M. T., "Cost Estimating by Computer," *Chem. Eng.*, p. 84 (May 29, 1972).

Bresler, S. A., and Kuo, M. T., "More Programs for Cost Estimating by Computer," *Chem. Eng.*, p. 130 (June 26, 1972).

Carroll, C. E., "Increase Your Plant Profitability by Decreasing Capital Costs," *Chem. Eng.*, p. 113 (November 27, 1961).

Champley, J. A., "A Business Approach to Capital-Budgeting Decisions," *Chem. Eng.*, p. 127 (September 22, 1969).

Dickens, S. P., and Douglas, F. R., "Off-site Investment and Working Capital," *Chem. Eng., Prog.*, 56(12), 44 (1960).

Drayer, D. E., "How to Estimate Plant Cost–Capacity Relationships," *Petro/Chem Eng.*, p. 10 (May 1970).

Enyedy, G., Jr., "Cost Data for Major Equipment," *Chem. Eng. Prog.*, 67(5), 73 (1971).

Enyedy, G., Jr., "Design of a COME Project Equipment Cost Data Sheet," *Am. Assoc. Cost Eng. Bull.*, p. 80 (June 1972).

Green, E. O., "Procedure for the Review of Cost Estimates," *Trans. Am. Assoc. Cost Eng.*, p. 164 (1972).

Guthrie, K. M., "Field-Labor Predictions for Conceptual Projects," *Chem. Eng.* p. 170 (April 7, 1969).

Hackney, J. W., "Estimating Methods for Process Industry Capital Costs," *Chem. Eng.*, p. 119 (April 4, 1960).

Hackney, J. W., and Robertson, W. E., "Cost Estimating Must Be Done Right," *Power Eng.*, p. 69 (November 1955).

Hirsch, J. H., and Glazier, E. M., "Estimating Plant Investment Costs," *Chem. Eng. Prog.*, 56(12), 37 (1960).

Hirschmann, W. B., "Has the Cost of Building New Refineries Really Gone Up?" *Chem. Eng. Prog.*, 67(8), 39 (1971).

Holland, F. A., and Brinkerhoff, R., "How to Scale Up Cost Estimations," *Chem. Eng.*, p. 97 (February 4, 1963).

Jelen, F. (ed.), *Cost and Optimization Engineering*, McGraw-Hill, New York, 1970.

Jones, L. R., "Building-Cost Escalation," *Chem. Eng.*, p. 170 (July 27, 1970).

Katell, S., and Duda, J. R., *Bibliography of Investment and Operating Costs for Chemical and Petrochemical Plants, January–December 1969*. Information Circular 8478, U.S. Department of the Interior, Bureau of Mines, Washington, D.C.

Layshook, J. L., "Estimating and Cost Control from Inception to Completion," *Am. Assoc. Cost Eng. Bull.*, p. 85 (September 1969).

Loring, R. J., "The Proposal Manager's Work," *Chem. Eng.*, 77(18), 98 (1970)

Nachod, J. E., "How Leasing Conserves Capital," *Chem. Eng.*, p. 150 (June 16, 1958).

Quigley, H. A., "Economics of Multiple Units," *Chem. Eng.*, p. 97 (August 29, 1966).

Schagrin, E. F., "How Much Do Minicomputer Control Systems Cost?" *Chem. Eng.*, 78(7), 103 (1971).

Schwartz, C. C., "Estimate Plant Costs Quickly, Accurately," *Oil Gas J.*, p. 156 (November 11, 1963).

Whittum, J. B., "Computerized Factored Chemical Plant Estimates," *Trans, Am. Assoc. Cost Eng.*, p. 170 (1972).

Wilson, J. D., "Want to Predict Cost Escalation?" *Hydrocarbon Process.*, 45(1), 157 (1966).

Yen, Y. C., "Estimating Plant Costs in the Developing Countries," *Chem. Eng.*, p. 89 (July 10, 1972).

References

1. American Association of Cost Engineers, *Cost Engineer's Notebook*, "Cost Engineering Terminology," Index No. AA-4.000, Rev. 1, May 1971.

2. T. G. Papavero, "Cost Control through Our 'Piping Efficiency Program,'" *Transactions*, American Association of Cost Engineers, Pittsburgh Meeting, 1969, p. 75.

3. J. E. McGinness, Jr., "Automated Take-off and Computerized Estimating," *Transactions*, American Association of Cost Engineers, Houston Meeting, 1968, p. 1–1.

4. M. F. Schewe, "Practical Aspects of Computerized Estimating," *Transactions*, American Association of Cost Engineers, Houston Meeting, 1968, p. 2–1.

5. H. C. Bauman, *Fundamentals of Cost Engineering in the Chemical Industry*, Reinhold, New York, 1964.
6. "Construction Cost Index," *Eng. News-Rec.*, (published weekly).
7. T. H. Arnold and C. H. Chilton, "New Index Shows Plant Cost Trends," *Chem. Eng.*, *70*(4), 143 (1963) (also published in regular issues of Chemical Engineering).
8. R. W. Stevens, Marshall and Stevens Equipment Index, *Chem. Eng.*, *54*(11), 124 (1947) (index values published monthly in Chemical Engineering).
9. R. B. Norden, "Development of a Cost Index," *Transactions*, American Association of Cost Engineers, Houston Meeting, 1968, p. 46–1.
10. W. L. Nelson, "Cost Indexes," *Oil Gas J.*, *63*(14), 185 (1965); and *63*(27), 117 (1965) (index published first issue each month).
11. W. L. Nelson, "Cost Indexes without Productivity Mean Little," *Transactions*, American Association of Cost Engineers, Houston Meeting, 1968, p. 45–1.
12. H. J. Lang, "Cost Relationships in Preliminary Cost Estimates," *Chem. Eng.*, *54*(10), 117–121 (1947).
13. H. J. Lang, "Simplified Approach to Preliminary Cost Estimates," *Chem. Eng.*, *55*(6), 112 (1948).
14. R. Williams, "Six-tenths Aid in Approximating Costs," *Chem. Eng.*, *54*(12), 124 (1947).
15. J. D. Chase, "Plant Cost vs. Capacity: New Way to Use Exponents," *Chem. Eng.*, *77*(4), 113 (1970).
16. K. M. Guthrie, "Capital and Operating Costs for 54 Chemical Processes," *Chem. Eng.*, *77*(6), 140 (1970).
17. D. E. Drayer, "Part 1: How to Estimate Plant Cost–Capacity Relationship." *Petro/Chem Eng.*, *5*, 10 (1970).
18. H. C. Bauman, "Factoring Costs of Chemical Plánt Installations," *Ind. Eng. Chem.*, *50*(8), 69A (1958).
19. M. S. Peters and K. D. Timmerhaus, *Plant Design and Economics for Chemical Engineers*, 2nd ed., McGraw-Hill, New York, 1968, p. 107.
20. B. G. Liptak, "Costs of Process Instruments," *Chem. Eng.*, *77*, 60 (September 7, 1970).
21. C. L. Williams, Jr., and R. D. Damron, "Which Cools Cheaper: Water or Air?" *Hydrocarbon Process. Pet. Refiner*, *44*(2), 139 (1965).
22. K. M. Guthrie, "Capital Cost Estimating," *Chem. Eng.*, *76*, 114 (March 24, 1969).
23. W. F. Wroth, "Factors in Cost Estimating," *Chem. Eng.*, *67*, 204 (October 17, 1960).
24. J. T. Gallagher, "Rapid Estimation of Plant Costs," *Chem. Eng.*, *74*, 89 (December 18, 1967).
25. J. Clerk, "Multiplying Factors Give Installed Costs of Process Equipment," *Chem. Eng.*, *70*, 182 (February 18, 1963).
26. K. M. Guthrie, "Rapid Calc' Charts," *Chem. Eng.*, *76*, 138 (1969).
27. J. S. Page, *Estimator's Manual of Equipment and Installation Costs*, Gulf, Houston, 1963.
28. J. S. Page and J. G. Nation, *Estimator's Electrical Man-Hour Manual*, Gulf, Houston, 1968.
29. J. S. Page, *Estimator's General Construction Man-Hour Manual*, Gulf, Houston, 1960.
30. J. S. Page and J. G. Nation, *Estimator's Piping Man-Hour Manual*, revised ed., Gulf, Houston, 1968.
31. J. S. Page, *Estimator's Man-Hour Manual on Heating, Air Conditioning, Ventilating and Plumbing*, Gulf, Houston, 1961.
32. J. E. Haselbarth, "Updated Investment Costs of 60 Types of Chemical Plants," *Chem. Eng.*, *74*, 214 (December 4, 1967).

33. J. M. Berk and J. E. Haselbarth, Cost Capacity Data Series, *Chem. Eng.*, *67*, 172 (December 12, 1960); *68* (January 23, 1961); *68*, 174 (February 20, 1961); *68*, 182 (March 1961).
34. *The Boeckh Building Valuation Manuals*, Vols. 1, 2 and 3. American Appraisal Co., Milwaukee.
35. "C. E. Plant Cost Index, New Ratios for Estimating Plant Cost," *Chem. Eng.*, *70*(20), 120 (1963).
36. L. R. Jones, "Estimating Cost Escalation." *Transactions*, American Association of Cost Engineers, San Francisco Meeting, 1970, p. 58.
37. K. M. Guthrie, "Costs," *Chem. Eng.*, *76*, 201 (April 14, 1969).
38. R. J. Loring, "Cost of Preparing Proposals," *Chem. Eng.*, *77*, 126 (November 16, 1970).
39. *Cost Engineer's Notebook*, American Association of Cost Engineers, 1971.
40. D. A. Bosworth, "Installed Costs of Outside Piping," *Chem. Eng.*, *75*, 132 (March 25, 1968).
41. G. Enyedy, Jr., "The Systems Approach to Estimating Equipment Costs," *Transactions*, American Association of Cost Engineers, Houston Meeting, 1968, p. 22.
42. C. G. Lindeman, "Reducing Construction Costs," *Transactions*, American Association of Cost Engineers, Pittsburgh Meeting, 1969, p. 292.
43. R. J. Johnson, "Costs of Overseas Plants," *Chem. Eng.*, *76*, 146 (March 10, 1969).
44. J. T. Gallagher, "Efficient Estimating of Worldwide Plant Costs," *Chem. Eng.*, *76*, 196 (June 2, 1969).
45. E. E. Ludwig, *Applied Project Management for the Process Industries*, Gulf, Houston, 1974, Chap. 6.

E. E. LUDWIG

Cost, Cash Flow Concepts

For an economic evaluation of a project or venture, it is necessary to identify the sources and represent the expected flow and disposition of the monies involved. To elucidate the money flow process for a going business operation, a diagram will be used that is analogous to a process flow diagram showing the input of raw materials, the various processing steps and the flows, recycle, and hold-up of intermediate material within the process, and finally the output of product. This way of looking at money transfers in a business is readily grasped by technical people. A money stream particularly useful in the economic evaluation of projects is *cash flow*, the sum of the depreciation and the net profit.

For the analysis of economic feasibility, the flow of money is idealized. It can be considered one-time, i.e., a lump sum at an instant such as the purchase of land or of equipment and the allocation of working capital to a new project. The alternative situation is regular payments as for wages, raw materials, services, etc. where the money flow is generally considered as continuous. Other regular schedules are sometimes employed, such as continuous but linearly decreasing or increasing costs. These idealized flows permit the use of simple models to describe the money flows for a project to be evaluated.

The various items of money flow will be looked at in more detail for the case of a going manufacturing business. The primary source of income necessary to maintain the health (indeed, the life) of the business is revenue from the sale of products. A large part of this income flows out as operating costs. Using the proposed analogy, just as we have a flow diagram for materials in a process, viz.,

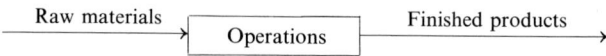

we can also have a similar chart for the flow of funds in a going operation:

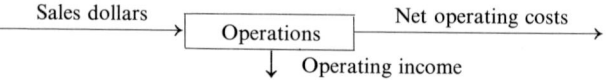

At this point it is instructive to review a checklist of operating cost items as may be found in Perry [1]. Net operating costs includes all payments made for other than capital investment, income taxes, stock dividends, and depreciation.

It is necessary to have a supply of funds, called working capital, to meet demands for current expenses, e.g., raw materials, wages, salaries, while waiting for receipts from the sale of products or services. This reservoir of money facilitates the business operation and insures smooth performance. From experience, each company works out the proper amount which it needs to accommodate comfortably the ups and downs in the business activity. Obviously, it is desirable to keep the amount of capital tied up this way to as low a figure as is practical. Thus we have:

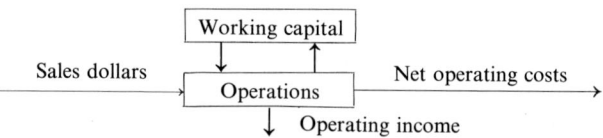

The working capital fund is constantly replenished. Over the long run the net flow of working capital into and out of the operations is zero, and none finds its way into operating costs or operating income.

Now we will consider the complete money flow diagram, Fig. 1. The excess of revenue over net operating costs is termed operating income. Depreciation is diverted from this stream, leaving gross profit. The amount diverted is determined by depreciation accounting, the standard procedure used in business for charging for capital expenditures over the period during which assets are in use. In other words, the outlay for a fixed asset is not charged to the operation when it is made but on a regular schedule over the life of the asset. Therefore, depreciation is handled as an item in the cost of manufacture and appears as such on cost sheets. Note, however, that it is an internal cost and is

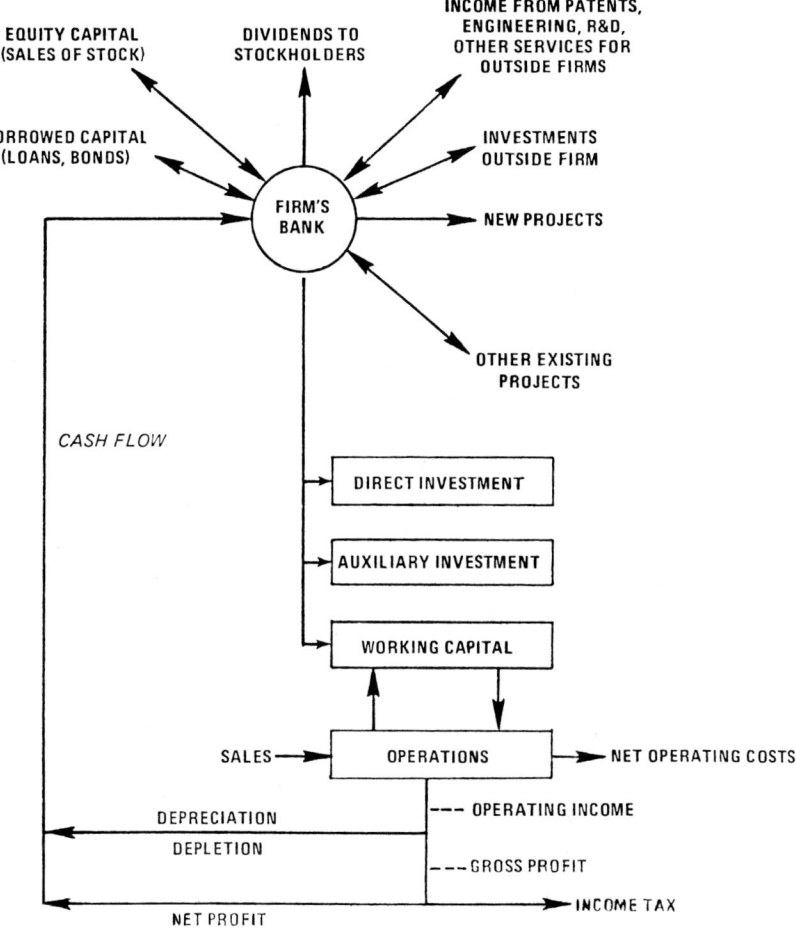

EQUITY CAPITAL
(SALES OF STOCK)

DIVIDENDS TO
STOCKHOLDERS

INCOME FROM PATENTS,
ENGINEERING, R&D,
OTHER SERVICES FOR
OUTSIDE FIRMS

BORROWED CAPITAL
(LOANS, BONDS)

INVESTMENTS
OUTSIDE FIRM

FIRM'S
BANK

NEW PROJECTS

OTHER EXISTING
PROJECTS

CASH FLOW

DIRECT INVESTMENT

AUXILIARY INVESTMENT

WORKING CAPITAL

SALES OPERATIONS NET OPERATING COSTS

DEPRECIATION --- OPERATING INCOME

DEPLETION

---GROSS PROFIT

INCOME TAX

NET PROFIT

FIG. 1. Money flow diagram. This demonstrates the source of *cash flow*.

returned to the business. This is the reason depreciation is shown in the money flow diagram as separate from the net operating costs which represent dollars leaving the company. Depletion is handled in a similar way; it takes into account the consumption of assets which are exhaustible material resources, such as the products of mines, forests, and oil and gas fields. For typical depreciation schedules and periods, see the *Depreciation, Obsolescence, Depletion, and Income Taxes* article.

The gross profit remaining after the diversion of depreciation and depletion is subject to federal and state income taxes. After these taxes are deducted, the net profit remaining, together with depreciation and depletion, is shown in the money flow diagram, Fig. 1, as streaming into the firm's bank. The combined stream is termed the *cash flow*. (Recall that *cash flow* in italics connotes the sum of depreciation and net profit.)

Cash transfers to and from the company, other than the revenues, net operating costs, and income taxes considered previously, are made directly to and from this fictitious bank. Funds in the bank are used to repay borrowed capital, loans and bonds, and to pay dividends on preferred and common stock. All of the net profit belongs to the owners (common stock shareholders) and, therefore, could be distributed as dividends. However, in most businesses roughly one-half of the net income is retained, and together with depreciation helps to meet the continuing capital needs of the firm. This retained net profit (termed retained earnings) obviously increases the owners equity, i.e., the value of each owner's share of the business.

In the previous discussion we assumed a going business for which the investment in plant facilities had already been made. Now we appreciate that this investment and the necessary working capital were provided by the firm's bank, and that the depreciation and retained earnings from the project are now available for new projects and other capital requirements. The complete money flow diagram (Fig. 1) shows the recycle of funds, the flow of funds into new projects, and other flows associated with the various activities of the firm's bank with outside entities. For example, the sale or leasing of patents and process know-how may be a very important source of funds.

We mentioned that the firm's bank disburses and receives funds other than those connected directly with operations. We can also look at the bank as the part of the firm that makes financial decisions concerning investments, expansions, and the like. This activity is similar to that of a commercial bank because it provides and allocates funds to competing projects. It decides what projects to approve and provides the necessary capitalization. All net profit and depreciation, i.e., *cash flow*, from the capitalized projects are then returned to the bank.

Generally, interest is focused on a given project and not the entire enterprise. The flow of funds into and out of a project during its life may be shown in a "cumulative cash position chart," Fig. 2. Zero time is taken as the time when the plant first begins to produce salable products. Prior to this, at a negative time, after the decision was made that the project had economic merit, the firm's bank secured the necessary capital and then made it available to the project as needed for the purchase of land, procurement and installation of equipment, investment in auxiliary facilities, and working capital. After time zero the revenues from sales exceed the operating expenses and the project begins to generate operating income. The cumulative cash flow is plotted throughout the life of the project and, at the end of the project life, adjustments are made for the recovery of land and working capital and also for salvage value, if any. All the foregoing items, together with their timing, appear on the cumulative cash position chart. Capital and other initial expenditures are shown below the horizontal base line in the negative region. *Cash flow* is positive and, therefore, accumulates in the positive direction and after a period of time shows a positive cash position. This graph indicates that the whole *cash flow* is used to retire the investment. Profit appears only after the investment is completely paid off.

The cash position chart proves to be useful in following the financial history of a project or individual facility. It also provides a powerful device to understand readily the common "criteria of profitability."

An example will illustrate the development of the money flow chart (Fig. 3)

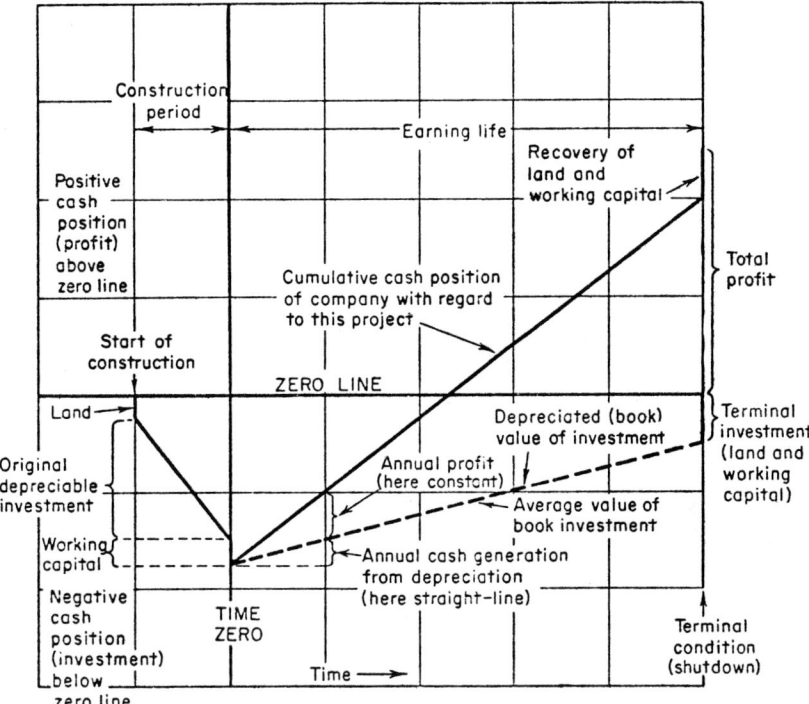

FIG. 2. Illustrative cumulative cash position chart [2].

and the cumulative cash position chart (Fig. 4). A chemical plant can be designed and constructed for $1,000,000. Operating income is estimated to be $600,000 annually. Assume 10-yr life, straight line depreciation, and 50% income tax rate. Land can be purchased for $100,000. Plant construction would take 2 yr and at the start of the operation $100,000 for working capital would be made available. For computing the annual depreciation rate, do not take into account salvage value at the end of the 10-yr period; however, consider that a net salvage value of $50,000 will actually be realized at the end of the tenth year which must then be handled as income.

For each year we have the following tabulation:

Depreciation	$1,000,000/10 = \$100,000/yr$
Gross profit	$600,000 - 100,000 = \$500,000/yr$
Income tax	$(500,000)(0.50) = \$250,000/yr$
Net profit	$500,000 - 250,000 = \$250,000/yr$
Cash flow	$250,000 + 100,000 = \$350,000/yr$

At the end of the tenth year the plant is fully depreciated. Therefore, the $50,000 salvage value is subject to a 50% income tax; hence $25,000 is the net recovery. In addition, land value and working capital are recouped, giving the following

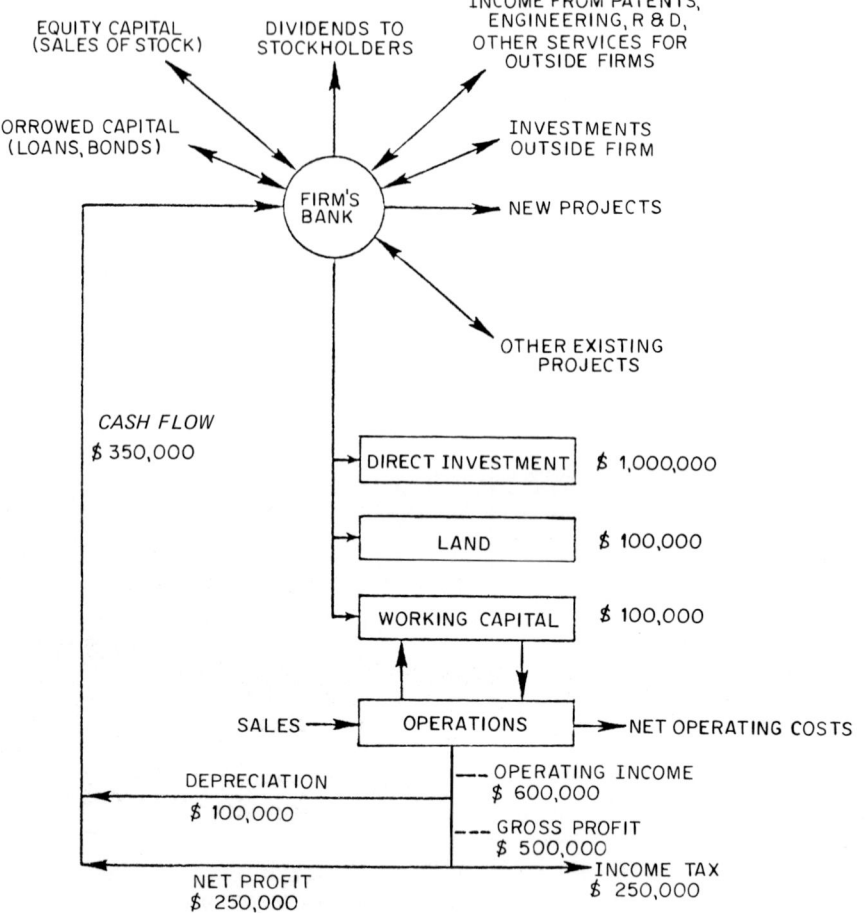

FIG. 3. Annual *cash flow* from operations.

total recovery at the end of the tenth year:

Net profit on salvage recovery	25,000
Recovery of land value	100,000
Recovery of working capital	100,000
Total recovery	225,000

The cumulative cash position table can be set up from information on cash flow and the investment expenditures with timing as follows:

Two years before start-up, purchase land	100,000
Build plant over 2-yr period	1,000,000
Assume equal expenditures per year	500,000
Provide working capital at plant start-up (reference time is 0)	100,000

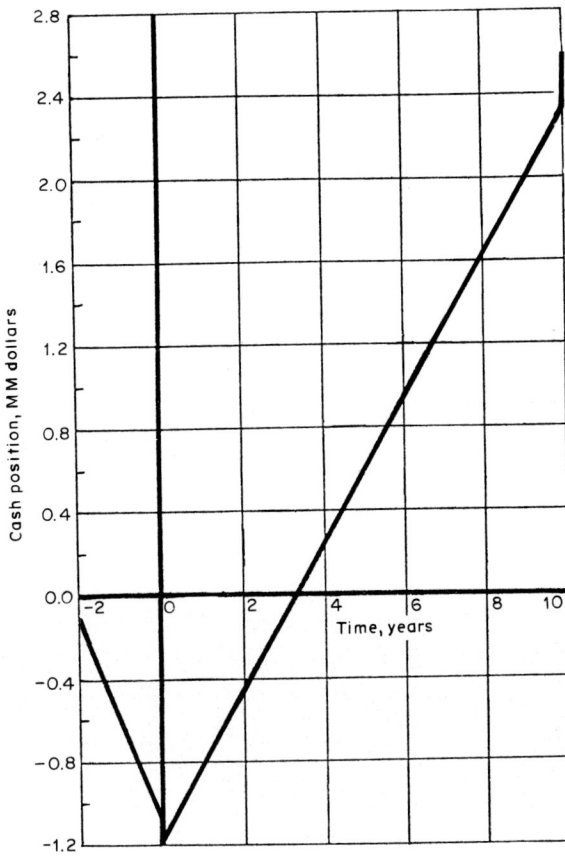

FIG. 4. Cumulative cash position chart for example.

TABLE 1 Cumulative Cash Position Table

Time Periods		Investment Summary	Cash Flow Summary	Cumulative Cash Position
Second year before start-up	At start	− 100,000		− 100,000
	During	− 500,000		− 600,000
First year before start-up	At start	—		− 600,000
	During	− 500,000		− 1,100,000
At time zero		− 100,000		− 1,200,000
First year			350,000	− 850,000
Second year			350,000	− 500,000
Third year			350,000	− 150,000
Forth year			350,000	200,000
Fifth year			350,000	550,000
Sixth year			350,000	900,000
Seventh year			350,000	1,250,000
Eighth year			350,000	1,600,000
Ninth year			350,000	1,950,000
Tenth year			350,000	2,300,000
At end of tenth year			225,000	2,525,000

It will be recalled that investment expenditures are negative and cash flow is positive.

The data in the cash position table (Table 1) are used to plot the cumulative cash position chart (Fig. 4).

This article is based on a chapter in Uhl and Hawkins [3].

References

1. R. H. Perry and C. H. Chilton (eds.), *Chemical Engineers' Handbook*, 5th ed., McGraw-Hill, New York, 1973, Table 25–28.
2. Ref. 1, Fig. 25–9.
3. V. W. Uhl and A. W. Hawkins, *Technical Economics for Engineers* (No. 5 in AIChE Continuing Education Series), American Institute of Chemical Engineers, New York, 1971.

A. W. HAWKINS

Cost Engineering (see also Economic Evaluation)

The subject "cost engineering" will be defined here as covering the nature of manufacturing costs and the assessment of project economic merit in the context of the chemical process industries. Engineers and scientists probably are most frequently interested in costs and economics in situations requiring the evaluation of new projects which are being considered. Although such cost engineering often may reasonably take a basis somewhat different from the cost and profit evaluations performed classically by accountants, the accountant's approach will be emphasized here for two reasons. First, accounting records of costs are very valuable sources of information for the estimation of similar costs in new circumstances. The structuring of the economic evaluation to be compatible with accounting procedures therefore makes it easier to prepare the evaluation. Second, and probably more importantly, good policy in an organization calls for a comparison of forecasted economics for a project to the actual economics experienced subsequent to installation and operation. This comparison can be performed with minimum explanations to management if done straightforwardly in the accounting structure familiar to management.

Before proceeding into details, it will be useful to review the overall economic position of the chemical process industries. The Bureau of the

Census, U.S. Department of Commerce, regularly publishes summary statistics for the United States manufacturing industries [1]. The industries are organized into a Standard Industrial Classification [2]. Table 1 shows a comparison of Chemicals and Allied Projects to the other manufacturing industries for the year 1971. In addition to data for value of shipments, the Annual Survey of Manufacturers also contains the following information for each industry: number of all employees and total payroll dollars, number of production workers, their man-hours worked and wages received, dollars of value added by manufacture, cost of materials, new capital expenditures, year-end inventories, fuels and electric energy used, book value of fixed assets, and geographical location. It can be seen from Table 1 that chemicals and allied products rank fourth among the manufacturing industries in dollar volume of shipments. Table 2 contains a breakdown of values of shipments of chemicals and allied products into product classes. The Survey of Manufactures provides still more detailed information for items within product class. In some cases the item is a specific product; for example, "chlorine" within "alkalis and chlorine." More generally, the item is generic; for example, "miscellaneous cyclic chemical products" within "industrial organic chemicals."

TABLE 1   Value of Manufacturing Industry Shipments in 1971 [3]

Industry, by Standard Industrial Classification	Shipments (billion $)
Food and kindred products	103.63
Tobacco manufactures	5.53
Textile mill products	24.03
Apparel and other textile products	25.02
Lumber and wood products	14.93
Furniture and fixtures	9.75
Paper and allied products	25.46
Printing and publishing	26.89
Chemicals and allied products	51.87
Petroleum and coal products	26.94
Rubber and plastics products, n.e.c.[a]	17.04
Leather and leather products	5.22
Stone, clay, and glass products	18.53
Primary metal industries	53.07
Fabricated metal products	42.03
Machinery, except electrical	55.56
Electrical equipment and supplies	49.17
Transportation equipment	86.92
Instruments and related products	12.28
Miscellaneous manufacturing industries	10.12
Ordnance and accessories	6.98
Total	670.97

[a]Not elsewhere classified.

TABLE 2   Value of Shipments of Product Classes within Chemicals and Allied Products in 1971 [4]

Product Class	Shipments (billion $)
Alkalis and chlorine	0.701
Industrial gases	0.667
Cyclic intermediates and crudes	2.078
Inorganic pigments	0.688
Industrial organic chemicals, n.e.c.[a]	6.815
Industrial inorganic chemicals, n.e.c.[a]	3.882
Plastics materials and resins	4.820
Synthetic rubber	1.153
Cellulosic manmade fibers	0.680
Organic fibers, noncellulosic	2.821
Biological products	0.415
Medicinals and botanicals	0.761
Pharmaceutical preparations	5.610
Soap and other detergents	2.522
Polishes and sanitation goods	1.355
Surface acting and finishing agents	0.516
Toilet preparations	3.920
Paints and allied products	3.141
Gum and wood chemicals	0.277
Fertilizers	0.892
Fertilizers, mixing only	0.530
Agricultural pesticides and other agricultural chemicals	0.935
Adhesives and gelatin	0.631
Explosives[b]	0.189
Printing ink	0.408
Carbon black (channel and furnace process only)	0.227
Chemical preparations, n.e.c.[a]	1.706

[a]Not elsewhere classified.
[b]Except government-owned, contractor-operated plants.

Recognition needs to be made of the fact that the Standard Industrial Classification is necessarily arbitrary, and much chemical processing also exists in industries other than "chemicals and allied products." Examples of such other industries would be petroleum and coal products, primary metals and rubber and plastics products. It will be impossible here to examine the cost and other economic characteristics of all the industries which employ chemical processing. This being so, the approach will be to analyze the basics of cost engineering and economic evaluation in the context of a few basic processes and situations, and to identify significant variations from those basics as seems appropriate.

A summary of the cost of manufacturing sulfuric acid is given in Table 3 to provide a vehicle for this discussion of cost engineering [5]. This summary shows two of the types of costs which are of primary inportance to cost engineering: investment or capital cost, and manufacturing cost. Investment cost will be treated in detail later in this section. The items comprising manufacturing cost will be considered first. Table 3 collects the individual costs into four rather classical categories, i.e., raw materials, utilities, operating expenses, and overheads, although the terminology may vary somewhat among different organizations.

Raw materials itemizes the materials which react to form the product. In Table 3, sulfur is the only raw material shown. Many products require more than one raw material, in which case each raw material is listed. The process for sulfuric acid illustrates a minor complexity which is frequently encountered. Oxygen and water are also required for the manufacture of sulfuric acid from

TABLE 3 Breakdown of Product Cost for 350,000 Short Tons/Year (1000 st/d) of 100% H_2SO_4

Investment cost for battery limits (B.L.) plant		$2,300,000	
Investment cost for offsites		300,000	
		$2,600,000	

	Annual Quantity	Unit Cost	Annual Cost	$ Cost/ST 100% H_2SO_4
Raw material				
Sulfur	116,700 tons	36$/ton	$4,200,000	
			$4,200,000	12.00
Utilities				
Power (B.L.)	14,000 MWh	1¢/kWh	140,000	
Power (offsites)	3,000 MWh		30,000	
Boiler feed water	500,000 tons	5¢/ton	25,000	
Process water	70,000 tons	2¢/ton	1,400	
Cooling water	700,000 tons	1¢/ton	7,000	
Steam (credit)	450,000 tons	1$/ton	(450,000)	
			(246,600)	(0.70)
Operating expenses				
Labor	1 man/shift		32,000	
Supervision	Part time		10,000	
Maintenance	4% investment		104,000	
			146,000	0.42
Overheads				
Payroll overhead	15% of labor supply		6,300	
General overhead	50% of operating expenses		73,000	
Depreciation, taxes	15% of investment		390,000	
			469,300	1.34
Total manufacturing cost			$4,568,700	13.06

sulfur. The oxygen may come from air or from an oxygen-rich stream. If the oxygen is consumed as air, or is produced as an oxygen-rich stream in an in-plant air liquifaction unit, the oxygen might not appear as a raw material, but its costs would instead be covered in the other components of the analysis. This is the case in Table 3, where utilities costs, for example, include the cost of filtering, drying, and compressing the air. Had the oxygen been purchased from an external source at some unit cost, then it would have appeared under the raw material section. Similarly, the water required to form the acid is not shown here as a raw material, but instead appears under the utilities section as process water. Another kind of complexity arises when a process produces a by-product. The usual practice is to show the by-product under the raw material section as a credit at whatever unit price the by-product will command. The total raw material cost is then the net for the materials consumed and the salable by-products. Still another kind of complexity related to raw materials exists for some processes. This complexity occurs when a material is consumed in the process, but at very low rates or at rates which are not inherently fixed to the process stoichiometry. Examples of these are catalysts which lose their activity over long periods of time and must be replaced, solid dessicants and adsorbents, and inhibitors used to stabilize a product. These kinds of material consumptions are frequently costed under a "supplies" category as an operating expense rather than as a raw materials item.

It is conventional in accounting practice to show negative quantities in tabulations by enclosing them in parentheses rather than by prefixing a minus sign. Since managers are accustomed to this practice, engineering cost analyses should preferably use the same convention.

The *utilities* shown in Table 3 for the sulfuric acid process include about all of the commonly-occurring utilities items, with one exception. That exception is steam which is required for heat or for motive power. Plant steam systems commonly have loops of steam available at several pressures, e.g., at 50, 250, and 600 lb/in.^{2} gauge. Various duties for steam in a process will optimumly call for one or another of these pressures. The unit cost of steam depends on the steam pressure. It is therefore desirable to itemize steam consumption at the pressures used in order to include these variations in steam unit costs. Table 3 shows that this process for sulfuric acid generates steam which is in excess of the acid plant's requirements. A credit appears for this excess, and the total utilities costs are thus reduced. It should be recognized that this credit is valid only if a customer exists to purchase the steam. If this plant were isolated so that there were no potential users of the steam, then the credit would be invalid. Even worse, cooling water (or air) would then have to be used to absorb the cooling duty equivalent to the steam being generated, and utilities costs would increase beyond the amount which would result simply from deletion of the steam credit.

The utilities section of the cost estimate should define units of measure for each item. There are no standard units of measure, but those most commonly used are probably kilowatt-hours (kWh) for electricity, thousands of pounds (Mlb) for steam, thousands of gallons (Mgal) for cooling water, thousands of standard cubic feet (MSCF) for natural gas, and millions of Btu's (MMBtu) for fuel oil and coal. Although the foregoing units and their abbreviations have been traditional in chemical and petroleum process industries, care must be taken in interpreting abbreviations. There has been a strong recent trend toward the use of M to signify millions or "mega." Thus the quantities for power in Table 3 are in megawatt-hours.

The *operating expenses* section of Table 3 shows items for labor, supervision, and maintenance. Labor requirements per shift are relatively difficult to estimate but, fortunately, operating labor is usually a small cost because of the characteristic of most chemical processes to have only one or a few operators controlling very large production units by means of highly instrumented and automated systems of equipment. Supervision is similarly a quite small cost item. Maintenance cost is also a difficult item to estimate, but it is a more significant cost item. It is usually estimated as a percentage of capital investment, which may range from 2 to 20%/yr. Maintenance cost will depend upon the severity of corrosion and other process conditions such as temperature and pressure, upon geographical location, and upon materials of construction chosen for the equipment. There is a trade-off between maintenance cost and capital cost as affected by the choice of materials of construction. In a given service, the use of carbon steel equipment may reduce capital cost but increase maintenance cost. The percentage of capital required for maintenance cost then will be increased by having a larger numerator and smaller denominator. Conversely for the same service, the use of stainless steel increases capital cost, decreases maintenance cost, and results in a percentage reflecting a smaller numerator and larger denominator. Observed or reported maintenance costs for a comparable process and materials of construction are the best basis for estimating this item. Table 3 does not show some other items which frequently are classified as operating expenses. One of these is supplies, to provide for lubricants, chart paper, workers' clothing and protective equipment, and other miscellaneous needs. Another operating expenses item might be a royalty paid for a licensed process per unit of product manufactured. Still another operating expense might be the cost of laboratory control analyses on raw materials and process and product streams. As mentioned earlier, costs of catalysts, inhibitors, etc. might also exist as items of operating expense.

The *overhead* costs of Table 3 encompass those manufacturing plant expense items which may be chargeable in full to the process being evaluated. The plant will have personnel and facilities for administration, industrial relations, security, cafeteria and recreation, health care, fire protection, purchasing, warehousing, receiving, shipping, accounting, production scheduling, general laboratory services, waste disposal, process engineering, and plant engineering functions. Insurance, local taxes, and community relations are other forms of overhead costs. If a plant produces only a single product, then the cost of all these overhead functions is chargeable to that product. However, if the product being considered is part of a multiproduct plant, then only some portion of these overhead costs must be allocated to that product. Actually, all of the overhead costs itemized in Table 3 do not fit this generally accepted definition of overhead cost. The first item, payroll overhead, is simply the employer's cost for fringe benefits such as social security, health and life insurance, workmen's compensation premiums, vacations, holidays, and sick leave. This payroll overhead stems from the operating labor and supervision costs in the operating expenses. The payroll overhead item thus might more appropriately have been included as part of operating expenses. The general overhead item in the overhead section apparently reflects this company's experience that a "fair" way to allocate plant overhead expenses is as 50% of the manufacturing unit's operating expenses. Companies have developed numerous different methods of charging plant overhead based upon correlations which they have observed in their particular circumstances. The third item shown in the overhead section of Table 3, capital charges, is not really a plant overhead item at all but instead is a "fixed cost" on the sulfuric acid plant. It consists of depreciation, property taxes, and insurance. The depreciation portion represents an expense which the government allows to be taken for the wearing out of the capital facilities in use and in time. More attention will be given to depreciation later. Even though depreciation on a manufacturing unit is directly assignable to that unit, it is customary in economic evaluations to show depreciation, or capital charges, as a part of overhead costs. This peculiarity results from interfacing and overlapping in some cost concepts which also will be examined later at length: variable costs versus indirect costs.

Several other aspects of Table 3 warrant discussion. One of these is the terms "battery limits" and "offsites." "Battery limits" defines a geographical area containing the process equipment which enters directly into the manufacture of a product. "Offsites" defines those service facilities which do not enter directly into the manufacture of a product, such as utilities, feed and product storage, shops, roads, fire protection systems, and communications systems. Another item of significance in Table 3 is the right-hand column, headed "$ Cost/ST 100% H_2SO_4." The contributions of individual cost items to total unit manufacturing cost of the product appear in this column. These contributions to unit cost are not really necessary to the presentation of estimated or actual economics, but they provide very helpful information to thought processes. For example, the cost of sulfur raw material for the process has already been shown as $4,200,000 for 350,000 tons/yr of 100% H_2SO_4. However, these numbers for the cost of sulfur do not register nearly so well in most individual thought processes as does the $12.00/short ton of 100% H_2SO_4 appearing in the right-hand column.

The use of annual costs in Table 3 circumvents a confusing point which arises when shorter time periods are used as the basis of the analysis. Economic analyses frequently are developed on the basis of 1 day. The reason for this is apparently to reduce the magnitude of numbers to quantities more easily managed in the mind. However, two different kinds of "days" are variously used, and the intended basis must be clearly stated. One of the kinds of days, the "stream day," is used when costs and material quantities reflect the process' being operated at its instantaneous rate. The other kind of day, the "calendar day," is used when costs and materials quantities reflect the process' being operated at its average rate. The average rate is less than the instantaneous rate because the process does not operate throughout the year. Equipment malfunctions and process upsets cause a reduction in productive operating time. Furthermore, it is conventional practice to shut down a process unit at planned intervals for "turnarounds" to do preventive maintenance. These turnarounds further reduce productive time. The resultant stream efficiency for processes may vary typically from 90 to 98%, where stream efficiency is 100 times the ratio of number of days per year of operation at the instantaneous rate divided by 365. The letters CD in an evaluation are used to denote a calendar day basis and the letters SD are used to denote stream day. The kind of day used in an evaluation should be clearly shown, and the stream efficiency taken is also valuable information to impart.

The effective practice of cost engineering requires an understanding of distinctions which exist for direct costs, indirect costs, variable costs, and fixed costs. These four classifications of costs exist primarily as two dichotomies, of which the first is direct costs versus indirect costs and the second is variable costs versus fixed costs. However, these two dichotomies also interface. For example, direct costs and indirect costs may be of either the variable or fixed type. Similarly, variable costs and fixed costs may be of either the direct or indirect type.

A direct cost is defined to be one which may logically be assigned in full to a particular product, while an indirect cost is one which is affected by the manufacture of a product but which is not easily assignable to that product. Obviously, in a single-product operation, all costs become direct costs. In multiproduct operations, however, numerous difficulties arise. One example of such difficulties is the situation where one raw material produces more than one product, as in the partial oxidation of butane. There, just about everything that can happen chemically does happen, and the products consist of an array of acids, aldehydes, ketones, alcohols, and ethers. It is impossible in such cases to assign raw material costs to individual products on anything but an arbitrary basis. Another example is a distillation column producing methanol as the overhead product and ethanol as the bottoms product. The operation of the column will incur costs for, among other things, cooling water for the condenser and steam for the reboiler. The costs of cooling water and steam can only be assigned to the methanol and ethanol on some arbitrary basis. There is no easy way out of such difficulties. A device that is sometimes used when one product is dominant is to assign all costs to that dominant product and to show the lesser products as credits either under the raw materials section or under a section of the economic analysis created especially to set forth the by-products. When an arbitrary basis for assigning costs must be used, one can select from among several rationales. Sometimes costs are assigned on the basis of weight or volume processed. In energy-intensive situations, costs are sometimes assigned on the basis of energy consumptions. Perhaps the most favored assignment of costs is on the basis of unit selling prices of the products. This approach can be unrealistic but it does offer the aesthetic appeal of minimizing the likelihood that total unit costs for some individual products will be greater than the unit selling prices for those individual products.

Indirect costs most typically arise in the service or administrative functions of a manufacturing plant. An example would be the process engineering or technical service department of a plant. The process engineers engage in a great variety of activities directed toward troubleshooting and upgrading plant operations. The volume of process engineering activities, and hence costs, will probably vary with the number of products produced by the plant. The allocation of such costs is difficult. If a process engineer works on a project involving, say, the plant steam system, the cost of his activities may be assigned to the cost of steam and hence ultimately to steam-using individual products. In any event, the cost of his activities has reached the individual products by a very indirect route.

The other cost dichotomy requiring understanding is the distinction between variable cost and fixed cost. A variable cost is one which increases or decreases linearly with increases or decreases of quantity of product manufactured. A fixed cost is one which remains constant, or essentially so, even though volume of product varies. In the chemical process industries, raw material costs and utilities costs are almost always variable costs. At the other extreme, items such as depreciation and plant general overhead usually are fixed costs. These costs continue regardless of whether a production unit is operated at capacity, at a fraction of capacity, or is shut down even for an extended period. There is a middle ground of cost items which cannot be classified so definitely, but which tend to be fixed costs. Among these are costs for production labor and supervision, maintenance, supplies, and laboratory. One or a few operators in and around a process unit's control house are going to be present no matter whether the unit is operating at capacity or at a fraction of capacity. Maintenance cost probably does vary with a unit's rate of production, but some maintenance costs continue even if the unit is shut down for a temporary period. Hence maintenance cost tends to be a fixed cost. The components of supplies cost also tend to be required in quantities independent of production rate. Laboratory costs in the form of periodic sampling and analysis similarly must be performed at fractional rates of capacity. Subject to all the hazards of any generalization, in the chemical process industries the costs of raw materials and utilities tend to be variable costs and all other usual costs tend to be fixed costs.

Depreciation on a product's manufacturing facilities is a fixed (and direct) cost which calls for special attention. Equipment for the manufacture of chemicals wears out as it is used. The federal government allows the owner of the equipment to apply as a deduction for income tax purposes a depreciation expense so that the original cost of the equipment is recovered in annual increments over the life of the equipment. Depreciation thus appears as an item on the manufacturing cost sheet. The dollar value of depreciation taken in a given year depends on the original capital investment in the equipment, the life of the equipment, and the kind of depreciation schedule chosen, of which there are several. Bulletin "F" of the Bureau of Internal Revenue, U.S. Treasury Department, tabulates estimates of the average useful lives of different kinds of equipment such as pumps and fractionating columns in different types of service, i.e., in the manufacture of alkalis, electrochemicals, specific acids, etc. These numbers of years for equipment lives are negotiable, and can be altered if adequate justification for the alteration is presented to the Internal Revenue Service. It is more common to speak of depreciation in terms of a rate than in terms of years of useful life. For example, a life of 11 years is equivalent to a rate of depreciation of (100/11), or approximately 9.09%/yr. The overall depreciation rate of a facility can be developed by compositing the rates for the individual items in the facility. Another approach is to use the depreciation ranges published by the Internal Revenue Service for capital assets classified by the service of the assets. Table 4 contains some of the published depreciation

TABLE 4 Depreciation Ranges in Years and Guideline Repair Allowance Percentages for Certain Classes of Assets [6]

Class of Asset	Depreciation Range in Years			Guideline Repair Allowance Percentage
	Lower Limit	Guideline Period	Upper Limit	
Petroleum refining	13	16	19	7.0
Manufacture of pulps from wood and other cellulose fibers and rags	13	16	19	4.5
Manufacture of paper and paperboard	13	16	19	4.5
Manufacture of chemicals and allied products	9	11	13	5.5
Manufacture of rubber products	11	14	17	5.0
Manufacture of miscellaneous finished plastics products	9	11	13	5.5
Manufacture of glass products	11	14	17	6.0
Manufacture of cement	16	20	24	3.0
Manufacture of primary metals:				
Ferrous	14.5	18	21.5	8.0
Nonferrous	11	14	17	4.5
Manufacture of textile mill products	11	14	17	4.5
Land improvements	—	20	—	—

ranges. The right-hand column of Table 4 gives a percentage of the capital investment which can be used to determine the maintenance expense item on the manufacturing cost sheet; a different percentage can be used if it can be justified to the Internal Revenue Service.

When money is spent which affects the operability of a manufacturing unit, the consideration must be faced about whether the expenditure is an operating expense and therefore currently deductible for income tax purposes, or increases the capital value of the facility and is therefore to be treated as depreciation in future years. In the view of the Internal Revenue Service, whether an expenditure is a repair or a capital improvement depends on whether it prolongs the life of an asset [6]. Many expenditures related to chemical processing plants offer flexibility in being declarable as current operating expense or as becoming part of the capital investment. For example, the design of a new plant may require substantial expenditures for engineering work for several years preceding the construction of a plant. A company can choose to "expense" those funds or to "capitalize" them. If expensed, the funds become part of the manufacturing cost and so reduce the profits (and taxes) for the years in which they occurred. If capitalized, they contribute to depreciation deductions in future years and so reduce the profits (and taxes) in the future years. It should be noted that the decision to expense or capitalize an expenditure does not affect the cumulative tax due the government, assuming that the company's profits keep it in the same tax bracket over time. The only effect is to have higher or lower profits (and taxes) in the current or future years. Company managements can use the decision to create more favorable cash flows, retained earnings, and dividends to stockholders to achieve a desired balance of objectives during a, say, 5-yr planning period.

While it can be seen from Table 4 that the amount of money allocated annually is dependent in part on the life of the facility, the annual amounts need not be uniform. This is because the owner of the facility is allowed by the government to choose a schedule by which the facility is to be depreciated. Three broad schedules are available: a uniform rate, an accelerated rate, and a retarded rate. Assuming again that a company's profits keep it in the same tax bracket over time, the choice of a depreciation schedule cannot serve to increase or decrease the cumulative tax paid to the government out of the project's earnings. The schedule chosen affects only *when* the varyingly sized increments are to be paid during the project's life. Table 5 summarizes the four most

TABLE 5   Summary of Four Different Depreciation Schedules

Definitions

C	original capital investment
D_n	depreciation amount for year n; $n = 1, 2, \ldots, N$
F	some multiplier; limited to ≤ 2
I	interest rate, a fraction
N	number of years of facility life
Q_n	"digit" for year n
T	terminal value of facility
U_n	undepreciated balance at beginning of year n

Uniform depreciation

$$D_n = \frac{C - T}{N}$$

Declining balance depreciation

$$D_n = (U_n)(F)\left(\frac{C - T}{NC}\right)$$

The schedule must change to linear depreciation at some year to get 100% percent depreciation

Sum-of-digits depreciation

$$D_n = \left[\frac{N - n + 1}{N(N + 1)/2}\right][C - T]$$

Sinking fund depreciation

$$D_n = (C - T)\left[\frac{I}{(1 + I)^N - 1}\right](1 + I)^{n-1}$$

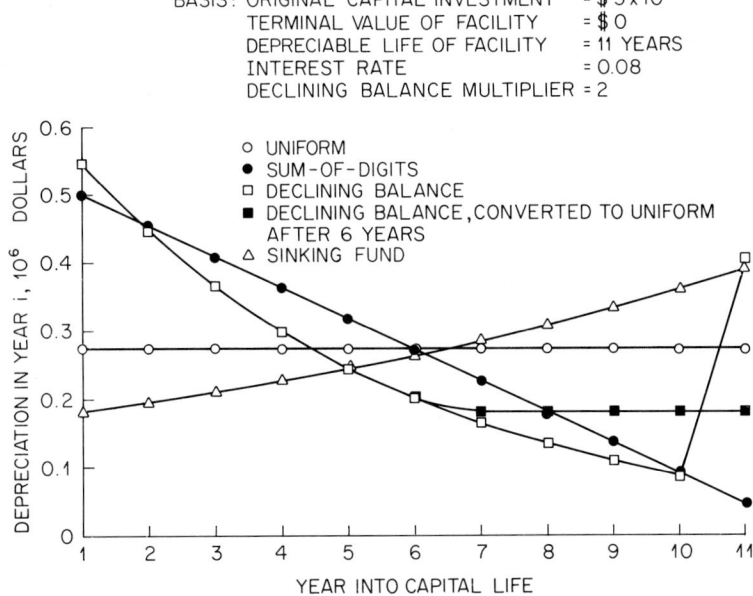

FIG. 1. Comparison of annual depreciation schedules.

common depreciation schedules. *Uniform* depreciation provides an annual amount which is the same in each of the years. *Declining balance* and *sum-of-digits* provide accelerated depreciation in which the annual amounts are large initially and decrease as time passes. *Sinking fund* depreciation provides retarded depreciation in which the annual amounts are small initially and increase as time passes. Figure 1 shows these characteristics for a hypothetical example.

Company managements thus can use a mixture of depreciation methods in a multiple set of new ventures to achieve a desired schedule of cash flows over a planning horizon. Cash flow may be viewed as the sum of after-tax profit and depreciation. The use of an accelerated depreciation method maximizes cash flow during the early years if all other economic parameters are constant. A larger amount for depreciation increases manufacturing cost and decreases before-tax and after-tax profit. But while after-tax profit is decreased by approximately one-half (depending on the tax rate being 48, 50, or 52% of before-tax profit in the particular year) of the depreciation, the full amount of depreciation is added back to after-tax profit to get cash flow. Thus the larger depreciation will create a larger cash flow.

Two factors tend to cause company managements to favor accelerated depreciation within the other constraints of its desired schedule of cash flows. One of these is the concept of risk and the other is the concept of the time value of money. Risk increases as time passes because of possible unforeseen changes in raw material availabilities and prices, product demands and prices, and the advent of new and superior technology. This situation makes it desirable to schedule cash flows to be a maximum during the early years of a project's life when uncertainty is a minimum. The time value of money conceives the present worth of a cash flow in year n to be the quotient of the year's cash flow divided by the sum of one plus a fractional interest rate I with the sum raised to a power equal to the value of the year into the future:

$$\text{Present worth of cash flow in year } n = \frac{\text{cash flow in year } n}{(1 + I)^n}$$

$$= 1, 2, 3, \ldots \tag{1}$$

The interest rate I is usually taken to be constant during the planning horizon. The denominator of this expression increases exponentially as time moves into the future and, for a given cash flow in the numerator, the quotient decreases. So the present worth of all cash flows is maximized by having the cash flows to be larger in the early years when the exponents n are smaller.

Although these two concepts favor the use of accelerated depreciation ultimately, some company managements prefer that all preliminary evaluations of project economics be made using linear depreciation. This procedure has the advantage of placing all project evaluations on the same basis for comparison, independent of the variations which appear when different depreciation methods and rates are used.

A very basic requirement in cost engineering is the estimation of the capital necessary to construct the plant for the manufacture of the product or products. This estimate of the fixed capital for a project is in itself an expensive activity. The expense for the estimate needs to be kept consistent with the purpose of the economic evaluation. In very "quick-and-dirty" evaluations, only a few minutes or hours may be invested in the estimate of fixed capital. One or more of several procedures might be used in this situation. These involve data available from the company's experience or from published values for such factors as capital cost per annual ton of product, the ratio of capital cost to annual sales, and the exponent β in the relationship

$$\text{Capital cost at size 2} = \text{capital cost at size } 1 \left(\frac{\text{size 2}}{\text{size 1}} \right)^\beta \tag{2}$$

Reference 3, for example, contains such values. A next class of capital estimate measured in terms of the effort and cost expended might be described as a "preliminary" estimate. In this, a process flow sheet is used to determine the size and cost of individual pieces of process equipment such as reactors, distillation columns, heat exchangers, pumps, compressors, and storage tanks. The purchase prices of these items are totaled. The costs of other equipment, such as foundations, piping, instrumentation, and buildings, are then estimated as fractions of the total for the process equipment. These fractions may be based on a company's experience or on values which have been published (e.g., Ref. 7). The sum of process equipment and its related piping, instrumentation, etc. may be considered to be the "plant physical cost." Fractions of this cost are then taken to cover engineering and construction expense, profit for the contractor, and an allowance for the contingencies which almost inevitably arise. The sum of these three costs plus the plant physical cost finally leads to the total capital cost of the plant. A third class of estimate of fixed capital might be called the "appropriation" estimate. Here the estimate will be the fixed capital which will be stated to top company management as being the actual cost of constructing the plant ready for operation. It is important that this estimate be of good accuracy because the company will have to live with the result after the plant is built. Therefore, detailed estimates are made of the cost of process equipment; of the pipe, valves, etc. required for piping; of the sensors, controllers, etc. required for instrumentation; and so on for every category of fixed capital cost for the plant. The appropriation estimate can be very expensive to prepare, but conceptually, at least, should result in the greatest accuracy.

There is a second kind of capital requirement in the operation of a plant. This is working capital, which is money that must be set aside for investment in inventories of raw materials, work-in-process materials, and products. Continuous operation of the plant depends upon raw materials always being available. In most situations this means that an inventory of raw materials will have to be maintained. These raw materials will have had to be purchased and placed in storage for the eventuality that normal supply schedules will be interrupted. This investment in raw materials is one component of working capital. A 30-day supply of raw materials is frequently taken as the basis of this component. In the operation of the plant there will be materials in various stages of completion between raw materials and finished products. The company will have invested all elements of manufacturing costs in these work-in-process materials, including raw materials, utilities, operating expenses, and overheads depending on the proximity of the given material to being a finished product. A second component of working capital is the investment necessary to maintain the inventory of these work-in-process materials. Thirty days of manufacturing costs are frequently taken as the basis of this component. The third component of working capital is the investment required to maintain a supply of finished product. There likely will be variation in the day-to-day sales and shipments of products. There also will be day-to-day variation in the production from the plant because of upsets and other interruptions. A supply of finished products helps to protect customer relations against failures to ship products on schedule. It is common practice to allow customers 30 days to pay for materials which they have received. A second 30 days of manufacturing costs are frequently taken as the basis of this third component of working capital. It should be apparent that a hypothetical company organized to build a plant for the manufacture and sale of a product would have insufficient capital to function if all of its capital just met the fixed capital requirement. Substantial additional capital is also required for working capital.

Cost engineering often is necessary in developing information for the following kind of decision: "We have an opportunity to sell an additional Q units of product at a price of Y cents per unit. Should we accept the opportunity?" The correct answering of this question makes it necessary to distinguish between average unit cost and incremental unit cost. Average unit cost is the quotient of total manufacturing cost at the increased production rate divided by the quantity produced at the new rate. Incremental unit cost is the net increase of cost incurred in increasing the production rate divided by the increase in production. The sulfuric acid economics of Table 3 may be helpful for demonstrating the difference between average cost and incremental cost. Suppose the plant is producing at 85% of capacity. If operating expenses and overheads are fixed costs while raw materials and utilities are variable costs, then total manufacturing cost at the 85% rate is $3,975,700/yr [= $146,000 + $469,300 + (0.85)($4,200,000 − $246,600)]. Production at the 85% rate is 297,500 tons/yr. The average cost at this rate is $13.36/ton [= $3,975,700/ 297,500]. Suppose also that the additional quantity of product in question is 10% of the plant's capacity, or 35,000 short tons/yr. The total manufacturing cost at the increased production rate would be $4,371,000/yr [$146,000 + $469,300 + (0.95)($4,200,000 − $246,600)]. The new production rate would be 332,500 tons/yr, for a new average cost of $13.15/ton. Suppose finally that the price offered is $13.00/ton net at the plant. On the basis of average cost, the decision would be to reject the offer because the price of $13.00/ton is less than the average manufacturing cost of $13.15/ton. But the incremental cost of the additional acid would be only $11.30/ton [= ($4,371,000 − $3,975,700)/ 35,000]. So, the offer actually should be accepted because it results in a plant profit of $1.70/ton [= $13.00/ton − $11.30/ton]. It should be noted that this incremental cost could have been more easily obtained from Table 3 by simply taking the net of the $12.00/ton raw material cost and $0.70/ton credit for utilities shown there.

In this discussion of incremental cost versus average cost, the average

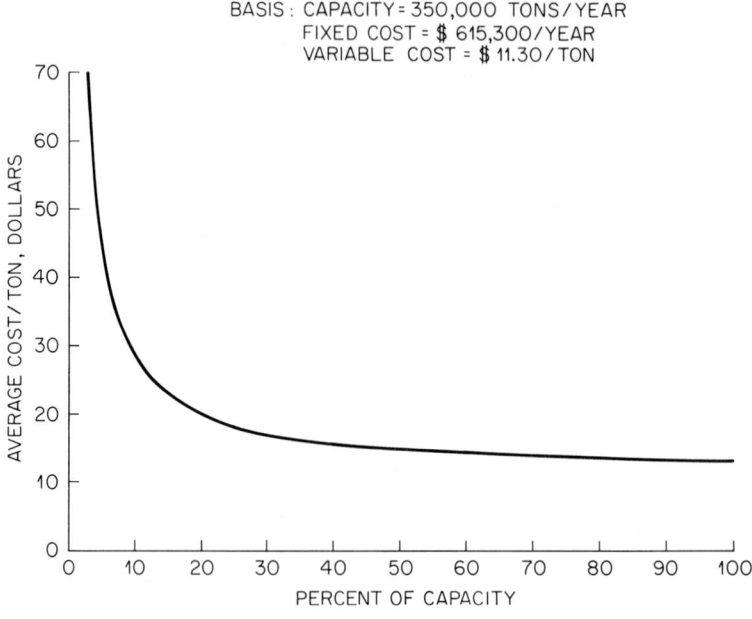

FIG. 2.   Effect of operating level on manufacturing cost.

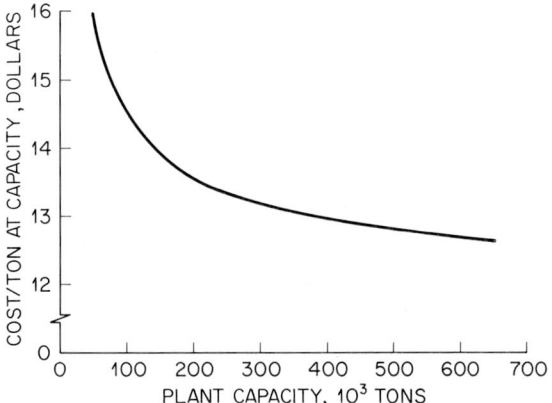

BASIS: C_2 = FIXED CAPITAL AT GIVEN SIZE
$$= (\$\ 2{,}600{,}000)(\frac{SIZE}{350{,}000})^{0.62}$$
VARIABLE COST = $11.30/TON
FIXED COST = $69,300/YEAR + $(C_2)(0.21)$/YEAR

FIG. 3. Effect of plant size on unit manufacturing cost.

manufacturing cost was shown to decrease from a value of $13.36/ton at 85% of capacity to $13.15/ton at 95% of capacity. This demonstrates another important concept in cost engineering: the variation of average cost with operating rate in a plant having a given capacity. When a plant is producing nothing, it nevertheless will be incurring its fixed costs. The first increment of product must absorb not only the variable costs but also the fixed costs. Therefore the average manufacturing cost for the initial unit of production will be extremely high. As production rate is increased, the fixed cost portion of total cost is spread over the greater quantities of production with the result that the average manufacturing cost continues to decrease. Figure 2 demonstrates this characteristic for the sulfuric acid plant of Table 3.

Just as a reduction in unit manufacturing cost can be achieved in a given plant by operating that plant nearer to its capacity, so also can unit manufacturing cost be reduced by building larger plants. Equation (2) gave a relationship between plant size and fixed capital cost. The exponent β in Eq. (2) is generally less than unity. This being so, the capital investment in plants to manufacture a given product increases less than linearly with plant size. The result is that the elements of manufacturing cost such as depreciation which are dependent on fixed capital become a smaller fraction of unit manufacturing cost. Therefore, lower manufacturing costs can be obtained as the size of the plant is increased so long as the exponent β is less than unity. Returning again to the sulfuric acid plant of Table 3 and using a value of $\beta = 0.62$ cited in Ref. 7 for contact sulfuric acid plants, it is possible to develop Fig. 3 which shows the effect of plant size on unit manufacturing cost at plant capacity. These economics for sulfuric acid are dominated by raw material costs. For some other products such as ethylene, variable costs are much less significant than are the fixed capital-dependent costs, and the variation of unit manufacturing cost with plant size is much greater than that shown for sulfuric acid.

To this point the discussion has centered upon costs which occur in the manufacturing plant. After the product leaves the plant, it may have to incur shipping costs depending on whether the selling price is effective at the plant or at the product's destination. Shipping cost has many parameters. One of these is the mode of shipment: pipeline, tanker, barge, rail, or truck. Even within a particular mode, the volume shipped affects costs. Another parameter is the distance involved. Still another is the identity, i.e., the nature, of the material. If shipment is to be by tank car or boxcar as is frequently the situation, published rates are available which can be used to develop approximate shipping costs:

1. Refer first to the "Uniform Freight Classification (No.)" of the Uniform Classification Committee, Chicago, Illinois to obtain a rating for the product and container. For example, page 512 of the Uniform Freight Classification 12 shows sulfuric acid in tank cars to have a class rating of 25.
2. Refer next to a regional freight association tariff to get a rate basis (standard mileage) and then use this with the class rating to determine a shipping cost per hundredweight (cwt) of product. For example, the Southern Freight Association Tariff 1011, dated 9/10/75, shows a rate basis from Baton Rouge, Louisiana to Mobile, Alabama of 219, and a rate basis from Baton Rouge, Louisiana to Norfolk, Virginia of 1150. The same tariff shows for a Class 25 rating that the shipping cost is $1.14/cwt for a rate basis of 219, and is $2.75 for a rate basis of 1150. These costs can be converted to 10.4¢/ton-mi for the 219-mi distance and 4.78¢/ton-mi for the 1150-mi distance.
3. Estimate an average shipping distance for the product from plant to customers. Use that distance with these kinds of cents/ton-mile costs to get a unit shipping cost for the product. Costs for shipping by rail determined in this manner tend to be conservative, i.e., larger than they actually may be. When very substantial shipments of a commodity have been made between two points over an extended period of time, it is possible for the shipper to obtain a reduced rate from the carrier which is an exception to the published rate. A number of other factors can affect rail shipping costs.

The cost of shipping products to customers obviously can become a significant cost item if distances are large. Although product shipping costs can be reduced by locating the plant near to customers, this may result in an increase of shipping costs for raw materials if the raw materials are not available in the vicinity of the customers. In terms of shipping costs only, the plant site should be selected to minimize the sum of raw material and product shipping costs. The cost of shipping raw materials or products increases as the method moves from pipeline to tanker to barge to rail to truck, depending on volume shipped and distance. Therefore, this consideration should be made important in selecting a plant site. Inventories can also be of major importance in the selection of a plant site. In general, the greater the distance from supplier to the plant or from plant to the customer, the larger should the inventory be. It is, however, prohibitively expensive to maintain sizable inventories of some commodities, e.g., those in the gaseous state. Other factors which need to be considered in the selection of a plant site are labor availability and attitude, the tax structure, the receptivity of the community, and the living conditions for the plant's employees. It must be kept in mind when estimating the total cost of a product that there exist costs in a company which are not immediately associated with the product or plant. Among these are the costs of marketing the product, the costs of the company's headquarters (administrative) staff, and the cost of the company's research organization. The overall cost of these functions is assigned to a product for purposes of economic evaluation, usually as a percentage of the sales revenue generated by the product based on the company's experience.

Having given attention to the costs required to deliver a product to the market place, it will next be necessary to identify a selling price for the product. The selling price may already be known in some specific situations in which negotiations are underway with a customer. At other times the selling price may have to be estimated from knowledge about the market place in general. One of the sources of information about product prices is the *Chemical Marketing Reporter*. Three government publications also may be useful:

U.S. Bureau of Census Current Industrial Report: *Industrial Gases*, Series M28C (year)—14

U.S. Bureau of Census Current Industrial Report: *Inorganic Chemicals*, Series M28A (year)—14

U.S. Tariff Commission Publication: *Synthetic Organic Chemicals, U.S. Production and Sales* (year)

Prices obtained from the *Chemical Marketing Reporter* tend to be higher than a product might command from buyers under high-volume contracts. The prices from the government publications suffer from being a year or so old by the time the publication is issued. They do, however, have an advantage in being representative because they are derived from the quotient of the United States totals for sales revenue and quantity sold. These government publications also are valuable as sources of information of total United States production of individual chemical products. Obviously, these four publications also are useful for information on raw materials. If better information is needed than these sources can provide, every sizable company has a market research function which can be asked for assistance.

There is one other economic factor which must be recognized before project profitability can be calculated, and that is the investment tax credit. This is an amount of money which the federal government allows the investor to deduct from the investor's income tax. It is viewed by the government as a stimulus to the economy. By it, a company can deduct from its income tax a prescribed percentage of the value of purchased items which are: (a) depreciable, (b) have a life greater than 2 years, and (c) are personal property which " . . . covers primarily machinery and equipment but excludes buildings, land, parking areas, docks . . . " [8]. The investment tax credit can be taken only one time for

TABLE 6 Calculations for Profit and Economic Merit of Plant Producing 350,000 tons/yr 100% H_2SO_4

	M$/yr
Sales revenue at $17.00/ton at plant	5950
Less SAR costs at 8% of sales	476
Net revenue	5474
Less manufacturing cost	4569
Gross profit	905
Less income tax at 48%	434
Net annual profit	471

Fixed capital $=$ $2,600,000

Working capital $= [30/365][\$4,200,000 + (2)(\$4,569,000)] =$ 1,096,000

             $3,696,000

Depreciation per year $= (0.11)(\$2,600,000) = \$286,000$

$$\text{Return on investment} = \frac{\$471,000}{\$3,696,000} \times 100 = 12.7\%$$

$$\text{Payout} = \frac{\$2,600,000}{\$286,000 + \$471,000} = 3.43 \text{ yr}$$

a given investment. Its amount is limited to the lesser of: (a) the income tax liability shown on the return or (b) $25,000 plus 50% of the tax liability in excess of $25,000.

The project profitability can now be determined. For the purpose of working in the context of the sulfuric acid economics being used, assume that the sulfuric acid has a price of $17.00/ton at the plant, that sales, administrative, and research (SAR) expenses are 8% of sales, that the income tax rate is 48% of gross profit, and that a linear depreciation rate of 11% applies. The upper portion of Table 6 shows the calculation of net profit, amounting in this situation to $471,000/yr. Although this profit places the project in a favorable light, profit is an inadequate measure of project merit. It needs to be related to the amount of investment which is made to earn the profit. In the example, the after-tax profit of $471,000/yr may be judged to be satisfactory for the fixed capital investment of $2.6 million, but probably would be judged as unsatisfactory had the fixed capital investment been 10 times as large, or $26 million.

Several measures relating annual profit to investment are in common use. These in general examine the economics of the project in a mature year when costs and sales have stabilized. One of these measures is percent return on investment (%ROI):

$$\%ROI = \frac{\text{annual after-tax profit}}{\text{fixed + working capital}} \times 100$$

Another measure which has been in long usage is payout period:

$$\text{Payout, years} = \frac{\text{fixed capital investment}}{\text{annual depreciation + after-tax profit}}$$

Payout here considers only fixed capital investment because of a philosophy that working capital can always be recovered at approximately its full value by selling the raw materials and product at their costs to the owner. It should be noted that these definitions of $\%ROI$ and payout are not universal. For example, some companies prefer to include depreciation in the numerator of the expression for $\%ROI$. The lower portion of Table 6 shows the calculation of ROI and payout for the sulfuric acid plant project.

Percent return on investment (or payout) is a measure of profit versus investment, but as a single number it lacks meaning unless it is referred to some kind of scale. Some companies use a decision table such as the following to evaluate a return on investment:

If risk of project is	Then minimum acceptable ROI is
Low	10%
Medium	15%
High	20%

The risk of a project may be categorized in the context of a second decision table:

For			
Marketing know-how	Manufacturing know-how	Product demand	Risk is
Weak	Weak	Weak	High
Strong	Strong	Strong	Low

A project having medium risk would show a mixture of strong and weak entries in the matrix. A company's marketing know-how will be weak or strong for a product depending upon whether the company has experience and an established position in the marketing of this kind of product. For example, a synthetic fibers company may find itself poorly prepared to market an agricultural chemical. Similarly, a company may have a strong or weak background in the manufacture of the product. Again, for example, an expansion of capacity for an existing product should place the company in a strong position for manufacturing technology. Product demand is measured in terms of the long-term high-volume demand for the product. Sulfuric acid would have a strong position, but a solid rocket fuel might not.

Percent return on investment and payout as measures of project merit have come to be supplemented by "Interest Rate of Return (IRR)" in recent years because they do not recognize the time value of money. During a project's life, revenues and expenditures are likely to vary from year to year. For example, several years may be required after a plant starts up to reach marketing and manufacturing levels at the plant's nominal capacity. Furthermore, the fixed capital investment usually is spread over the several years it may take to design and construct a plant. The time value of money thus becomes significant. Suppose that $1.00 can be invested at an interest rate of 6%, compounded annually. Then at the end of the first year the investment will have increased to a value of $1.06, at the end of the second year to a value of $1.1236, etc. The first two columns of Table 7 generalize this relationship. The converse must also be true. An amount of money existing at the end of a future year will have a reduced present worth. The third column of Table 7 quantifies this relationship. The IRR is that value of i (as a percentage) which satisfies the following expression:

$$0 = \sum_{n=1}^{N} \frac{CF_n}{(1 + i)^n}$$

Table 8 demonstrates the trial-and-error calculation of the interest rate of return for the sulfuric acid project. The basis for the calculations in Table 8 is:

1. 40% of the fixed capital is expended in Year 1 and 60% is expended in Year 2.
2. The plant starts up at the beginning of Year 3.
3. The full amount of working capital is expended in Year 3.
4. Plant life is 17 years with zero salvage value on the fixed capital equipment.
5. Plant produces and sells at 70% of capacity in Year 3, and at 100% of capacity thereafter.
6. Investment tax credit is applied to other company profits, and is 7% of $2,300,000 of the fixed capital.
7. Cash flow = net profit + depreciation + investment tax credit − fixed capital−working capital.

TABLE 7   Effects of Time Value of Money upon Investments and Annual Cash Flows at Interest Rate i

Year	Value of I at End of Year[a]	Present of Worth of CF_n Occurring in Year[b]
1	$I(1 + i)^1$	$CF_1/(1 + i)^1$
2	$I(1 + i)^2$	$CF_2/(1 + i)^2$
3	$I(1 + i)^3$	$CF_3/(1 + i)^3$
...	...	...
N	$I(1 + i)^N$	$CF_N/(1 + i)^N$

[a] I = an investment made at the beginning of the first year. CF_n = a cash flow occurring during the nth year.

TABLE 8   Determination of Interest Rate of Return for Sulfuric Acid Plant Project

Item	1	2	3	4–11	12	13–17
Fixed capital	(1040)	(1560)	—	—	—	—
Working capital	—	—	(1096)	—	—	—
Net revenue	—	—	3832	5474	5474	5474
Manufacturing cost	—	—	(3198)	(4569)	(4569)	(4569)
Gross profit	—	—	634	905	905	905
Income tax	—	—	(304)	(434)	(434)	(434)
Net profit	—	—	330	471	471	471
Depreciation	—	—	286	286	26	—
Investment tax credit	64	97	—	—	—	—
Cash flow	(976)	(1463)	(480)	757	497	471

Trial 1

$$\text{Assume } i = 0.15; \ \sum_1^{17} \frac{CF_n}{(1.15)^n} = 350 \neq 0$$

Trial 2

$$\text{Assume } i = 0.17; \ \sum_1^{17} \frac{CF_n}{(1.17)^n} = 90 \neq 0$$

Trial 3

$$\text{Assume } i = 0.178; \ \sum_1^{17} \frac{CF_n}{(1.178)^n} = -1 \simeq 0 \text{ and IRR} = 17.8\%$$

The interest rate of return for this project is 17.8%.

To this point the analysis of economics has been performed as if the numerical value of each parameter is known with certainty, so that the system is *deterministic*. When such certainties do not exist, modern cost engineering employs probabilistic techniques to obtain a better assessment of economics. This is a *stochastic* treatment of the economic system. Using the sulfuric acid plant as the example, suppose that uncertainty exists about the selling price of the sulfuric acid. Suppose further that on some basis, the following probabilities are assigned for the price of the sulfuric acid in a given year:

Probability is	*That selling price will be*
0.3	$16/ton
0.4	$17/ton
0.2	$18/ton
0.1	$19/ton

The procedure for the determination of IRR would then be, for each year in which sulfuric acid is sold, to generate a price for the sulfuric acid in that year as a function of a random number (RN) obtained from a table of such or from within a computer program:

If	*The selling price is*
$0 \le \text{RN} < 0.3$	$16/ton
$0.3 \le \text{RN} < 0.7$	$17/ton
$0.7 \le \text{RN} < 0.9$	$18/ton
$0.9 \le \text{RN} < 1$	$19/ton

Having selling prices for each of the years in a planning horizon, the IRR can then be determined for the "case." Many cases, say 500, are examined. There will result a frequency distribution of IRR's. This distribution can then be examined to make judgments such as the following:

There is an $X \%$ probability that IRR will be less than 15%; and/or
There is a $Y \%$ probability that IRR will lie between 15% and 20%; and/or
There is a $Z \%$ probability that IRR will be greater than 17%

The power of the stochastic treatment increases as the number of parameters possessing uncertainty increases. Thus, if uncertainties also existed on the values of fixed capital, raw material cost, and sales volume, then a stochastic simulation would be the only means of getting a good assessment of project economics.

Reference

1. U.S. Bureau of the Census, U.S. Department of Commerce, *Annual Survey of Manufactures*, U.S. Government Printing Office, Washington, D.C.
2. Statistical Policy Division, Office of Management and Budget, *Standard Industrial Classification Manual, 1972*, U.S. Government Printing Office, Washington, D.C.
3. Ref. 1, 1970–71, p. 11.
4. Ref. 1, 1970–71, pp. 63–66.
5. J. M. Connor, "The Economics of Sulfuric Acid Manufacture," *Chem. Eng. Prog.*, *64*(11), 59–65 (November 1968).
6. Internal Revenue Service, U.S. Department of the Treasury, *Tax Information on Depreciation, 1974 Edition*, U.S. Government Printing Office, Washington, D.C.
7. F. A. Holland, F. A. Watson, and J. K. Wilkinson, *Chem. Eng.*, *81*(7), 71–76 (April 1, 1974).
8. Internal Revenue Service, *Tax Guide for Small Business*, U.S. Government Printing Office, Washington, D.C.

O. L. CULBERSON

Cost, Equipment and Materials*

Contracting for the engineering, procurement, and construction of petrochemical plants and oil refineries has always been a cyclical industry. Heavy work loads come and go every 3 to 5 years. But the cycle that began in 1972–1973 was amplified beyond anything in memory.

The industries of the supplier firms who provide our materials and equipment experience the same cycles we do. In fact, this is one of our problems. Manufacturers are never sure of the volume of goods they will be asked to produce. So, they are reluctant to risk capital expenditures to expand capacity. When the surge comes, they are usually caught without sufficient capacity.

Boom/Collapse, 1974

This point was demonstrated in early 1974. Spiraling demand with insufficient productive capacity resulted in shortages and delays. Prices rose and delivery times extended at high rates, boosting project costs coincidentally.

The boom collapsed suddenly. Deferrals and cancellations of capital expansion programs were widespread. Almost in lock-step, orders for equipment and materials evaporated.

Price Response

In previous cycles, these events would have led to quick downward response in both prices and deliveries. But not this time. The response has been a mixed bag. Prices for a number of items continue to rise. Some prices have dropped, but just a few. Delivery schedules for most items have been shortening, but not for all.

The nature of our industry seems to be in transition to some different format. The old rules may no longer apply. The recession has not affected some of our suppliers, but it appears to be putting others out of business. Overall price performance seems a little strange for a recession.

V-Notch Recession

One theory is that the recovery has started and suppliers are holding prices up, hoping for a resurgence in demand. Performance of the leading indicators

*This material was originally presented as a speech to the Engineering and Construction Contracting Committee, American Institute of Chemical Engineers, on October 13, 1975. It has not been updated.

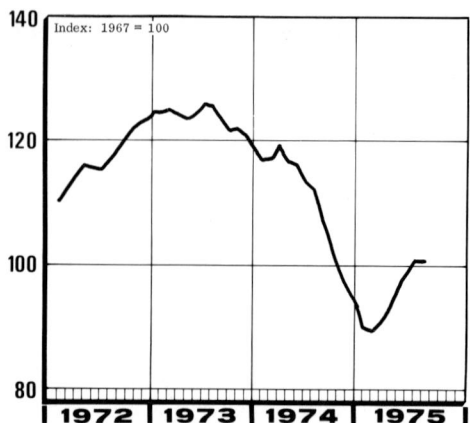

FIG. 1.   Leading indicators. Source: Commerce Department Reports.

might support this theory. Figure 1 depicts the trend of a Department of Commerce composite index of 10 leading indicators. Note the V shape. Some of the economists predicted a V-notch recession. That is, a short recession with a rapid rebound to recovery. Perhaps that is what is happening.

Another Department of Commerce statistic is the new order rate reported by manufacturers of durable goods. Figure 2 shows an unreasonably sharp climb in mid-1974, followed by a catastrophic drop that wiped out the gains of two previous years. The rise in 1975, so far, is frighteningly steep. Does this mean we can expect another sharp drop later this year?

Industrial output, as reported by the Federal Reserve Board, swings along the same shape—but a bit more moderately (see Fig. 3).

People who follow economic trends watch performance of indices like these. But these indices are of little value unless they track closely the performance of the prices of materials of interest.

There is actually little correlation between these publicly reported indices and the actual supply/demand and price/delivery performance in our industry.

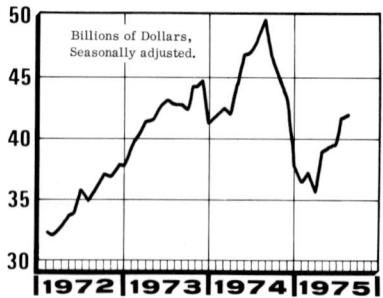

FIG. 2.   Durable goods orders. Source: Commerce Department Reports.

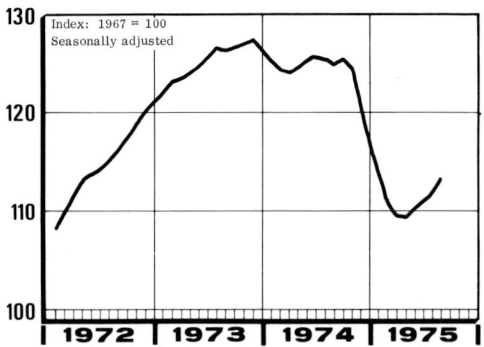

Index: 1967 = 100
Seasonally adjusted

1972   1973   1974   1975

FIG. 3.   Industrial output. Source: Federal Reserve Board Reports.

For this reason, we maintain a continuing effort to track the trends of specific commodities we buy. Charts showing these trends follow.

In preparing these remarks, I asked the leaders of a number of our major supplier firms for their views on price changes. A summation of their views indicates the mixed situation we find today.

Pricing

One question put to these industry leaders was: How do your current prices compare with those at the peak of your 1974 load?

Half said prices were up more than 10%.
A quarter said they have dropped.
A quarter said prices have stabilized.

We also asked about the trend of prices for the balance of 1975.

Half said prices would rise more than 10%.
A quarter said they would drop.
The rest predicted they would stabilize.

Capacity

Another question was: Do you believe the capacity of your industry is sufficient for the demand you foresee for, say, the next 3 or 4 years?

No, said one quarter, there will be a shortage.
Yes, said one quarter, capacity is adequate.
Half said it depends on demand.

The genesis of this question was the observation that in the midst of recession, the prices and deliveries of some items have not dropped one iota. Notably, big centrifugal gas compressors, nuclear valves, and others. This suggests that even though demand has dropped, the supply—that is, the capacity to produce—is really insufficient. Yet our discussions with industry leaders failed to secure outright agreement that a capacity shortage does exist.

A subtle factor may exist here. No manufacturer wants to encourage competition to build up its capacity. Perhaps those who replied that capacity sufficiency depends on demand were really recognizing that a buildup in demand would create a shortage. From this, one can only conclude that three-fourths of our respondents see shortages if normal demand for their products returns.

Recalling the chart of new orders for durable goods, it looks like the move to a shortage economy has already started. How soon this trend will bring us back to 1974-type shortages does indeed depend on how soon full demand develops.

Those of you involved in the new projects in the Middle East are well aware of the scope of the demand these jobs present to industry worldwide. It's staggering. Orders for compressors were released in the past few weeks that run well over $100 million. Even at today's prices, that's a lot of compressors. The companies involved say they can handle them plus additional business. That remains to be seen.

Shortly, other new Middle East jobs will be buying more compressors, pumps, valves, bubble columns, heat exchangers, and even instrumentation in gargantuan quantities. The demand levels will come close to the capacity of the industries to produce, if they don't actually exceed them. This places a responsibility on buyers and sellers alike to adopt better measures for evaluating capacity, and for entering into orders and contracts with really realistic schedules and pricing.

Another sign of the transitional times we are living in is that some items have been in short supply even in the depths of the recession. Of interest to us are castings and forgings, mainly those that our suppliers make or buy for compressors, pumps, valves, and other similar products.

The casting industry and the forging industry have had a capacity shortage for some years now. Capital investment to expand capacity has been in short supply. The return on investment of foundries and forge shops has diminished to very low levels. Besides the rising costs of the production equipment, foundries and forge shops have had to channel huge sums into pollution-control equipment. Many small firms have looked at the costs of this equipment and decided there was no payout, so they closed up shop instead.

The healthy, forward-looking firms have complied with EPA and OSHA regulations, and they are still producing. But they are the rarity, and their costs for compliance have really jumped. One major Eastern foundry raised prices 100% during 1974, 25% more so far in 1975.

The new regulations are improving the environment, and that is important.

Yet we should all be aware of the impact on costs and schedules of these new controls. It may be possible to go too far too fast and, possibly, strangle the basic industries that provide essential products to our industry.

Besides environmental controls, there are two other culprits that discourage capital investment. One is inflation. The other is cost of money. Both of these are largely affected by governmental actions. Without detailing this argument, it seems imperative for industry leaders to study the effects of factors limiting capital expansion and to take an activist role aimed at minimizing, or correcting, their deleterious effects.

Now, let's look at the price trends of equipment and materials used in petrochemical plants and oil refineries. The charts show price trends for the prior year, since October 1974, and our guess for the future year, to October 1976. Price is reflected in a percentage, with 100 being the level in January 1974.

Centrifugal Compressors

Note the steady rising trend in Fig. 4. Demand is close to supply and may actually exceed capacity. Pricing reflects the rising costs manufacturers face for both labor and materials.

In our talks with manufacturers, a recurrent theme was the problems they face with castings and forgings. Because of critical specification requirements,

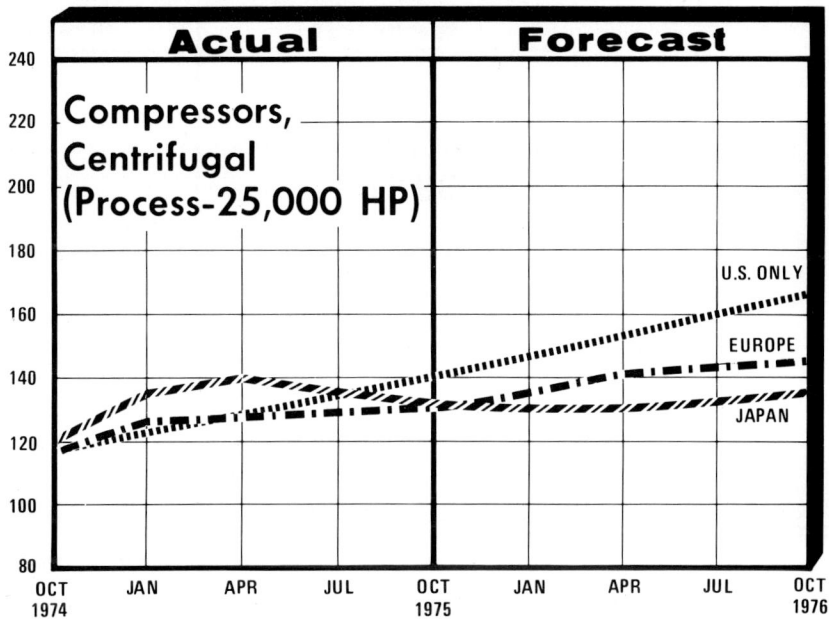

FIG. 4.   Price trends. Price base: January 1974 = 100.

manufacturers are often limited to one or two sources. This often results in price gouging, about which manufacturers can do very little. They need to have the parts promptly to meet schedules they have committed to.

Moreover, a number of the small foundries and other subsuppliers have been backed into costly labor agreements. One form is called COLA, cost-of-living-adjustment. Once COLA is built into the labor agreement, the supplier's labor cost rises with the cost-of-living index named in the agreement. This then becomes a fact of life. Then, when the contract expires, the union presses for a substantial wage agreement piled on top of the COLA formula. In shortage cycles there is no option but to agree, and, of course, this is highly inflationary.

The compressor industry doesn't feel the current recession. Backlog remains high and rising, and so are lead times. Deliveries of 130 weeks seem likely by year-end.

Pricing overseas hasn't risen this year. In fact, Japanese prices have declined.

Circulating Water Pumps

This is another industry whose performance is controlled by the casting/forging supply crisis. This led to a more than 50% price hike during 1974. Figure 5 shows prices of large circulating water pumps. These have risen steadily. Given

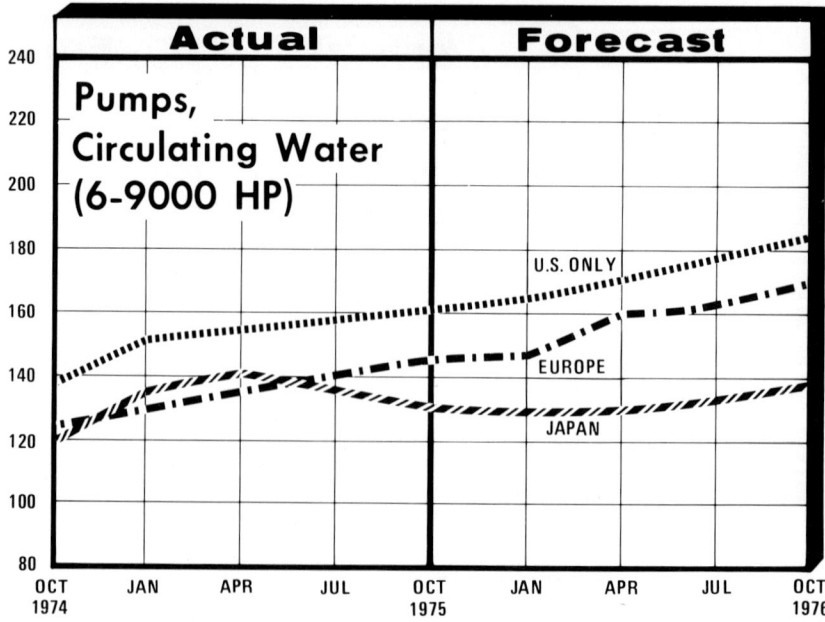

FIG. 5.   Price trends. Price base: January 1974 = 100.

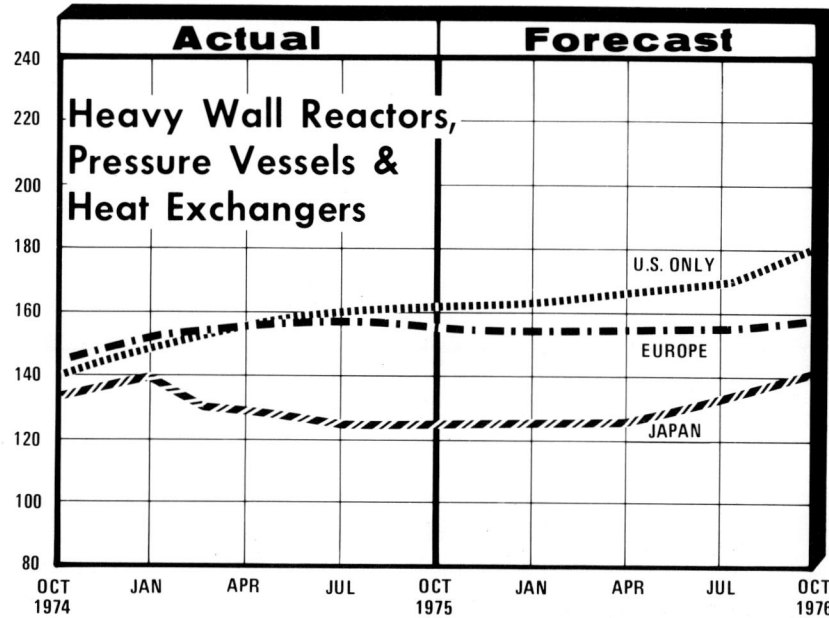

FIG. 6.   Price trends. Price base: January 1974 = 100.

the current shortage in industry capacity, it seems unlikely that the price trend will level out even well into the foreseeable future.

Heavy Wall Reactors

This trend line is a mixture of a number of fabricated steel products as indicated in Fig. 6. Like all averages, the trend line does not necessarily reflect the situation for individual products.

Heavy wall reactors are on a continually rising trend. There are only a few top quality shops that build reactors for our industry. Several of these shops have added capacity recently and they hope they have enough for the foreseeable future.

Backlogs have declined, resulting in improved deliveries. Lead times peaked around 40 months. They are down to around 20. And the Japanese lead times are often shorter, reflecting their surplus shop space.

How long this will continue is hard to guess. The industry leaders I talked to expressed concern that there will again be severe shop overloads if, as one of them said, "everything we expect to take place in the next 3 or 4 years actually happens."

Run-of-the-mill pressure vessel shops are less loaded. Prices at some of these shops have actually dropped, probably as much as 10% so far this year. There is not enough work immediately available for these prices to stabilize for the rest of this year.

Heat exchangers fall somewhere in between. One manufacturer reported the following: "Current price structure is somewhat higher than at the peak price of 1974, but is somewhat lower than it might have been because of the reduction in prices due to a soft market. It is estimated that prices at this time have risen approximately 5% over mid-1974. Our peak backlog, in about mid-1974, was approximately 18 to 20 months. At this time, this backlog has dropped to approximately 14 to 16 months."

European prices nearly parallel domestic prices, but Japanese prices have dropped. This doesn't mean they are actually lower. Remember, these curves are based on an index of 100 equals January 1974 pricing. At that time, Japanese prices were higher than domestic. Today it is a standoff.

Fabricated Structural Steel

Fabricators of structural steel were among the first to feel the decline in 1974. Bookings began to fall off rapidly in late summer. However, most of the major shops have had enough backlog that they have kept pretty busy into 1975. Bookings have picked up this year, but they are nowhere near the levels of 1973 and 1974.

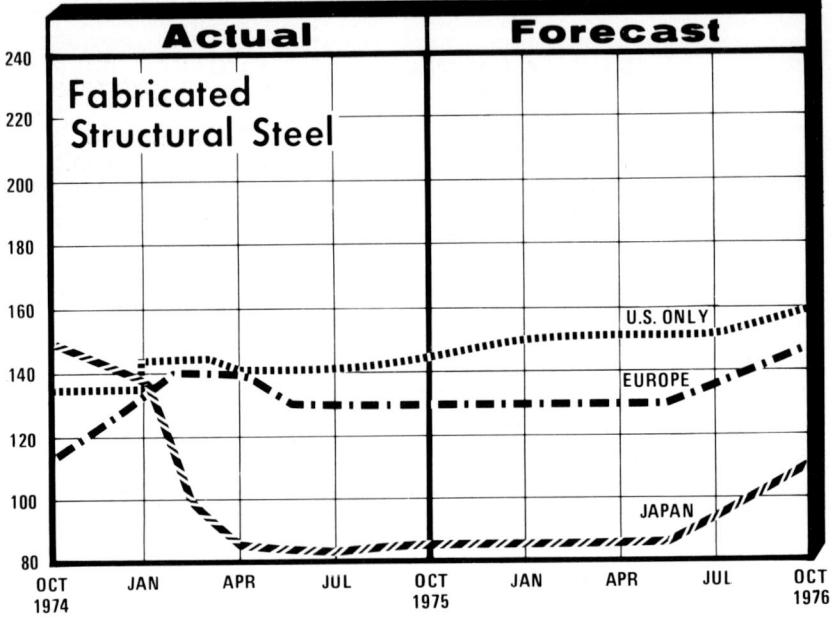

FIG. 7.   Price trends. Price base: January 1974 = 100.

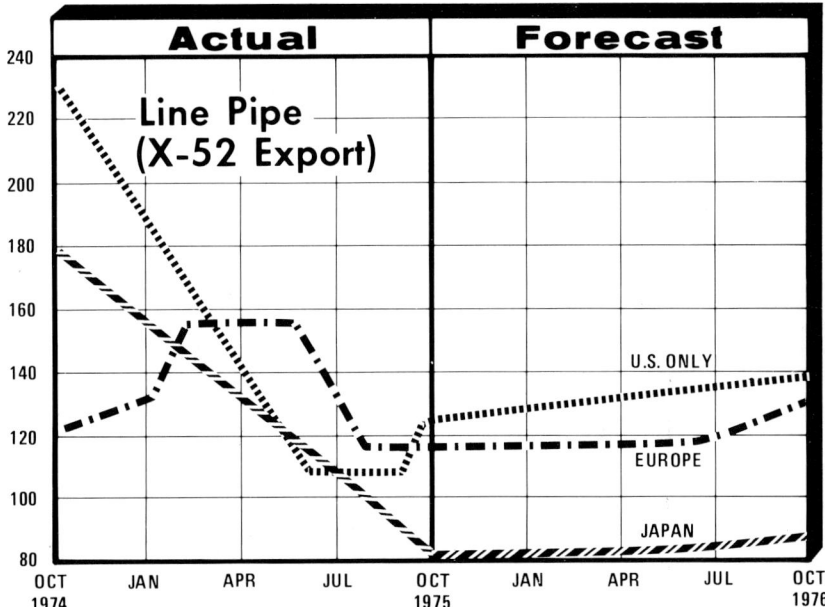

FIG. 8.   Price trends. Price base: January 1974 = 100.

The result of this reduced workload was a softening of the market, but as can be seen from the curve in Fig. 7, it did not soften very much. There have been spot bargains, of course, but on the whole, price levels have risen. Fabricators cite their rising costs for both material and labor to explain why prices continue to rise. Generally, the industry feels their profit margin is insufficient.

As to the question of capacity, the top quality fabricators are predicting serious shortages in the next 3 to 4 years. These are the ones that are still enjoying a reasonable work load due to last year's heavy backlog.

Lead times vary widely depending on size and complexity of the structure. But deliveries of steel shapes from the mills are reported to be "the best we have had in a number of years." In addition, many fabricators are carrying large inventories. Depending on what is ordered, deliveries of fabricated shapes can be had quickly.

Line Pipe

Line pipe was one of the commodities that shot out of control price-wise in early 1974. Prices went from around $350 a ton to as high as $1000 a ton, but then the bubble burst just as rapidly. As can be seen on the price curve of Fig. 8, the market returned to reasonable levels quite rapidly. For the coming year a gradual increase reflecting cost increases is foreseen.

Lead times for line pipe have also dropped from about a year at the peak of demand to around 3 months now.

Flanges and Weld Fittings

Prices jumped about 70% on the average during 1974 as shown in Fig. 9. This was in large part due to the increased prices of materials charged by the steel mills plus the extreme shortage conditions that existed.

As demand disappeared and distributor stocks began to build up, pricing from distributors softened. Prices will probably remain soft for the balance of the year. A gradual increase from then on is expected, because of rising costs.

Valves

Valves increased 100% during 1974, as indicated in Fig. 10. This year, valve pricing in general has been flat. There have been a few price increases announced but, generally, the competitive situation has not allowed them to take effect.

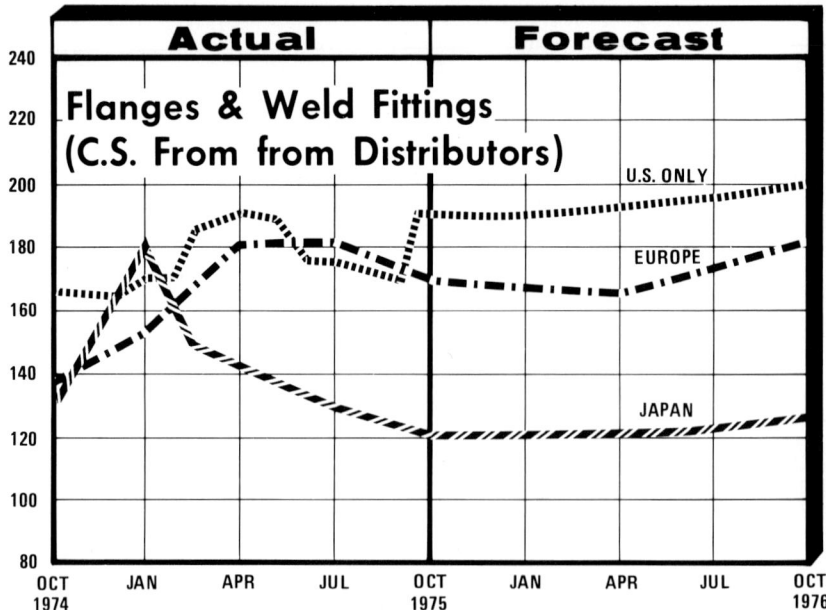

FIG. 9.   Price trends. Price base: January 1974 = 100.

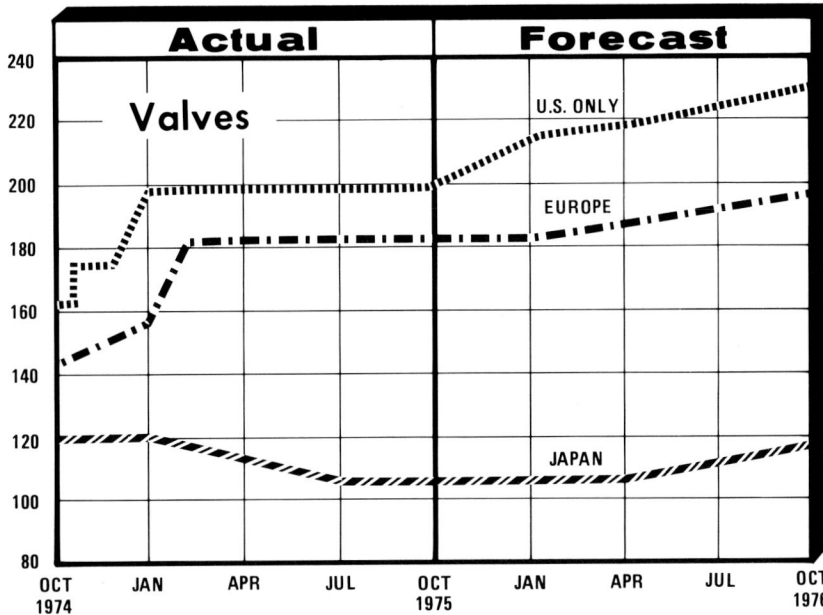

FIG. 10.   Price trends. Price base: January 1974 = 100.

For the balance of this year, manufacturers are predicting a 10% price rise. However, this may turn out to be wishful thinking. It does not appear there is enough market strength to support an increase right now. Sometime next year, as the workload builds up, the trend of pricing is once again expected to begin to rise.

Lead times for standard cast steel valves have dropped significantly. They are currently running about 3 to 5 months. Special valves, like those for nuclear service, are still in the range of 18 to 24 months. There has not been enough dip in demand to reduce this.

Electrical Equipment

Electrical equipment, which rose nearly 50% in 1974, has been a rather quiet performer price-wise this year (see Fig. 11). The manufacturers do not expect any pick up until perhaps the third quarter of 1976. Then, like other manufacturers, they are expecting a heavy boost in workload.

Much of this will come from the new work in the Middle East. The preliminary bid requests for these jobs are underway now. It does look like there is sufficient capacity to handle the initial orders, but next summer should begin to present some problems both as to price and availability.

Instrumentation

Pricing of instrumentation, in general, has leveled off as indicated in Fig. 12. An increase hasn't been seen in about 2 months. There have been firm price quotations on short delivery instrumentation. Short delivery includes schedules as long as 8 months.

The manufacturers of instruments talk much like other suppliers. They expect a heavy workload sometime early next year. They expect panic to set in around mid-1976. By early 1977, they expect that the industry will be back where it was during the toughest days of 1974.

Control Valves

Control valves are normally grouped with instrumentation in the engineering and procurement activities in the industry. However, recent price performance of control valves is different from instrumentation. Pricing is not stable and it has gone up. It is expected that control valve prices will rise more later this year. In this respect, control valves really should be compared with standard cast steel valves because the manufacturers face relatively similar problems.

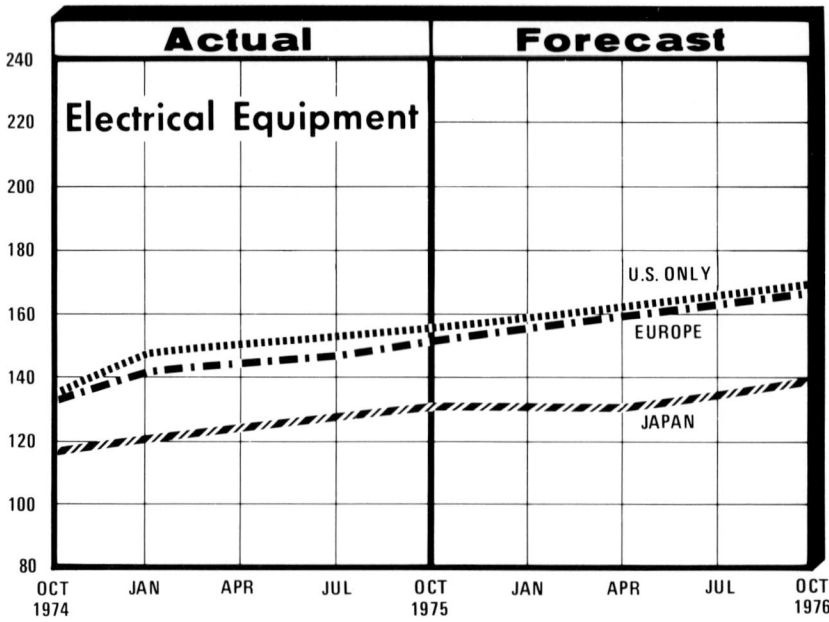

FIG. 11. Price trends. Price base: January 1974 = 100.

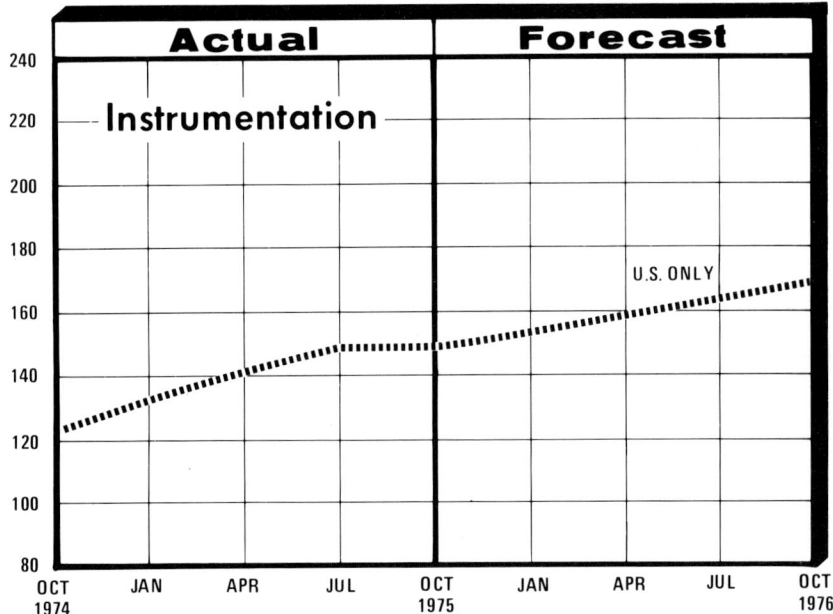

FIG. 12.   Price trends. Price base: January 1974 = 100.

Summary

Price advances during 1975 have eased off from the furious pace of 1974. In some areas there has actually been a return to a buyers' market. Buyers have been counseled, however, not to expect this to last very long. Within the next few months, demand should pick up and, in most lines, it will exceed supply. This will cause a return to the era of rampant price increases, long deliveries, and shortages. The question is: How soon?

Whether this bulge in business happens or not depends on a number of factors. Availability of capital is perhaps the major problem that may delay recovery. Considering this, demand may not increase until well into 1976.

On the opposite hand, virtually all the manufacturers we have talked to expect that a heavy workload is coming shortly. These trend charts tend to reflect this expectation.

Among the forecasts considered to be of importance is the outlook for steel for 1976. A recently completed study forecasts industry shipments of about 95 million tons during 1976, compared to a level of 80 to 82 million tons in 1975. Steel shipments were reduced this year by both the recession and inventory liquidation.

Domestic steel consumption in 1975 was around 101 million tons, including imports and inventory liquidations. The study forecasts an increase in consumption of about 5 million tons during 1976. This should be achievable with current industry capacity.

Some forecasters have been predicting a return of the shortages of 1974. Our view is that the moderating influences that have cooled 1975 will apply well into 1976. Shortages are coming when the recovery takes full effect, but probably not until after 1976.

JOHN E. EGGLESTON

Cost Indexes

Indexes for Finished Steel

Costs for various types of finished steels, like costs for most types of equipment, have advanced steadily, comparable in some instances with the increases in labor costs.

Table 1 shows quarterly values for the Nelson Cost Indexes for composite and seven individual types of finished steels.

The cost-increase trend is clearly shown. All index values are based on 1946 = 100.

The composite cost index for finished steel increased from 335.9 to 545.1 during the 4-yr period, 1973–1976. Low-alloy bars increased from 353.9 to 577.5.

Plate, structural, seamless tubes, and tubing increased in about the same proportion.

Sheet and strip steel had the smallest increases, from 208.1 to 312.2, and from 204.6 to 307.7, respectively.

Itemized Refining Cost Indexes

Cost indexes (Table 2) may be used to convert prices at any date to prices at other dates by ratios to the cost indexes of the same date. Item indexes are published each quarter (first weekly issue of January, April, July, and October). In addition the Nelson Construction and Operating Cost Indexes are published in the first issue of each month of *The Oil and Gas Journal*.

W. L. NELSON

TABLE 1 Indexes for Finished Steel

Year and Quarter	Composite	Bars (low alloy) 3120	Plate (carbon)	Sheet AISI 302	Strip AISI 430	Structural (carbon)	Tubes seamless 2–2¼	Tubing AISI 304
1973								
1st	335.9	353.9	379.4	208.1	204.6	404.2	355.8	265.5
2nd	339.7	358.5	379.4	218.5	208.5	404.2	369.2	275.5
3rd	340.2	359.8	379.4	220.5	210.5	404.2	369.2	275.5
4th	343.0	359.8	379.4	220.7	210.5	404.2	369.2	275.5
Year	339.7	358.0	379.4	217.0	207.1	404.2	365.9	273.0
1974								
1st	354.9	362.2	392.4	227.8	218.1	411.7	390.6	275.9
2nd	410.6	401.5	443.7	272.0	256.2	457.4	390.6	296.5
3rd	476.7	441.1	507.8	306.6	279.8	560.5	423.3	306.0
4th	484.3	451.3	512.8	311.3	284.1	593.0	423.3	313.1
Year	431.6	414.0	464.2	279.4	259.6	505.7	407.0	297.9
1975								
1st	496.0	466.0	539.9	300.4	297.3	633.5	471.9	321.6
2nd	492.7	466.0	539.9	299.8	297.3	604.8	471.9	321.6
3rd	491.4	487.9	527.7	288.5	297.3	604.8	471.9	423.9
4th	510.4	514.7	557.2	277.9	278.0	634.1	479.1	464.6
Year	497.6	483.7	541.2	291.7	292.5	619.3	473.7	389.2
1976								
1st	512.4	514.7	557.2	288.0	287.2	634.1	479.1	464.6
2nd	513.4	514.7	557.2	288.0	294.7	634.1	497.9	464.6
3rd	539.0	548.9	596.5	305.3	307.7	678.3	507.8	483.4
4th	545.1	577.5	606.3	312.2	307.7	678.3	507.8	483.4
Year	527.5	539.0	579.3	298.4	299.3	656.2	498.2	474.0

TABLE 2 Itemized Refining Cost Indexes

	1954	1962	1972	1973	1976	August 1977
Operating cost (based on 1956 = 100.0)						
Power, industrial electrical	98.5	106.7	131.2	138.7	237.6	282.4
Fuel, refinery price	85.5	100.9	152.0	170.2	350.5	383.8
Gulf cargoes	85.0	95.8	130.4	198.8	480.1	481.8
N.Y. barges	82.6	84.8	169.6	202.9	478.1	543.8
Chicago low sulfur	—	—	—	—	472.3	486.6
S. Pedro bunkers	84.3	108.5	168.1	189.7	487.1	520.8
Oklahoma, northern shipments	60.2	87.4	128.1	154.8	528.1	605.9
Natural gas at wellhead	83.5	152.2	190.3	206.6	459.3	673.1
Inorganic chemicals	96.0	105.2	123.1	127.8	238.2	247.3
Acid, hydrofluoric	95.5	89.4	144.4	144.4	208.7	212.4
Acid, sulfuric	100.0	100.0	140.7	144.9	245.4	240.5
Ammonia	103.0	109.6	63.9	68.3	139.3	139.3
Platinum	92.9	80.4	121.1	148.9	165.4	163.7
Sodium carbonate	90.9	104.3	119.4	119.4	207.8	223.6
Sodium hydroxide	95.5	115.7	136.2	149.8	346.9	328.0
Sodium phosphate	97.4	97.7	107.0	113.5	274.6	297.8
Organic chemicals	100.0	93.9	87.4	88.1	213.2	217.3
Furfural	94.5	95.8	137.5	151.6	363.9	391.7
MEK, tank-car lots	82.6	103.3	87.5	87.5	159.7	166.7
Phenol	90.4	77.1	47.1	46.2	161.0	141.5
TEL	102.8	91.8	101.6	101.6	165.8	187.4
Operating labor cost (1956 = 100)						
Wages and benefits	88.7	121.1	210.0	221.2	314.3	342.7
Productivity	97.1	131.8	197.0	214.9	216.2	227.4
Skilled const. (1946 = 100)	174.6	241.5	499.9	532.3	657.6	708.3
Common labor (1946 = 100)	192.1	284.9	630.6	683.4	862.6	933.2
Refinery const. (1946 = 100)	183.3	258.8	545.9	585.2	729.4	787.0

Equipment or materials (1946 = 100)						
Bubble trays	161.4	202.0	324.4	340.8	464.4	507.3
Building materials (nonmetallic)	143.6	164.5	212.4	219.3	313.3	—
Brick—building	144.7	171.5	252.5	270.0	365.2	431.6
Brick—fireclay	193.1	242.5	322.8	341.1	459.5	496.3
Castings, iron	188.1	218.3	274.9	308.2	524.8	552.7
Clay products (structural, etc.)	159.1	193.9	342.0	255.2	338.5	381.9
Concrete ingredients	141.1	167.5	218.4	225.4	331.4	344.2
Concrete products	138.5	154.9	199.6	211.9	285.6	307.7
Electrical machinery	159.9	189.4	216.3	220.2	287.2	302.9
Motors and generators	157.7	166.4	211.0	219.8	330.3	347.1
Switchgear	171.2	222.6	271.0	275.1	393.2	421.6
Transformers	161.9	160.8	149.3	154.3	220.1	240.2
Engines (combustion)	150.5	183.5	233.3	238.3	348.3	381.4
Exchangers (composite)	171.7	183.6	274.3	313.7	478.5	453.6
Copper base	190.7	169.2	266.7	307.0	431.4	414.4
Steel	156.8	194.7	281.9	320.3	521.2	480.5
Fractionating towers	151.0	185.8	278.5	295.5	401.9	439.8
Hand tools	173.8	239.8	346.5	360.1	516.3	560.1
Instruments (composite)	154.6	314.8	328.4	338.0	464.2	484.1
Pressure recorder (pneu.), PR	168.0	217.0	299.0	317.6	515.4	557.4
Pressure recorder (elect.), PR	162.8	265.9	422.2	437.7	542.7	558.0
Pressure recorder cont. (pneu.), PRC	163.5	205.1	312.0	321.2	486.0	518.4
Pressure recorder cont. (elect.), PRC	160.2	248.9	450.3	465.8	572.0	589.5
Pressure gauge only, PI	108.6	124.9	223.1	227.3	274.2	287.9
Flowmeter (mech.), FP	161.6	277.8	348.6	383.7	545.8	571.9
Flowmeter (pneu.), FR	161.4	235.4	345.3	358.1	590.8	626.5
Flowmeter (elect.), FR	162.3	245.5	358.1	361.5	427.9	442.1
Flow control (pneu.), FRC	160.3	202.7	306.2	313.0	477.4	506.5
Flow control (elect.), FRC	159.1	241.0	415.9	424.6	510.0	526.6
Temp. recorder (pneu.), TR	157.3	200.8	318.8	330.9	510.2	542.2
Temp. recorder (elect.), TR	160.8	266.7	288.3	291.4	405.7	413.1
Potentiometer, 6-point, TR	167.2	240.8	243.0	243.0	270.4	283.2
Temp. control (pneu.), TRC	154.7	189.2	296.3	302.9	450.2	476.0
Temp. control (elect.), TRC	160.7	256.4	387.0	396.0	512.4	524.0
Control valve (elect.)	161.9	196.9	304.2	306.7	416.0	428.2

(*continued*)

TABLE 2 (*continued*)

	1954	1962	1972	1973	1976	August 1977
Insulation (composite)	198.5	188.5	272.4	261.6	428.0	461.7
Lumber (composite)	197.8	198.2	353.4	456.3	515.2	637.8
Southern pine	181.2	179.8	303.9	376.1	432.3	563.4
Redwood, all heart	238.0	192.6	310.6	392.0	527.6	637.2
Machinery						
General purpose	159.9	207.8	278.5	289.2	431.2	463.0
Construction	165.9	225.9	324.4	335.6	513.0	557.0
Oilfield	161.9	198.1	269.1	281.3	460.4	504.6
Paints—prepared	159.0	186.5	231.8	240.3	342.5	361.2
Pipe						
Clay sewer	140.1	180.9	207.4	211.8	271.7	299.9
Black iron	195.0	258.2	346.9	360.8	567.1	628.1
8-in. line	182.7	247.7	319.9	332.9	558.3	620.3
Pumps, compressors, etc.	166.5	222.5	337.5	346.9	538.6	580.4
Steel, finished	187.1	242.4	330.6	339.7	527.5	591.5
Bars (low alloy) 3120	198.7	246.0	349.4	358.0	539.0	632.9
Plate (carbon)	187.0	254.3	365.5	379.4	579.3	657.1
Sheet, AISI 302	177.0	204.4	225.9	217.0	298.4	364.1
Strip, AISI 430	169.0	187.6	221.2	207.1	299.3	341.1
Structural (carbon)	193.4	268.8	386.7	404.2	656.2	678.3
Tubes, seamless 2–2¼	199.4	282.6	347.3	365.9	498.2	569.7
Tubing, AISI 304	180.0	209.6	265.5	273.0	474.0	558.7
Tanks and pressure vessels	147.3	181.5	246.4	262.5	367.4	405.4
Tube stills	123.0	90.4	125.3	130.0	191.1	208.3
Valves and fittings	197.0	254.6	350.9	367.8	595.7	629.7
Nelson Refinery (Inflation Index)	179.8	237.6	438.5	468.0	615.7	662.6
Nelson Refinery Operation (1956)	88.7	105.8	118.5	125.7	209.3	229.3
Nelson Refinery Process Operation (1956)	88.4	105.8	147.0	168.0	267.1	286.1

Cost, Operating Expenses Estimation

Introduction

Operating expenses are continuing expenses incurred in day-to-day plant operation. The most difficult aspect of project evaluation is the estimation of operating expenses. There are many components of operating expenses but the techniques for estimating these components are at best suspect and subject to great error. Whatever techniques have been developed are usually proprietary, and as a result, the open literature on this subject is rather meager.

In preparing an operating cost estimate, it is always wise to seek the advice of production personnel who have had operating experience on processing units similar to the one for which the estimate is being prepared. This is particularly true for the direct cost portion of the manufacturing cost estimate.

Manufacturing Cost Estimation

Terminology

The manufacturing cost is the expense involved in keeping a plant in daily operation. In order to clarify some of the terminology, the words "manufacturing" or "production" cost may be used interchangeably, and will mean those costs necessary to make a product and prepare it for shipment. "Operating" costs include the manufacturing cost plus other cost elements such as selling and distribution costs, and indirect costs of company, research, sales, administration, finance, and other overhead items.

The manufacturing cost sheet included as Fig. 1 should be helpful to refer to as this terminology is discussed.

Direct Costs

These are directly associated with the manufacture of a product, such as labor, utilities, and maintenance. Raw materials are separate items.

Indirect Costs

These costs are not directly related to production, such as depreciation, taxes, and plant indirect expenses. The indirect costs continue whether the manufacturing unit is operating or not.

Date:
By:

MANUFACTURING COST SHEET

LOCATION: DEPT NO: ____ PRODUCT:
Mfg. Capital   Design: per Yr. (Hr.)
Yields: Prod: per Yr. (Hr.)

RAW MATERIALS	UNIT	QUANTITY	$/UNIT	$/YEAR	c/
GROSS R. M. COST					

BY-PRODUCT (CREDIT)

TOTAL CREDIT					

NET MATERIAL COST

DIRECT EXPENSE	UNIT	QUANTITY	$/UNIT		
Steam	MM Btu				
Electricity	KWH				
Water—CT	M Gal				
—Filt.	M Gal				
—Sea	M Gal				
Fuel Gas	MM Btu				
Comp. Air	MCF				
Steam Condensate	M Gal				
TOTAL UTILITIES					
Labor	....				
Supervision					
Payroll Charges	....				
Repairs—M & E					
—Inst.	....				
Factory Supplies					
Laboratory	....				
Product Control					
Technical Service	....				
Royalty					
	....				
Container Expense					

TOTAL DIRECT CONVERSION EXPENSE

TOTAL DIRECT MFG. COST (DIRECT EXPENSE plus NET MATERIAL)

INDIRECT EXPENSE
  Depreciation
  Factory Indirect Expense

TOTAL INDIRECT CONVERSION COST

TOTAL MANUFACTURING COST

*Indicates negative quantity or credit.

FIG. 1

Variable Costs

Variable costs vary directly with the production rate. Examples of these items are raw materials, utilities, packaging, and shipping. The cost of these items per unit of production tends to be constant, regardless of the number of units produced.

Fixed Costs

The fixed costs are not affected by the production rate, e.g., depreciation, taxes, insurance. The cost per unit of production for indirect costs increases as production decreases.

Semivariable Costs

These costs decrease somewhat as production decreases but not in direct proportion. Maintenance, labor, overhead, and supervision are examples of these costs. Generally, the semivariable costs will be lower at lower production rates but do not diminish to zero at zero production. Some companies put labor and maintenance in the "semivariable" category or may call these "regulated," "budgeted," or "allocated" costs.

Raw Materials

Estimates of the quantities of the raw materials consumed in a process are obtained from material and energy balances. Although many authors will suggest the use of *Chemical Marketing Reporter*, *Chemical and Engineering News*, *Chemical Week*, etc. as sources of raw material costs, these should be used as a last resort. Each company enters into long-term agreements with suppliers at special contract prices for raw materials. Price quotations will be expressed as FOB vendor's plant or some other basing point. Care must be taken to be sure that freight charges are included. Freight information can be obtained from the company's traffic department or from the railroads, trucking companies, or barge lines in the local area.

Transferred raw materials are those obtained by transfer from another operating unit or division of the company. The *transfer price* is established by company policy and may be either the current market price or manufacturing cost plus the transportation costs, depending upon how the company keeps its books.

Prices of raw materials can vary considerably depending on the form in which they are received. For example, delivered prices of caustic soda, 50% caustic in tank cars, is considerably less than anhydrous flake caustic in drums. The estimator should make certain that the most economical type is used consistent with the manufacture of a quality product. In some cases a lower grade raw material may be used at a considerable cost saving. If a change in raw materials is contemplated, the research department should be consulted to ensure that raw material quality will not affect the product.

By-product materials are handled in the same fashion as raw materials. The prices of by-products can be estimated from the market prices of these materials less purification, packaging, selling, and transportation costs. If the by-products are intermediate products for which no market exists, they may be credited at their net value to downstream or subsequent operations at a cost value equal to their value as a replacement.

Direct Costs

For the production of a chemical product, the direct manufacturing costs require the greatest time and the most careful consideration in making an estimate.

Utilities

The amount of each utility required is obtained from the material and energy balances. A company's standard manufacturing cost sheet for the plant in which the new unit is to be located is the best source of cost information. However, the estimator should check with the power house superintendent or the company's utilities supervisor to discuss future costs as well as the incremental utility demands which the new project will require. When large incremental uses are involved, it may be necessary to tie in with the local utility as a standby utility source and use the local utility's cost figures. It is especially important to analyze utility costs for large increments, since unit costs often decrease substantially as demand increases. For small increments of utility consumption, the cost per unit usually may be assumed constant.

With the current energy shortage, the cost of all utilities is soaring. Utility cost figures need to be reviewed continuously. In the Southwest United States, typical utility costs as of 1976 are:

Steam	$0.80–1.50/Mlb
Electricity	$0.80–1.50/CkWh
Natural gas	$1.00–2.50/MMBtu
Cooling water	$0.02–0.10/Mgal
Process water	$0.20–0.50/Mgal

Labor

The safest way to establish the operating labor requirements is to layout shift, week-end, and vacation coverage for the process. The local labor contract is the best source of cost information. It is also a good idea to get the reaction of what future demands might be in the next labor negotiations to forecast future labor rates. This information may be obtained from the plant employee relations personnel. Labor rates have been steadily increasing on what seems to be an accelerating basis. Average rates should account for shift differential and overtime. In the Gulf Southwest a reasonable average operating labor cost is about $15,000 to 17,000 per man per year.

Direct labor charges, while not a major part of the total manufacturing cost, are used to estimate other items. Therefore, it is necessary to have reliable data. Typical labor rates for chemical workers are found in Table 1.

If the estimate is for exploratory purposes, direct labor requirements may be estimated from the number of processing steps and the production rate by Wessel's method [1]:

$$\frac{\text{Operating man-hours}}{\text{Tons of product}} = t \left[\frac{\text{no. of process steps}}{(\text{capacity in tons/d})^{0.76}} \right]$$

where t is determined by the process as follows:

Batch operation with a maximum of labor, 23
Operations with average labor requirements, 17
Well-instrumented continuous process operations, 10

The above equation accounts for the improvement of labor productivity as throughput increases, and it can be used as a guide to extrapolate known man-hour requirements from one plant to another of different capacity.

Haines [2] has developed man-hour requirements using elemental time values for novel projects.

TABLE 1   Average Hourly Earnings of Chemical Workers, July 1976

	$	% Annual Increase since 1970
Texas	7.35	8.5
Oklahoma	7.21	10.1
Louisiana	7.59	9.8
Alabama	5.82	10.3
Michigan	6.52	7.1
Indiana	6.37	8.0
Illinois	6.01	7.7
Ohio	5.91	7.4
Pennsylvania	5.61	8.2
New Jersey	6.20	8.3
New York	5.73	8.2
Connecticut	5.62	6.9
Massachusetts	5.55	6.6

Supervision

For quick estimates, and in the absence of other information, supervision can be assumed to average between 20 and 30% of operating labor. The upper limits should be used for batch or complex processing. Some estimators consider supervision as a fixed expense at the level required for 100% capacity operation. If the supervision position can be identified, then actual salaries can be used.

Payroll Costs

This item includes workmen's compensation, pensions, group insurance, paid vacations and holidays, social security, unemployment taxes, profit sharing, and an ever-expanding list of fringe benefits. A company will usually have a set figure which is revised frequently by the accounting department as more benefits are added. A rough estimate of 35 to 50% of the labor plus supervision cost is reasonable [2].

Maintenance

Most plants maintain records of maintenance frequency for various equipment items and the costs involved. Plant accounting can usually supply these cost items. For preliminary estimating purposes, however, an average figure of 6 to 10% of the fixed capital investment per year is reasonable. The higher figure should be used where the process has a large number of rotating equipment items, or where the process is operating under extreme conditions of temperature and/or pressure. A low figure of 6 to 8% would be typical for processes operating at or near ambient conditions. It should be noted that the above percentages include maintenance labor and supplies.

Operating Supplies

These are supplies such as filter cloths, brooms, mops, and instrument charts, exclusive of other items listed on the manufacturing cost sheet. Company records are again the best source of information, if they are available. If not, 6% of the operating labor is a reasonable figure.

Laboratory

Laboratory expenses will be a function of the amount of quality control desired and the nature of the process. These costs also are increasing rapidly and a good estimate would be $25 to $30 per laboratory hour. An alternate method for a complex process is to use 10 to 20% of the operating labor cost.

Royalties

This is a difficult item to estimate since the royalty or licensing agreement may involve many complexities. An approximation should be used in the absence of other data. Certain royalty payments and patent purchase costs may be considered under general expense as they may be regarded as a replacement for a research and development expense.

Clothing and Laundry

This item appears on some cost sheets, especially for departments manufacturing toxic materials. A reasonable figure to use is 10% of operating labor costs.

Technical Service

With some companies this is a budgeted item to spread the cost of maintaining a technical service group of engineers who are concerned with small departmental engineering jobs, process troubleshooting, or process improvement. A flat amount, say $20,000 per year per man, is a reasonable figure. (Note: These last two items are sometimes included in a term called "other direct costs" which might amount to from 1 to 5% of the total direct costs.)

Indirect Costs

The indirect costs for a manufacturing process can be divided into two main categories: depreciation and plant indirect expenses.

Depreciation

For the manufacturing cost sheet this is generally estimated by the straight line method over the useful life of the project. Accelerated depreciation may be used for tax and economic evaluation purposes. According to the IRS guidelines, chemical plants may be depreciated over an 11-yr period and petroleum refineries over a 16-yr period.

Plant Indirect Expenses

The plant indirect expenses include such items as local taxes, insurance, and local plant service expenses. Plant service expenses are those associated with railroad spurs, plant roads, fire protection, cafeteria, employee safety, parking lots, etc.

The company's accounting department usually has a percentage factor developed for each plant location. In the absence of such information, Hackney

[3] presents a method for order-of-magnitude figures based upon an investment factor and a labor factor.

	Investment Factor, %/yr	Labor Factor, %/yr
Heavy chemical plant—large capacity	1.5	45
Power plants	1.8	75
Electrochemical plants	2.5	45
Cement plants	3.0	50
Heavy chemical—small capacity	4.0	45

Example: A small electrochemical unit has the following costs:

Fixed capital investment	$1,000,000
Total labor	$30,000/yr

Calculate the plant indirect expense.

Investment portion = $1,000,000 (0.025)	$25,000
Labor portion = $30,000 (0.45)	13,500
Total plant indirect expense	$38,500

Loading, Packaging, and Shipping Costs

These costs are best estimated from other similar company operations, modified for the case(s) studied. These costs are easily measured and most companies have good cost control systems which can provide actual up-to-date cost information.

The cost of containers can be obtained from suppliers or from Perry's *Chemical Business Handbook* [4] with the proper escalation factors. If the plant is large enough, there is a materials handling specialist who can advise the estimator on these costs.

Transportation costs are best obtained from the company's transportation specialist or from local railroads, truck companies, or barge lines. Large volume, long haul (over 250 miles) rates are as follows:

Transportation Means	$/ton mile (1976) [5]
Pipeline	0.003–0.005
Barges and tankers	0.004–0.010
Railroads	0.02 –0.05
Trucks	0.03 –0.06

In general, distribution charges are segregated from other manufacturing costs to show the effect of the costs of various containers and methods of shipment.

Total Manufacturing Cost

The sum of all the preceding items constitutes the total manufacturing product cost. General overhead expenses are separate as discussed in the following section.

General Overhead Expenses

The general overhead expenses include the costs of operating sales offices, staff engineering departments, research facilities, and administrative offices. Also, the finance cost which includes bond interest may be included. Usually a percentage of sales is charged to a project for these expenses. Some companies add these costs to the total manufacturing cost to obtain the total operating expenses.

The Accounting Department can usually provide the percentage figure for this item. It may vary from 6 to 15% of the net annual sales figure, depending on the chemical sold and the amount of customer service provided.

Total Operating Expense

The total operating expense is the sum of the total manufacturing product cost and the general overhead expense.

Summary

The accuracy of the total operating expense estimate is dependent upon the amount of time which can be devoted to this activity as well as the quality of the information available.

In the presentation of a manufacturing cost estimate, costs should be computed for various operating levels; for example, 0, 50, 75, and 100% of capacity. This information is valuable in the profitability analysis of the project.

Bibliography

Berry, R. M., "Estimate Manufacturing Costs," *Chem. Eng.*, *67*, 123–128 (1960).
Dickens, S. P., "How To Estimate Operating Costs," *Hydrocarbon Processing Pet. Refiner*, *40*(7), 133–139 (1961).

Holland, F. A., Watson, F. A., and Wilkinson, J. K., "Manufacturing Costs and How to Estimate Them," *Chem. Eng.*, *81*, 91–96 (1974).

Keim, C. R., "Meaningful Production Costs," *Chem. Eng.*, *78*, 184–189 (1971).

Liebson, I., and Trischman, C. A., "Spotlight on Operating Cost," *Chem. Eng.*, *78*, 69–74 (1971).

Liebson, I., and Trishman, C. A., "How To Cut Operating Costs," *Chem. Eng.*, *78*, 92–95 (1971).

Weinberger, A. J., "Calculating Manufacturing Costs," *Chem. Eng.*, *70*, 81–86 (1963).

Wobus, R. S., "Estimation of Direct Operating Labor Requirements," *Chem. Eng. Prog.*, *53*, 581–585 (1957).

References

1. H. E. Wessel, *Chem. Eng.*, *59*, 209 (1952).
2. I. Haines, *Chem. Eng. Prog.*, *53*, 556–562 (1957).
3. J. W. Hackney, *Chem. Eng.*, *68*, 179–184 (1961).
4. J. H. Perry (ed), *Chemical Business Handbook*, McGraw-Hill, New York, 1954.
5. G. M. Davis, Private Communication, 1977.

JAMES R. COUPER

Cost, Plant Utility

In the recent past, because utility costs did not have the same impact on the total operating expense in a utility plant as they do in today's high-energy cost era, those costs were merely contingent to total operating expense. However, with the rapid escalation in capital and energy costs in recent years, utility costs have become a major consideration in financial evaluation calculations such as payout and return on investment. Typical escalations from 1970 to 1984 for fuel costs are shown in Fig. 1. This figure illustrates the rapid escalation in fuel costs during the period 1970 to 1975 and the projected trend through 1984.

Because of the dramatic fuel cost increases shown in Fig. 1 and the concurrent rise in both labor and capital costs, utility generation costs have also increased. In this article we are suggesting a simple calculation method to arrive at utility costs. In addition, we present graphical illustrations for steam generation costs which have been calculated using this method.

Table 1 presents a range of the typical fuel, water, and labor cost factors at an American Gulf Coast location in the first quarter 1975, which were used in the computation of the steam generation costs. Also listed in Table 1 are typical investment ratio factors to use when scaling the cost of a utility plant from one size to another.

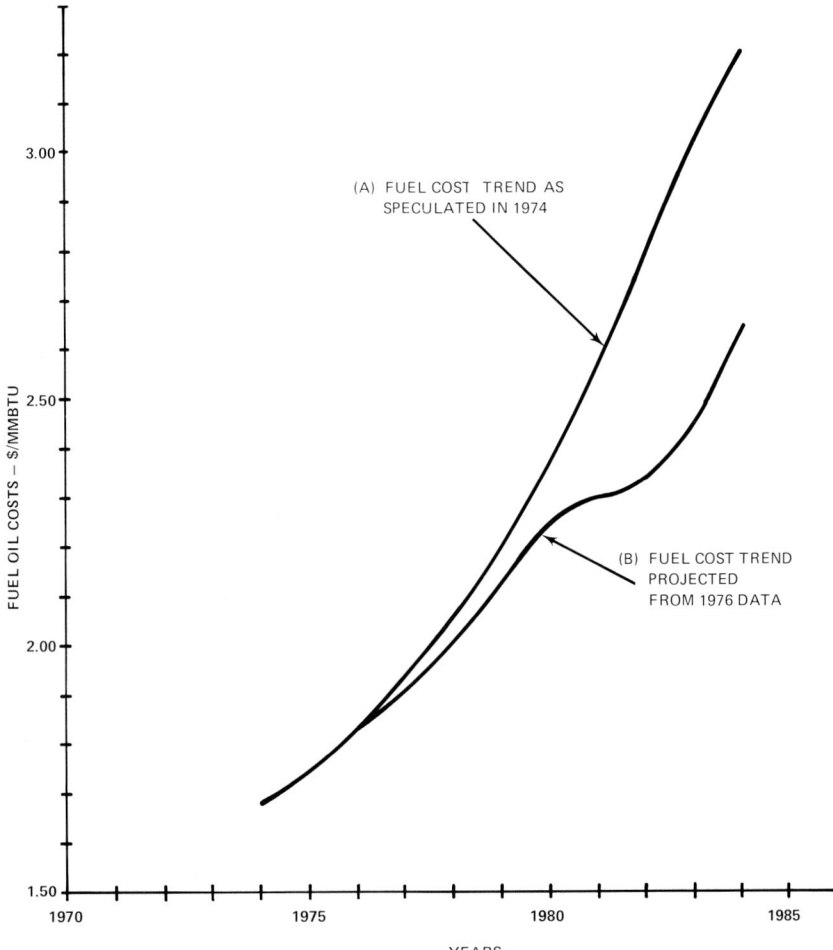

FIG. 1. Fuel cost trends.

Listed in Table 2 are the additional parameters which were used to calculate total plant manufacturing costs; these include typical basic efficiencies and fixed costs.

We have defined steam "manufacturing costs" as the sum of all operating costs plus fixed costs. Steam "value" is defined as the dollar amount per unit of production which pays out a plant in either 2 years before taxes (PBT) or 5 years after taxes (PAT). The following simple procedure defines the method which can be used to calculate these "values" for most utilities systems:

$$\text{Cash accumulation} = \text{net profits* + depreciation}$$

$$= 0.5X - 0.5M + D \tag{1}$$

*Tax at 50% (actually 52% in 1975).

TABLE 1   United States Gulf Coast Cost Factors

	Cost Factor
Plant investment (1975)	
Steam generation plant	0.7
Cooling water system	0.6
Power generation plant	0.8
Fuel costs (1975)	
Gas at Gulf Coast wellhead	
(includes gathering system costs):	
Existing contracts, \$/MMBtu	0.85
New contracts, \$/MMBtu	1.50
Fuel oil, \$/MMBtu at 6.25 MMBtu/bbl	1.75
Gas transmission costs, ¢/100 mi	3.5
Plant fuel gas, \$/MMBtu	1.75
Purchased power (Mid-Continent United States), ¢/kWh	2.8
Labor rates (1975)	
Foremen, \$/h	8.00
Operators, \$/h	7.00
Helpers, \$/h	6.25
Chemists, \$/h	6.50
Labor burden, % of direct labor	30
Plant general, % of total labor + burden	40
Maintenance, % of plant investment	
(60% labor, 40% materials)	3
Water costs (1975)	
Raw water, ¢/1000 gal	40
Cooling water treatment, ¢/1000 gal	5
Boiler feedwater treatment	See Fig. 2

TABLE 2   United States Gulf Coast Cost Factors

Efficiencies	
Boiler efficiency,[a] % (HHV)	80
Pumps (500 hp and larger), %	80
Turbines (500 to 2000 hp), %	60
Turbines (larger than 2000 hp), %	75
Motors (500 hp and larger), %	90
Station heat rate (power generating plant), Btu/kWh	10,000
Overall efficiency (power generating plant), %	35
Fixed costs	
Depreciation allowance, % of depreciable capital	10
Interest, % of total capital investment	5
Taxes and insurance, % of total capital investment	3.5
Income tax, % of gross profit (1975)	52

[a]85% can also be used with somewhat higher boiler investments.

where X = "value" of commodity, in $/yr
       M = manufacturing costs (including depreciation allowance), $/yr
       D = depreciation, $/yr

Cash accumulation required for 5-yr payout = 0.2 (capital investment)   (2)

Combining Eqs. (1) and (2):

$$X = 0.4 \text{ (capital investment)} + M - 2D$$

Therefore, for a 5-yr payout after taxes:

$$\text{Steam value, \$/yr} = 0.4 \text{ (capital investment)} + M - 2D$$

and for a 2-yr payout before taxes:

$$\text{Steam value, \$/yr} = 0.5 \text{ (capital investment)} + M - D$$

The above procedure can also be applied to arrive at cooling water operating costs and electric power costs, which are explained later.

Steam Generation Costs

Fuel costs vary with plant location. In the first quarter 1975, fuel oil could be purchased in the Gulf Coast for about $1.75/million Btu. Recent trends for fuel gas purchase at Gulf Coast wellhead indicate costs between $1.25 and $1.5/million Btu under new contracts and costs between 85 and 90¢/million Btu with existing contracts. The trend with time is steadily upward.

Using the Gulf Coast as a base point, transmission costs to other areas can be estimated by adding 3.5¢ to the fuel cost for each 100 miles of transmission. Thus typical fuel oil prices at Gulf Coast United States plant sites are $10/bbl and may reach as high as $17/bbl at Northeastern United States plant sites depending on the crude source and the percent sulfur in the fuel oil.

Fuel consumption costs are a function of boiler efficiencies. Based on higher heating value (HHV), an efficiency of 80% can be safely assumed for a typical process plant boiler. Higher efficiencies (on the order of 85%) can also be achieved. The calculation for total fuel consumption is based on 105% of required steam output to cover blowdown and power requirements in the boiler plant itself. However, continuous blowdown for high-pressure boilers will vary depending upon the type of boiler feedwater (BFW) treating employed.

Boiler capacities in the range of 500,000 to 1,000,000 lb/h, including all auxiliaries such as deaeration and water treating, require the average operating labor force described in Table 3. Labor requirements not scaled directly with plant size tend to remain constant over a fairly wide range of plant capacities.

Chemical costs vary with water conditions, plant size, plant water balance,

TABLE 3 General Computation

Base year	1975
Steam pressure, lb/in.^{2}gauge	600
Plant capacity, lb/h	500,000
Plant investment, $	5,800,000
Plant investment, $/lb/h	11.6
Fuel cost, $/MMBtu	1.75
Boiler efficiency, %	80

Manufacturing costs	$/yr
Operating costs	
Operating labor	
1 foreman, 8.00 $/h, 24 h/d	69,100
3 operators, 7.00 $/h, 24 h/d	181,500
2 helpers, 6.25 $/h, 24 h/d	108,000
1 chemist, 6.50 $/h, 40 h/wk	13,400
30% labor burden	111,600
Subtotal	483,600
Fuel consumption	11,700,000
Boiler feed water treating (chemicals included)	487,000
Maintenance, 3% of plant investment[a]	174,000
Plant general, 40% of total labor[b]	235,200
Total operating costs	13,080,200
Fixed Charges	
Depreciation 10% of investment	580,000
Taxes and insurance, 3.5% of investment	203,000
Interest, 5% of investment	290,000
Total fixed charges	1,073,000
Total manufacturing costs	14,153,200

	$/1000 lb
Steam generation costs	3.28
Steam value[c]	3.55
Steam value[d]	3.82

[a]60% labor and 40% materials.
[b]Total labor includes maintenance labor.
[c]Including 5-yr payout after income taxes (PAT).
[d]Including 2-yr payout before taxes (PBT).

and the quality of condensate which is returned as boiler feedwater. Figure 2 typifies actual and future costs of boiler feedwater for various plant capacities. For example, with 80% condensate return and a value of 50¢/1000 gal for makeup untreated raw water, the makeup water costs for a 1-million-lb/h steam plant will be about $100,000/yr. However, using 1975 as base year, with clarified or pretreated water, the makeup water value becomes 40¢/1000 gal and the

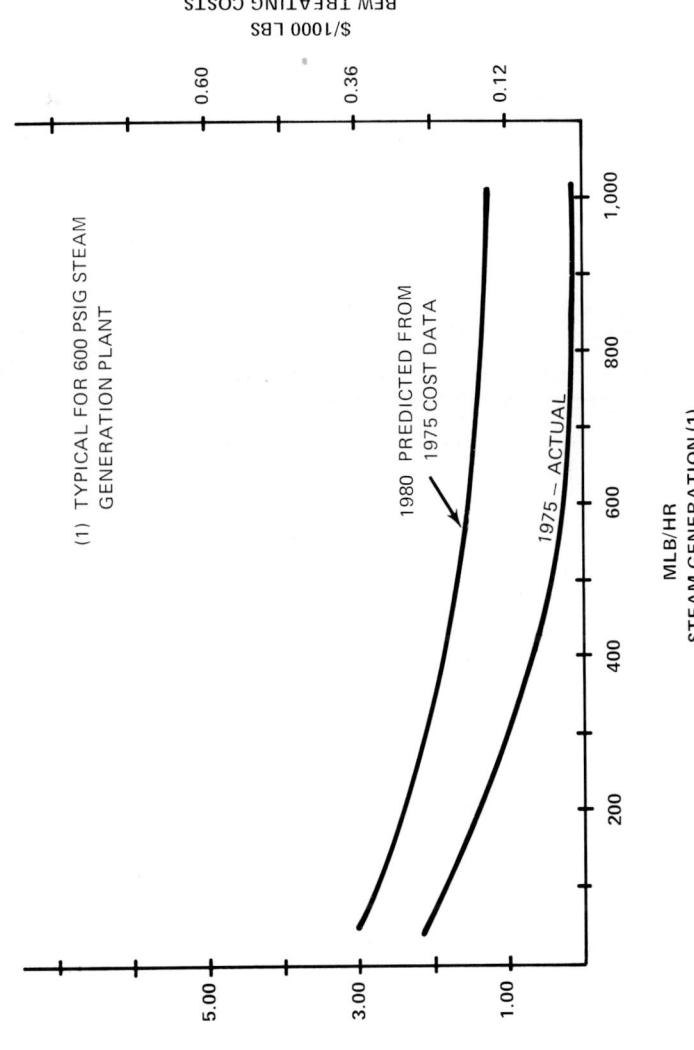

FIG. 2. Boiler feedwater treating costs.

makeup water costs for a 1-million-lb/h steam plant becomes $140,000/yr. At 50% condensate return, the cost for treated water makeup increases to more than $350,000/yr.

In the first quarter 1975, annual interest rates averaged 9 to 9.5% of the unpaid balance over the life of a typical long-term loan, which is equivalent to about 5% a year of total capital investment. Other fixed charges, which include *ad valorem* taxes and plant insurance, will run about 3.5% of the total capital investment per year. Table 3 presents a typical calculation of steam values, which were calculated using a fuel oil cost of $1.75/million Btu.

Figure 3 was developed using the described method presented earlier. The steam "values" and steam manufacturing cost can be determined for 1975 using 1975 fuel cost, and estimated for the year 1984 using a higher fuel cost of $2.76/million Btu. Figure 4 presents the steam plant cost index to be used for scaling plant investment.

Figure 5 manifests the upward trend of steam manufacturing costs caused by rising plant investment and labor costs.

Utility generation costs are mainly determined by the cost of fuel and the size of the utility plant. Steam generation plant sizes vary from 50,000 to 300,000 lb/h for a package installed boiler and from 200,000 to 1 million lb/h for a field erected boiler. Figure 6 illustrates a significant decrease in steam generation costs per unit increase in the size of a package boiler compared to those of a field-erected boiler.

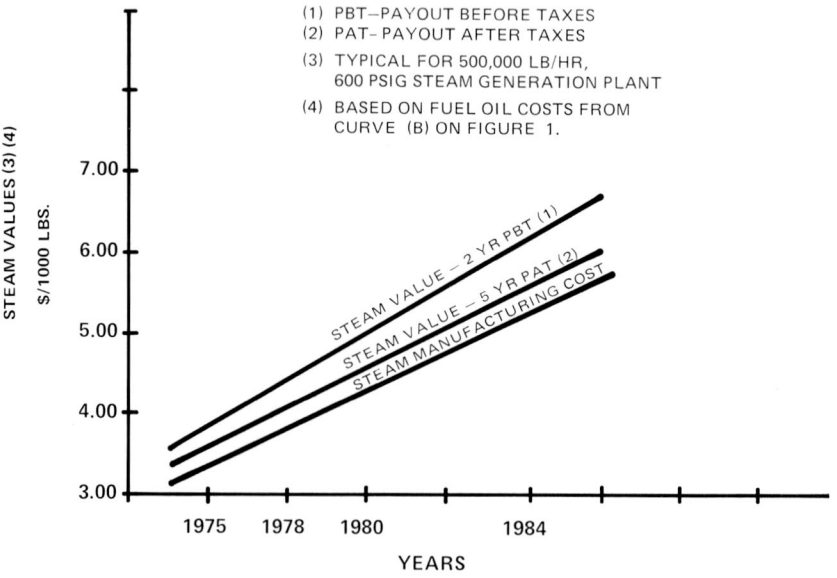

FIG. 3.   Yearly steam values.

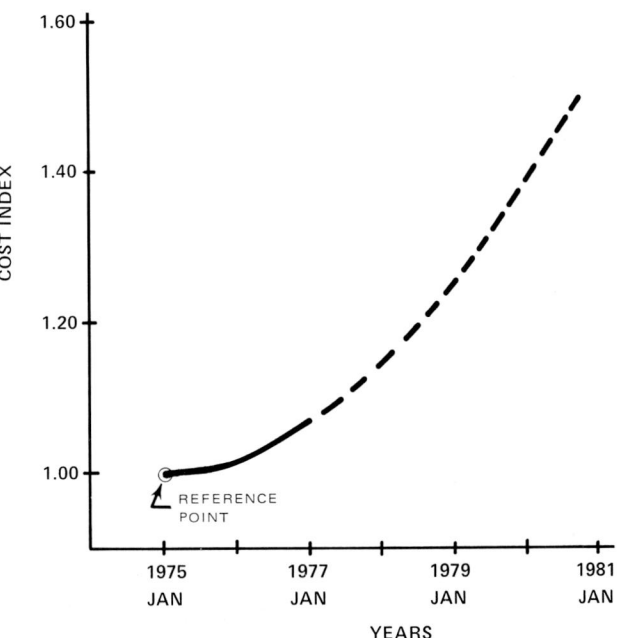

FIG. 4.   Steam plant cost index.

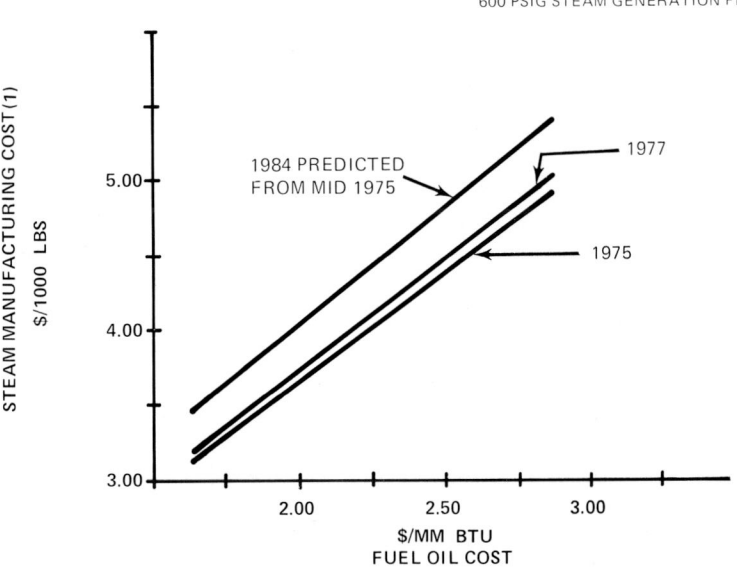

FIG. 5.   Steam manufacturing costs.

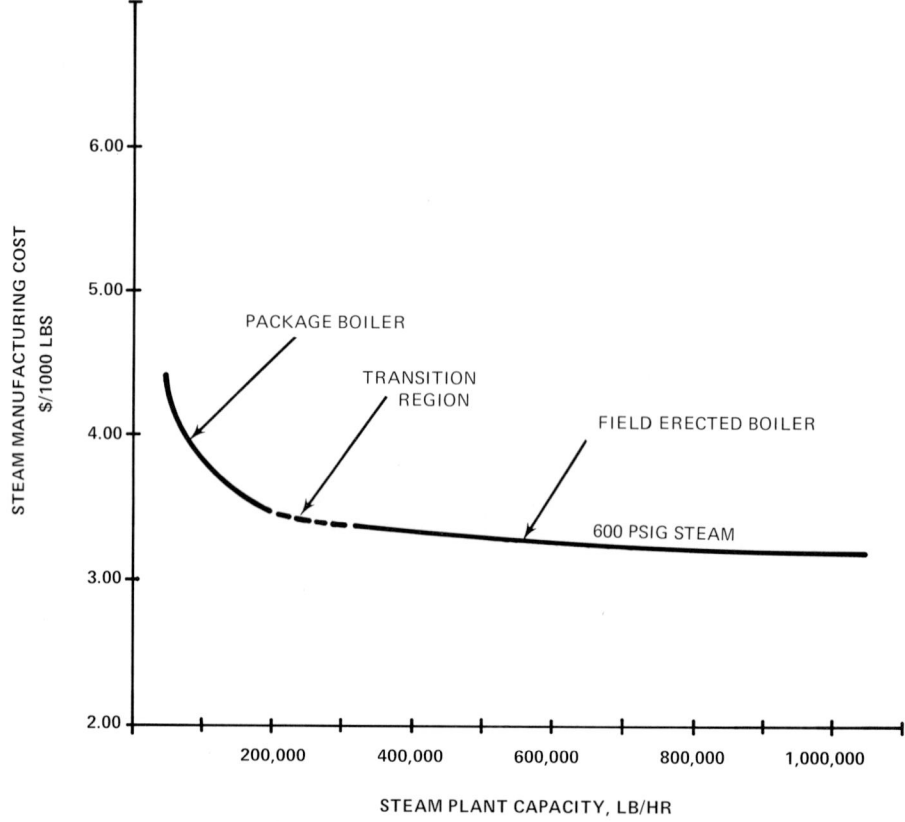

FIG. 6.   Steam manufacturing costs.

Cooling Water Costs

The method for arriving at cooling water operating costs is the same as for steam generation. To establish operating costs, start by first estimating the differential pressure of the system and then calculating the differential pressure of the cooling water pump. A good initial estimate for typical systems ranges betweeen 50 and 70 lb./in.^{2}. Usually, cooling water pumps have lowhead and high capacity, and therefore will have efficiencies in the range from 80 to 85%.

   Using the estimated differential pressure and pump efficiency, the pump brake horsepower can be calculated. A steam turbine drive with a spare motor drive has been assumed, because this arrangement represents the highly dependable and flexible system preferred by most clients. In addition, for large refinery and/or chemical plant complexes, there is usually a need for low-pressure steam, and the cooling water pumps afford a good application for back-pressure turbines to obtain low-pressure steam. However, total motor drive is frequently used in areas having highly reliable electric supply sources.

Therefore, the type of driver that is selected in each case should be based on a careful evaluation of all parameters.

For instance, if a steam turbine drive is the choice, estimate the steam required per brake horsepower (usually called the "water rate") and charge for this steam at established steam costs; if a motor drive is the choice, estimate the motor horsepower using pump brake horsepower and a motor efficiency of 90 to 93%, and charge for power at electric power costs.

One operator and one helper a day should be sufficient for most cooling water systems. Again, 30% for labor burden and 40% of the total operating labor plus maintenance labor must be added for supervision and overhead expenses, or commonly called plant general expense.

Chemical costs vary with water conditions. However, about 5 to 25¢/1000 gal provides a good range for a simple water treating system using: (1) acid for pH control, (2) chlorine for algae control, (3) chromate for corrosion control, and (4) phosphate for scale control.

Raw water costs will vary from 40 to 50¢/1000 gal or more depending upon location and availability. Makeup raw water required in a cooling system will be about 5% of the circulation rate. This includes losses resulting from evaporation (about 1%/10°F temperature range), drift loss, and blowdown.

Maintenance for the cooling water system is about 3% of plant investment. A depreciation allowance can be taken as 10% of depreciable capital investment. *Ad valorem* taxes and plant insurance are estimated at 3.5% of total capital investment, and average annual interest is 5% of total capital investment.

The total of the above items is the operating cost. The "value" and manufacturing cost for this commodity are calculated by the same method as the costs for steam.

Electric Power Costs

The method of calculating electric power costs is also the same as for steam and cooling water, if power is to be generated rather than purchased. Usually, it is cheaper to purchase power if it is available and dependable. Since most utility companies have relatively large installations, their unit costs for each kilowatt of power will be correspondingly lower than the cost of on-plot kilowatt generation in a process facility.

Two operators, for about $\frac{1}{3}$ time, should be able to operate most process plant power generation systems. Again, 30% to cover labor burden and 40% of the total operating labor plus maintenance labor must be added for plant general expense.

To estimate fuel costs, use a "station heat rate" of 10,000 Btu/kWh. This means that 10,000 Btu/h of fuel is consumed for each kilowatt output. This also assumes a condensing turbine drive for the generator and includes boiler efficiency, turbogenerator efficiency, and heat losses. Maintenance, depreciation allowance, *ad valorem* taxes and insurance, interest, and values for electric power are calculated using the same formula used to determine steam costs.

This article has indicated the impact that rising fuel costs and plant investment have on the cost of steam generation. It also outlines a method which can be used to determine the utility values for other utility systems such as cooling water and electric power. The utility values thus obtained by this method can be used as a tool when energy conservation measures are implemented and/or alternate plant arrangements are compared on an economic basis.

This material appeared in *Chemical Engineering Progress*, November 1977.

CANDACE C. JOHNNIE
DARSH K. AGGARWAL

Cost—Process Equipment (Updating)

One of the most significant economic factors in the evaluation of project proposals is the need to secure data regarding the cost of process equipment. For small projects the only requirement may be for costs of what otherwise would be relatively insignificant items. When a larger job scope is involved, not only will the individual equipment items be larger and more expensive, but there will be substantially more components to deal with as well. In view of this range of potential needs, an adequate cost estimation procedure should provide for the incorporation and use of data over as wide an extent as possible.

Many chemical engineers involved in design work gradually accumulate a substantial file of these types of cost data for such process equipment as a result of involvement in a variety of projects. However, there are unfortunately two major problems generally associated with the subsequent effective use of such information in circumstances not directly related to the particular project or economic estimate that was originally considered. The first of these difficulties is caused by the fact that the later potential application where the data is needed will most frequently involve apparatus of a significantly different size or capacity. The second problem results from the influence of general inflationary trends on the cost of all types of process equipment, so that file data related to purchase prices rapidly become outdated, even for equivalent sized items.

Generalized Concepts of Cost Estimation

When faced with the requirement of securing or developing an estimated dollar value for the future purchase price of a piece of process equipment or for the expected capital investment of an entire chemical plant, the project engineer usually has several alternatives with which to complete the task. In each such case, depending on the availability of time and/or funds, it is possible to choose from a spectrum of methods extending from an order-of-magnitude approach (with a probable accuracy of about $\pm 30\%$) to a detailed contractor's estimate (with an accuracy of $\pm 5\%$). Although the intermediate techniques are identified by a variety of names [6] and basic assumptions, in general they involve a trade-off of expense and time against accuracy. As an additional complicating factor, it should also be recognized that in the preliminary phases of a project design the capacity factors themselves may also include substantial degrees of uncertainty. However, since the data for costs are obviously needed before the evaluation of project feasibility may proceed, the trade-offs are made and the engineer, in order to complete the procedure that has been selected, may then call upon the available resources. These most frequently involve individual files, corporate files or correlations, vendor quotations, and information published in the open literature, all of which must then be reviewed in detail. As might be expected from such a diversity of sources, such data are frequently presented or stored in incompatible forms that make rapid and/or convenient use of the values almost an impossibility. In addition, the problems of capacity difference and time dependence mentioned earlier may arise in forms that add substantially to the difficulty of use.

Much work [2, 4, 5, 7] has been done in developing procedures or systems that are directed toward overcoming these problems. In general, most of these approaches are designed to allow for input that will compensate for size factors of different types and for time variations. The methods suggested are obviously much more useful over an extended range and for a longer time span if they also provide the possibility of incorporating additional private data into their general format. This potential inclusion of supplementary values is particularly crucial for designers involved in specialty operations or processes not generally covered in trade journals or other open sources.

Scale-up Factors

A variety of techniques are available in the literature for correlating the estimated purchase cost of assorted types of process equipment and/or components with a capacity-type factor related to the operation or performance of the item. Such correlations have been applied to costs of design components as simple as a single valve and to systems as complex as entire processing plants. The capacity factor for these presentations may be specified on the basis of geometry (pipeline diameter for a valve), energy requirements (horsepower for a motor), design parameters (transfer area for a heat exchanger), product capacity (tons per day for manufacturing facilities), or other similar characteristics.

Thus tabulations of costs at a specific time for each capacity of interest could be made available, with optional aspects of the equipment being included as subheadings or additional parameters. However, because of the extensive space requirements for even a cursory coverage of equipment categories, this approach has limited application. Its major advantage is in the capability to use a high degree of precision in the listings, although it should be noted that any unjustified precision may in effect be counterproductive by implying more knowledge than has actually been used in formulating the table.

The most frequent procedure used in this type of consideration is the plotting, on log-log coordinates, of the various available cost values at a particular time versus the appropriate capacity factor. Hence the resulting interpolation of such values will give acceptable accuracy if sufficient data from reliable sources are available to produce the plot. It also provides for more effective utilization of space for summarizing a broad scope of equipment parameters. This is usually done by presenting a family of curves on the same plot for each category of equipment items in a general classification. In the majority of cases this relationship may be adequately represented by a straight line, or a family of parallel lines, over a reasonable range. If the straight-line representation is considered to be a satisfactory correlation of the data in the area of interest, it is generally more convenient to use an analytical expression of the form

$$\text{Cost, desired capacity} = (\text{cost, known capacity}) \left(\frac{\text{desired capacity}}{\text{known capacity}} \right)^n \quad (1)$$

where n is a constant dependent on the specific type of equipment, but frequently is assumed to be approximately 0.6 for preliminary estimation purposes.

As mentioned previously, the major difficulty with this approach is that all of the data points used to produce the plot must be based on a known and common time of purchase. However, it is also obvious that if the 0.6 exponent is satisfactory for these generalized production purposes, a single data point is sufficient to establish the location of the curve on such a plot.

Time-Dependence Factors

The practice of including escalation factors in cost estimation calculations is a well-established procedure in the chemical process industries. For short-term work such as up-dating a slightly obsolete quotation for a specific piece of equipment, an "add-on" percentage may be adequate. In the case of intermediate projections, extrapolation by plotting recent data will frequently provide sufficient precision. For longer range corrections the typical method is to use one of the published price ratio values such as the CE Plant Cost Index or the Marshall and Swift Equipment Index. The details of these systems are well described in the literature [1, 8] and generally involve an expression of the form

$$\text{Cost, present time} = (\text{cost, previous date}) \left(\frac{\text{index, present time}}{\text{index, previous date}} \right) \quad (2)$$

Historical information regarding these ratios in the form of tables and/or plots covering extended time intervals is available in the literature. The CE Index uses 1957–1959 costs as the base, and the M&S Index uses 1926 as its basis.

Generalized Calculations

Since it will rarely be the case that a process design estimate will involve only up-dating or only size changes, it is advantageous to develop a more universal calculational procedure that will encompass both aspects. This could be done on a sequential basis using the two procedures individually, but this approach tends to be time consuming. Therefore, by combining the ratio functions which were described earlier into a single expression, the following may be secured:

$$\text{Cost, desired capacity at present time} = (\text{cost, known capacity at previous date})$$

$$\left(\frac{\text{desired capacity}}{\text{known capacity}}\right)^n \left(\frac{\text{index, present time}}{\text{index, previous date}}\right) \qquad (3)$$

Because this is a relatively simple calculation, little trouble is encountered if the appropriate data values are readily available. However, if the necessary numbers are scattered through either the open literature or the design engineer's files in comparatively unorganized form, the use of this relationship may become a very time-consuming task. This is particularly the case if n is not known and therefore must be determined or estimated. In view of these difficulties, short-cut methods that would avoid the need to use the equation are of considerable interest to engineering staff members involved in cost estimation.

A nomograph has been developed [3], (Fig. 1) to overcome these problems by taking advantage of the following generalized concepts:

1. The X–Y grid for the equipment category provides a concise method of tabulating a wide variety of components and systems. It should be noted that shifts along the X axis correspond to the use of various values of n. In fact, if n is known for the material or classification, it will simplify the tabulation of such data.
2. The time-index line allows the use of whichever of the two values is known for a given application. This avoids the need to determine the index value if only the date of a quotation is available. It may also facilitate the projection of future trends if a relatively short time-span is involved.
3. Up-dating is convenient by merely adding supplemental dates opposite the appropriate index numbers as the information becomes available.
4. New equipment categories may be added by evaluating and tabulating the X–Y coordinates using two known cost values at two specified dates (or index numbers if known).

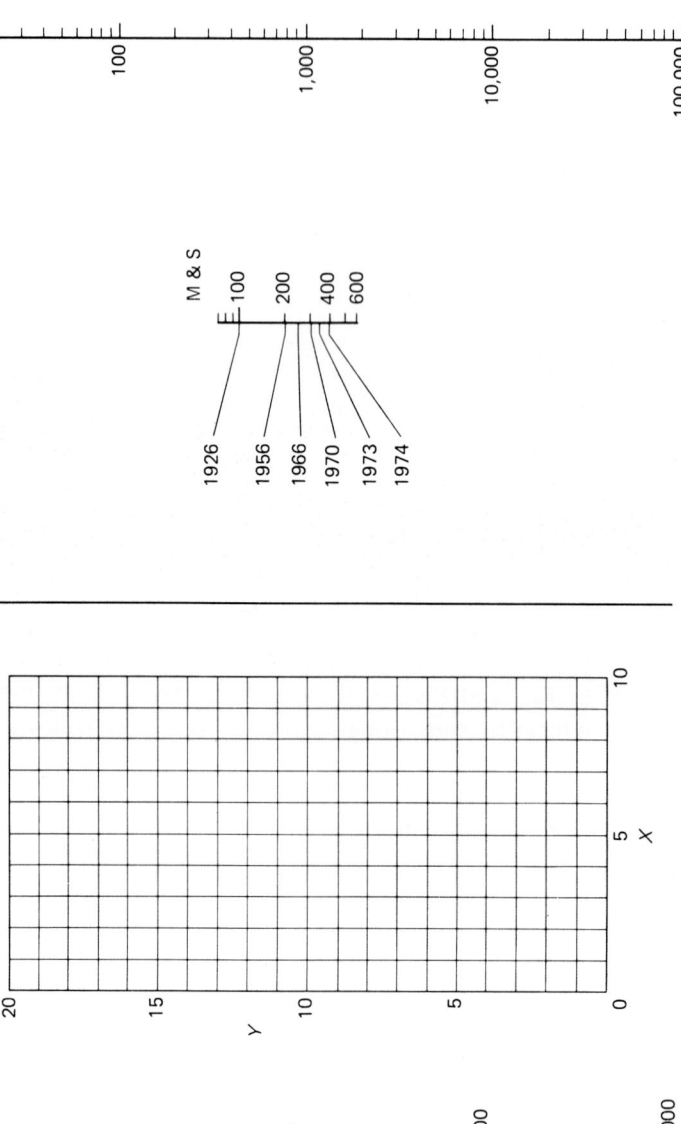

FIG. 1. A nomograph for correcting process and equipment cost data for both elapse of time and difference in size or capacity.

Because of the generalized nature of the correlations that form the basis of the nomograph, several precautions should be observed when using it in its current form or in adding supplemental data by back-plotting. Although this procedure is only intended to be used for approximate estimations, sufficient care must be used in order to secure adequate results. As pointed out by several authors [4, 5, 7], the straight-line representation is only valid over a limited range, and care should be used in extrapolating much beyond the tabulated values. Improved technology over a period of time may reduce the required investment needed to perform a given function, and in fact may decrease at a rate more rapid than increases due to inflationary factors, thus reducing the validity of the projection if such conditions exist. As additional items are incorporated into the grid, the descriptive categories used to classify them should be specific enough to provide adequate accuracy in selecting and interpreting the cost factors, but should not be excessively detailed so as to be relatively useless for the generalized estimates for which the method is intended.

Applications

The general techniques that would be involved in the use of the nomograph are directed toward meeting two needs: the estimation of the cost of an item already included in the listing, and the addition of new data to the grid. These may be accomplished as described below.

The more common case will be that of projecting the current cost from existing data, and this may be done by completing the following steps:

1. Refer to the alphabetized items in Table 1, and select the appropriate category for the desired equipment or process. If the necessary capacity is within the specified range, note the values of X and Y.
2. Construct a line from the point on the capacity scale corresponding to the equipment size (with units as specified in Table 1) through the X, Y point on the grid to the reference line.
3. Construct another line from this point on the reference line through the current index value on the M&S scale (or the year point, if desired) to the cost scale. The intersection will indicate the cost of the equipment. Note that a few values (as listed in Table 1) are given in thousands of dollars.

Although less frequently used, it is also of considerable importance to add new data to the grid for future use, and this may be done by completing these steps:

1. For each two sets of capacity–index–cost data available for an additional equipment category, construct a line from the cost scale through the M&S value to the reference line, and another line from there to the capacity scale.

TABLE 1 Nomograph Coordinates

Equipment, Process, etc.	Capacity		Cost Units	Coordinates	
	Range	Units		X	Y
Agitator, turbine	2–60	hp	$	6.4	14.4
Alumina, from bauxite	120–330	ton/d	$1000	6.8	13.8
Ammonium phosphate	270–750	ton/d	$1000	6.4	10.6
Blender, ribbon	10–239	ft³	$	6.2	14.0
Blender, sigma arm	4–70	hp	$	5.4	16.4
Blower, centrifugal	10–250	ft³/s	$	5.8	15.5
Boiler, 125 lb/in.²	100–500	lb/min	$1000	2.7	13.0
Butadiene, from butane	100–350	ton/d	$	6.8	16.8
Centrifuge, solid-bowl, continuous	15–50	in., diameter	$	1.8	16.2
Compressor, air, 125 lb/in.²	10–1000	ft³/min	$	7.7	15.8
Condenser, barometric, carbon steel	1000–9000	gal/min, water	$	4.1	8.8
Conveyor, 24 in. belt	20–250	ft, length	$	4.2	13.5
bucket, 10 in. × 6 in.	12–140	ft, length	$	6.0	13.4
18 in. roller	4–25	ft, length	$	3.3	10.2
Crusher, ball mill	10–800	ton/d	$	4.2	15.3
cone	40–300	ton/d	$	3.4	15.4
Crystallizer, vacuum, batch	500–7000	gal	$	7.4	14.3
Cyclone, single-stage	1,000–10,000	SCF/min	$	6.4	8.1
Dryer, pan, carbon steel	20–100	ft²	$	5.8	15.7
rotary, vacuum	50–300	ft²	$	6.3	16.3
spray	9–20	ft, diameter	$1000	2.9	10.9
Dust collector, cloth bag	30–200	ft³/s	$	4.4	13.0
Evaporator, vertical tube	100–10,000	ft²	$	4.5	12.0
Fan, centrifugal	50–350	ft³/s	$	3.8	9.5

Equipment	Size range	Units			
Filter, vacuum drum, continuous	3–90	ft²	$	7.1	17.0
plate and frame, cast iron	10–600	ft²	$	5.7	11.3
Gauge, pressure	3–12	in., dial	$	2.5	11.0
Heat exchanger, shell and tube, carbon steel	20–1000	ft²	$	6.2	10.8
316 SS	20–700	ft²	$	6.5	13.1
Heaters, immersion, electric	25–1000	kW	$	7.3	9.3
Kettle, cast iron, jacketed	20000	gal	$	7.3	13.5
glass-lined	200–3000	gal	$	7.4	10.5
Kneader, double arm	30–400	gal	$	6.0	15.8
Mixer, propeller	2–50	hp	$	5.5	13.5
Pressure vessel, 150 lb/in.² gauge	10–3000	gal	$	6.4	12.0
Pulverizer	500–10,000	lb/h	$	7.4	10.5
Pump, reciprocating, 316 SS, 25 ft head	10–60	gal/h	$	5.6	12.1
rotary, bronze	10–800	gal/min	$	6.6	10.9
Reactor, glass-lined, jacketed	20–3000	gal	$	6.7	13.6
agitated, jacketed	70–1000	gal	$	6.3	12.1
Refrigeration, mechanical, 20°F	4–100	tons	$	4.9	16.4
Rotameter, 316 SS	1–3	in., NPT	$	2.8	14.9
Screen, vibrating, single-deck	3–70	ft²	$	4.8	13.4
Tank, vertical	200–10,000	gal	$	6.0	7.8
horizontal	3–40	10³ gal	$	5.0	14.9
Tower, cooling	20–200	gal/s	$	4.7	16.6
distillation, bubble cap	20–100	in., diameter	$/ft	2.3	11.5
packed	10–150	in., diameter	$/ft	3.2	10.6
Tray, bubble-cap	3–9	ft, diameter	$/tray	1.3	14.7
Vacuum pump	10–3000	ft³/min	$	6.4	11.6
Valve, control, 316 SS	1–5	in., NPT	$	2.5	16.2
globe	1–3	in., NPT	$	4.9	8.8
plug, 316 SS	1–3	in., NPT	$	4.9	11.2

2. The intersection of the two reference–capacity construction lines will provide the X and Y coordinates for that equipment category, which may then be tabulated and used for later evaluations of costs.

References

1. Arnold, T. H., and Chilton, C. H., *Chem. Eng.*, *70*(4), 143 (1963).
2. Hazelbarth, J. E., *Chem. Eng.*, *74*(25), 214 (1967).
3. Hoerner, G. M., Jr., *Chem. Eng.*, *83*(11), 141 (1976).
4. Lang, H. J., *Chem. Eng.*, *54*(10), 117 (1947).
5. Mills, H. E., *Chem. Eng.*, *71*(6), 133 (1964).
6. Nichols, W. T., *Ind. Eng. Chem.*, *43*(10), 2295 (1951).
7. Peters, M. S., and Timmerhaus, K. D., *Plant Design and Economics for Chemical Engineers*, 2nd ed., McGraw-Hill, New York, 1968.
8. Stevens, R. W., *Chem. Eng.*, *54*(11), 124 (1947).

GEORGE M. HOERNER, Jr.

Cost, References

It is very helpful for a young engineer to study the literature on the cost estimation of chemical plants, processes, and equipment. For that reason we have prepared a comprehensive cost reference list. The list is not exhaustive but contains sufficient references to enable the designer to make his primary estimates.

At the end of the cost references the reader will find an index alphabetized in terms of chemical plants, chemical processes, and chemical equipment.

This compilation appeared in the *Chemical Age of India*, *24*(12), 867–882 (1973) and *25*(1), 9–26 (1974).

1. *Chemical and Process Engineering*, "Project News," *46*(4), 221 (1965).
2. *Chemical and Process Engineering*, "Process Costs—Acetic Acid," *47*(10), 49 (1966).
3. *Chemical Age of India*, "New Acetic Acid Plant at Hyderabad Goes into Production," *19*(12), 1128 (1968).
4. *Chemical Age of India*, "The Mysore Acetate and Chemicals Co. Ltd.," *20*(7), 639 (1969).
5. K. M. Guthrie, "Capital and Operating Costs for 54 Chemical Processes," *Chem. Eng.* (*N. Y.*), *77*(13), 140 (1970).

6. *Chemical and Process Engineering*, "Project News," *46*(1), 47 (1965).
7. *Chemical and Process Engineering*, "Project News," *46*(9), 506 (1965).
8. *Chemical and Process Engineering*, "Process Costs—Wuff Acetylene," *47*(2), 71 (1966).
9. R. B. Stobaugh, "Acetylene—How, Where, Who—Future," *Hydrocarbon Process.*, *45*(8), 125 (1966).
10. *Chemical Age*, "Acetylene from Methane," *99*(2882), 17 (1968).
11. U. Lorber, H. Reimann, and F. Rottmayr, "Acetylene Recovered from Ethylene Feedstock," *Chem. Eng. (N. Y.)*, *78*(17), 83 (1971).
12. R. E. Sanders, "Acrylic Fibers—Process Survey," *Chem. Process Eng.*, *49*(9), 100 (1968).
13. *Chemical and Process Engineering*, "Project News," *50*(4), 13 (1969).
14. R. E. Sanders, "Acrylic Fibers—Process Survey. 2," *Chem. Process Eng.*, *51*(9), 107 (1970).
15. *Chemical and Process Engineering*, "Acrylonitrile—Process Costs," *48*(6), 85 (1967).
16. *Chemical Age of India*, "New Processes," *19*(3), 227 (1968).
17. *Chemical Age of India*, "New Processes," *19*(12), 1195 (1968).
18. H. A. Sommers, "The Chlor-Alkali Industry," *Chem. Eng. Prog.*, *61*(3), 94 (1965).
19. K. S. Chari and B. G. Nadig, "Some Design Features of Diaphragm Cells for Caustic/Chlorine," *Chem. Age India*, *19*(1), 17 (1968).
20. G. Seelmann, F. Glos, and O. Wagner, "Optimisation of an Alkaline Chloride Electrolysis Plant with Mercury Cells," *Chem. Age India*, *21*(4), 434 (1970).
21. *Chemical and Process Engineering*, "Project News—1," *51*(5), 11 (1970).
22. J. D. Adhia, "Some Recent Developments in Chloralkali Industry," *Chem. Age India*, *23*(6), 483 (1972).
23. R. K. Gupta and V. Mehta, "Selection of Electrolytic Cells for Caustic Soda Chlorine Manufacture," *Chem. Age India*, *23*(6), 491 (1972).
24. N. R. Nandi and J. De, "Choice of Economic-Size Mercury Cells for Chlorcaustic Industry in India," *Chem. Age India*, *23*(6), 513 (1972).
25. F. A. Peters, P. W. Johnson, and R. C. Kirby, "Cost Estimate of Bayer Process for Producing Alumina," *U.S. Bur. Mines, Rep. Invest.*, *6730*, (1966).
26. G. Almasy et al., "Optimization Calculations of Ammonia Synthesis Loops," *Magy. Kem. Lapja*, *1965*(7–8), 353.
27. R. Habermehl, "Improved Catalysts Reduce Costs," *Chem. Eng. Prog.*, *61*(1), 57 (1965).
28. F. L. Applegate, Jr., "Ammonia Is Easy to Handle," *Chem. Eng. Prog.*, *61*(1), 66 (1965).
29. *Chemical and Process Engineering*, "Project News," *46*(2), 105 (1965).
30. O. J. Quartulli, "Check List for High Pressure Reforming," *Hydrocarbon Process.*, *44*(4), 151 (1965).
31. S. A. Bresler and G. R. James, "Questions and Answers on Today's Ammonia Plant," *Chem. Eng. (N. Y.)*, *72*(13), 109 (1965).
32. J. B. Allen, "Ammonia Manufacture," *Chem. Process Eng.*, *46*(9), 473 (1965).
33. *Chemical and Process Engineering*, "Project News," *46*(10), 576 (1965).
34. A. Cappelli, G. B. Ferraris, and M. Dente, "Design Optimization of an Ammonia Synthesis Unit. I—Optimization of the Reactor," *Chem. Ind.*, *48*(96), (1966).
35. A. Cappelli, M. Dente, G. B. Ferraris, and G. Colombo, "Design Optimization of an Ammonia Synthesis Unit. II—Economical Optimization of the Process," *Chem. Ind.*, *48*(702), (1966).

36. *Chemical and Process Engineering*, "Project News," *47*(1), 40 (1966).
37. N. D. Gopinath, "Modern Developments in Synthesis Gas Manufacture via Steam–Naphtha Reforming Process," *Chem. Age India*, *17*(3), 154 (1966).
38. *Chemical Age of India*, "Coke Oven Gas as a Feedstock for Ammonia Production," *7*(3), 196 (1966).
39. *Chemical and Process Engineering*, "Project News," *47*(12), 78 (1966).
40. *Chemical and Process Engineering*, "Project News," *48*(1), 66 (1967).
41. *Chemical and Process Engineering*, "Project News," *48*(2), 112 (1967).
42. *Chemical and Process Engineering*, "Project News," *48*(3), 94 (1967).
43. *Chemical Age of India*, "GSFC—Project Makes Commendable Progress," *18*(3), 179 (1967).
44. R. E. Blanco, J. M. Holmes, R. Salmon, and J. W. Ullmann, "Ammonia Costs and Electricity," *Chem. Eng. Prog.*, *63*(4), 46 (1967).
45. G. Honti, "Milestone in Ammonia Synthesis Technology," *Chem. Process Eng.*, *48*(4), 68 (1967).
46. *Chemical Age of India*, "Coromandel Fertilisers Limited," *18*(12), 887 (1967).
47. R. J. Morse, "Catacarb CO_2, Cuts Costs, Enjoys Big Growth," *Oil Gas J.*, *66*(17), 184 (1968).
48. *British Chemical Engineering*, "Project News," *13*(7), 927 (1968).
49. L. Axelord, R. E. Daze, and H. P. Wickhan, "The Largest Plant Concept," *Chem. Eng. Prog.*, *64*(7), 17 (1968).
50. J. A. Finneran, N. J. Sweeney, and T. G. Hutchinson, "Startup Performance of Large Ammonia Plants," *Chem. Eng. Prog.*, *64*(8), 72 (1968).
51. *British Chemical Engineering*, "Cost Data File," *14*(3), 361 (1969).
52. K. C. Khurana, "Trends in Steam–Hydrocarbon Reforming with Special Reference to Side-Fired Reforming Furnaces," *Chem. Age India*, *20*(9), 777 (1968).
53. *British Chemical Engineering*, "Cost Data File," *14*(10), 1459 (1969).
54. H. Grosskopf and H. Birnbaum, "Ammonia Synthesis Loop Design and Optimization," *Chem. Age India*, *21*(4), 459 (1970).
55. *Chemical and Process Engineering*, "Project News," *52*(6), 9 (1971).
56. E. Goeke, *Ammonia Production on the Basis of Coal and Lignite—Technical and Economic Aspects*, UNIDO Symposium, September 21–October 1, 1971.
57. P. D. Becker, H. Hiller, G. Hoghesand, and A. M. Sinclair, "Heavy Fuel Oil as Ammonia Plant Feed Stock," *Chem. Process Eng.*, *52*(11), 59 (1971).
58. K. S. Viswanathan and S. K. Mukherjee, "Ammonia Production Based on Various Raw Materials," *Chem. Age India*, *22*(12), 975 (1971).
59. E. Goeke, "Ammonia Production on the Basis of Coal and Lignite—Technical and Economic Aspects," *Chem. Age India*, *22*(12), 969 (1971).
60. *Chemical Age of India*, "Southern Petrochemical Industrial Corporation Limited," *23*(1), 41 (1972).
61. *Chemical and Process Engineering*, "Projects," *53*(7), 9 (1972).
62. R. Habermehl, "Process Optimising Industrial Catalysts," *Indian Chem. J.*, *3*(1), 155 (1972).
63. *Indian Chemical Journal*, "Inauguration of Ammonium Bicarbonate Plant," *3*(5), 57 (1968).
64. N. N. Udwadia and D. J. Mehta, "Ammonium Chloride and Light Magnesium Carbonate from Magnesium Chloride," *Chem. Age India*, *5*(2), 33 (1972).
65. A. J. Payne and P. G. Glikin, "Ammonium Nitrate," *Chem. Process Eng.*, *49*(4), 65 (1968).
66. *British Chemical Engineering*, "BCE Process Scan—Ammonium Nitrate, Hoechst—Uhde Process," *14*(7), 917 (1969).

67. *Indian Chemical Journal*, "HOC Goes the Whole Hog, Journalists Get a Preview of the Shape of Things to Come . . . ," *4*(11), 24 (1970).

68. *Popular Plastics*, "The Curtain Rises over a Giant Complex," *15*(5), 33 (1970).

69. S. D. Gupta, S. Prakash, and S. C. Joshi, "Process for Preparation of Antioxidant for Fats and Oils," *Chem. Age India*, *22*(8), 539 (1971).

70. K. S. Chari, N. M. Singh, and R. K. Tawney, "The Manufacture of Benzene Hexachloride in India," *Chem. Age India*, *19*(9), 814 (1968).

71. *Chemical Age of India*, "New Process," *19*(10), 944 (1968).

72. K. M. Rao, K. P. Patel, M. R. Oza, K. Seshadri, and D. S. Data, "Process for Recovery of Borax from Tincal," *Chem. Age India*, *17*(12), 1014 (1966).

73. *Chemical Age*, "Butadiene Extraction," *98*(2559), 6 (1968).

74. T. Reis, "Butadiene Extraction—Process Survey," *Chem. Process Eng.*, *51*(3), 65 (1970).

75. M. S. Kothari, "Process Design and Economics of the Production of Caffeine from Tea Waste," *Chem. Age India*, *17*(1), 41 (1966).

76. M. S. Kothari, "Technoeconomic Evaluation of the Production of Caffeine from Tea Waste," *Chem. Age India*, *17*(9), 699 (1966).

77. H. V. Shah and D. J. Mehta, "Calcium Silicate Filler in Non-Black Rubber," *Chem. Age India*, *19*(12), 1167 (1968).

78. *Chemical Age of India*, "Notes and News," *19*(5), 323 (1968).

79. *Chemical Age of India*, "Processes," *19*(6), 476 (1968).

80. *Chemical and Process Engineering*, "Project News," *49*(7), 11 (1968).

81. M. Taverna and M. Chiti, "Compare Routes to Caprolactam," *Hydrocarbon Process.*, *49*(11), 137 (1970).

82. *Chemical and Process Engineering*, "New Projects," *53*(5), (1972).

83. *Chemical Age of India*, "UCI's Carbon Black Plant Goes on Stream," *19*(1), 6 (1968).

84. J. A. Thurby and O. Sintnai, "Carbon Tetrachloride by Chlorination of Char," *Br. Chem. Eng.*, *17*(4), 319 (1972).

85. M. S. Kothari, "Technoeconomic Evaluation of Small Scale Production of Cement by Vertical Shaft Kiln Process," *Chem. Age India*, *17*(7), 541 (1966).

86. C. A. Miller, "New Cost Factors Give Quick and Accurate Estimates," *Chem. Eng. (N. Y.)*, *72*(19), 226 (1965).

87. F. Hine, "Economic Studies on Amalgam-Type Chlorine Cell," *Electrochem. Technol.*, *6*(1–2), 69 (1968).

88. H. A. Sommers, "Chlorine Industry Economics in 1970," *Chem. Age India*, *21*(11), 1013 (1970).

89. *Chemical Age of India*, "New Plant for Production of Chloromethanes," *19*(1), 31 (1968).

90. *British Chemical Engineering*, "Process Scan," *14*(12), 1660 (1969).

91. *Chemical Age of India*, "Herdillia Chemicals Limited—Report," *19*(2), 73 (1968).

92. *British Chemical Engineering*, "Project News," *15*(10), 1281 (1970).

93. F. A. Dufau, F. Eschard, A. C. Haddad, and C. H. Thonon, "Cyclohexane from Benzene, A Description of I. F. P. Process," *Chem. Age India*, *17*(5), 417 (1966).

94. *Chemical and Process Engineering*, "Project News," *47*(6), 294 (1966).

95. S. Field, and M. H. Dalson, "Economics of Making Cyclohexane," *Hydrocarbon Process.*, *46*(5), 169 (1967).

96. I. S. Rao, "Technology of DDT Manufacture—Recent Improvements," *Chem. Age India*, *19*(9), 811 (1968).

97. *Petroleum and Hydrocarbons*, "Notes & News—I.A.E.C. Demineralizing Plant for U.A.R.," *4*, 177 (1970).

98.   *Chemical and Process Engineering*, "Project News," *47*(9), 88 (1966).

99.   R. H. Jebens and D. I. Dykstra, "Advances in Sea Water Distillation Process," *Chem. Eng. Prog.*, *61*(8), 68 (1965).

100. L. S. Galstaun and E. L. Currier, "The Metropolitan Water District Desalting Project," *Chem. Eng. Prog.*, *63*(1), 64 (1967).

101. F. C. Standiford and H. F. Bjork, "Large Plants for Salt Water Conversion," *Chem. Eng. Prog.*, *63*(1), 70 (1967).

102. R. F. Detman, "Combination Process in Large Desalting Plants," *Chem. Eng. Prog.*, *63*(1), 80 (1967).

103. A. J. Barduhn, "Desalination by Crystallization Processes," *Chem. Eng. Prog.*, *63*(1), 98 (1967).

104. J. D. O'Toole, W. H. Comtois, and W. H. Stinson, "Characteristics of the PWR as an Energy Source in a Power-Desalting Application," *Chem. Age India*, *18*(3), 187 (1967).

105. H. A. Sindt, I. Spiewak, and T. D. Anderson, "Costs of Power from Nuclear Desalting Plants," *Chem. Eng. Prog.*, *63*(4), 41 (1967).

106. *British Chemical Engineering*, "Project Briefs," *12*(12), 1833 (1967).

107. *Chemical and Process Engineering*, "Project News," *49*(1), 13 (1968).

108. *British Chemical Engineering*, "Project News," *13*(10), 1371 (1968).

109. *British Chemical Engineering*, "Project News," *14*(2), 155 (1968).

110. E. Bahari, "Desalination Processes and Costs," *Chem. Process Eng.*, *50*(3), 71 (1969).

111. *British Chemical Engineering*, "Project News," *14*(6), 759 (1969).

112. *British Chemical Engineering*, "Project News," *14*(9), 1175 (1969).

113. *British Chemical Engineering*, "Project News," *14*(10), 1340 (1969).

114. *Chemical Process Engineering*, "Project News—1," *50*(10), 13 (1969).

115. *British Chemical Engineering*, "Project News," *14*(11), 1517 (1969).

116. J. E. Browning, "Desalting Process Costs Estimated," *Chem. Eng. (N. Y.)*, *77*(164), (1970).

117. *British Chemical Engineering*, "Project News," *15*(11), 1407 (1970).

118. *Chemical and Process Engineering*, "Desalination Costs," *51*(12), 5 (1970).

119. *Chemical Engineering*, "30th Inventory of New Processes and Technology," *78*(3), 74 (1971).

120. *Chemical and Process Engineering*, "CPE Processes and Licences—Desalination by Reverse Osmosis," *52*(11), 13 (1971).

121. *Desalting Digest*, *9*(12), 2 (1972).

122. *Chemical Engineering*, "32nd Inventory of New Processes and Technology," *79*(3), 74 (1972).

123. *British Chemical Engineering*, "Project News," *13*(3), 313 (1972).

124. R. E. Zimmermann, "Economics of Coal Desulphurization," *Chem. Eng. Prog.*, *62*(10), 61 (1966).

125. C. G. Cortelyou, R. C. Mallatt, and H. H. Meredith, Jr., "A New Look at Desulphurization," *Chem. Eng. Prog.*, *64*(1), 53 (1968).

126. N. P. Chopey, "Taking Coals Sulphur Out (Desulphurization, Part 4)," *Chem. Eng. (N. Y.)*, *79*(15), 86 (1972).

127. R. A. Duckworth, "Synthetic Detergent Powders—Process Survey," *Chem. Process Eng.*, *51*(4), 63 (1970).

128. S. D. Gupta and R. K. Srivastava, "Scheme for the Manufacture of Household Detergent Powders in the Small Scale Sector," *Chem. Age India*, *22*(5), 303 (1971).

129. *Chemical Age of India*, "De Nora Process for Dicalcium Phosphate Production," *20*(2), 155 (1969).

130. *Chemical Age of India*, "ATIC Industries Ltd.," *23*(1), 9 (1972).
131. C. Chapman, "Electrodialysis for a Developing Country," *Chem. Process Eng.*, *48*(4), 62 (1967).
132. C. E. Wood, "Costing and Control of Material in Electroplating Operations," *Plating*, *52*(4), 322 (1965).
133. K. V. Lad, J. R. Sandhavi, G. D. Bhat, and K. Seshadri, "Epsom Salt from Sels Mixtures," *Chem. Age India*, *19*(1), 33 (1968).
134. *Chemical and Process Engineering*, "Project News," *51*(3), 11 (1970).
135. E. H. Peters, "Ethylene, Organic Chemical Building Block," *Chem. Eng. Prog.*, *62*(6), 87 (1966).
136. R. B. Stobaugh, "Ethylene: How, Where, Who—Future," *Hydrocarbon Process.*, *45*(10), 143 (1966).
137. *Canadian Chemical Processing*, "Montreal—Swing to Naphtha Feedstock," *50*(11), 55 (1966).
138. S. B. Zdonik, E. J. Green, and L. P. Hallee, "Ethylene Worldwide—1, 2, 3," *Oil Gas J.*, *64*(48), 62 (1966); *64*(49), 108 (1966); *64*(51), 75 (1966).
139. P. Bonche and M. Carougeau, "Aspects Economiques de la Production d'Olefines," *Inst. Fr. Pet. Ann. Combust. Liq., Rev.*, *22*(2), 235 (1967).
140. W. Perry, "How to Make Ethylene Economically from Sugar-Cane Alcohol," *Chem. Eng. (N. Y.)*, *74*(18), 100 (1967).
141. *Chemical Age of India*, "The Dianor Process for the Production of Ethylenes, Ethylene Dichloride, Vinyl Chloride Monomer," *18*(11), 835 (1967).
142. *British Chemical Engineering*, "Project News," *13*(2), 175 (1968).
143. R. A. Duckworth, "Ethylene—Process Survey," *Chem. Process Eng.*, *49*(2), 67 (1968).
144. W. P. Hegarty, "Evaluating the Incremental Project: An Illustrative Example," *Chem. Eng. (N. Y.)*, *75*(19), 158 (1968).
145. P. R. Nayak, "India's 3rd Ethylene-Based Petrochemical Complex," *Indian Chem. J.*, *3*(4), 18 (1968).
146. L. Marshall and H. B. Zasloff, "Gas Oil Feedstocks for Ethylene Production," *Chem. Eng. Prog.*, *65*(10), 65 (1969).
147. R. G. Craig, C. E. Towler, and M. L. Raczynski, "Hydroprocessing of Ethylene Plant by Products (Aufarbeitung von Neben produkten der Aethylenherstellung)," *Erdoel Kohle, Erdgas, Petrochem.*, *22*(10), 613 (1969).
148. S. N. Rothman, "Ethylene Plant Optimization," *Chem. Eng. Prog.*, *66*(6), 37 (1970).
149. P. H. Spitz, "Chlor-Alkalis in the Caribbean Basin—2," *Chem. Eng. Prog.*, *66*(9), 20 (1970).
150. *Chemical Engineering*, "Giant Plants under Fire," *79*(6), 54B (1972).
151. A. J. Gambro, K. Menz, and M. Abraham, "Optimize Ethylene Complex," *Hydrocarbon Process.*, *51*(3), 73 (1972).
152. K. V. Satyanarayana and M. K. Raval, "A Plant for Ethylene Dibromide," *Pet. Hydrocarbons*, *3*(2), 66 (1968).
153. R. Landau, D. Brown, A. Saffer, and J. V. Porcelli, "Ethylene Oxide Economics—The Impact of New Technologies," *Chem. Eng. Prog.*, *64*(3), 27 (1968).
154. R. B. Stobaugh, G. C. Ray, and R. A. Spinck, "Ethylene Oxide: How, Where, Who—Future," *Hydrocarbon Process.*, *49*(10), 105 (1970).
155. C. D. Miserlis and J. R. Ghublikian, "Continuous Hydrolysis of Fats," *Chem. Eng. Prog.*, *62*(7), 114 (1966).
156. S. D. Vaidya and J. G. Kane, "A Twitchell Fat Splitting Unit in the Co-operative Sector," *Chem. Age India*, *19*(10), 924 (1968).

157. *Chemical Age of India*, "New Processes," *20*(8), 745 (1969).
158. *Chemical Engineering*, "New.Plants and Facilities," *72*(9), 111 (1965).
159. *Chemical Engineering*, "New Plants and Facilities," *72*(21), 191 (1965).
160. *Chemical and Process Engineering*, "Project News," *46*(11), 624 (1965).
161. *Chemical and Process Engineering*, "Project News," *47*(3), 147 (1966).
162. *Chemical Engineering*, "New Plants and Facilities," *73*(9), 151 (1966).
163. R. Horn and N. D. Fouser, "Production of 18–46–0 Fertilizer—Comparison of Processes," *Chem. Eng. Prog.*, *62*(7), 109 (1966).
164. *Chemical and Process Engineering*, "Project News," *47*(8), 76 (1966).
165. *Chemical and Process Engineering*, "Project News," *47*(10), 98 (1966).
166. *Chemical Engineering*, "New Plants and Facilities," *73*(22), 119 (1966).
167. *Chemical and Process Engineering*, "Project News," *47*(11), 82 (1966).
168. *Chemical Age of India*, "The Story of Fact," *18*(3), 170 (1967).
169. *Chemical Engineering*, "New Plants and Facilities," *74*(8), 187 (1967).
170. *Chemical and Process Engineering*, "Project News," *48*(4), 98 (1967).
171. B. C. Jain, "Our Food Problem and Fertilizers," *Chem. Age India*, *18*(5), 324 (1967).
172. *Chemical and Process Engineering*, "Project News," *48*(5), 86 (1967).
173. *British Chemical Engineering*, "Project Briefs," *12*(7), 1027 (1967).
174. J. Shah, "U.S. Newsletter—Elemental Phosphorus in Fertilizer Manufacture," *Chem. Age India*, *18*(7), 482 (1967).
175. *British Chemical Engineering*, "Project Briefs," *12*(8), 1177 (1967).
176. *Chemical and Process Engineering*, "Project News," *48*(8), 9 (1967).
177. *Chemical Engineering*, "New Plants and Facilities," *74*(21), 221 (1967); *75*(8), 147 (1968).
178. W. C. Scott and J. A. Wilbanks, "Fluid Fertilizer Production," *Chem. Eng. Prog.*, *63*(10), 58 (1967).
179. *Chemical Age of India*, "Madras Fertilizers Ltd.," *18*(11), 854 (1967).
180. *British Chemical Engineering*, "Chemical Processes in Europe," *12*(11), 37 (1967).
181. *Chemical and Process Engineering*, "Project News," *48*(11), 9 (1967).
182. *Chemical and Process Engineering*, "Project News," *48*(12), 9 (1967).
183. *British Chemical Engineering*, "Project News," *13*(1), 27 (1968).
184. *Chemical Age of India*, "Gorakhpur Unit Near Completion," *19*(2), 92 (1968).
185. I. K. Shankar, "High Protein Fertilizers—A Step Towards Self-Sufficiency," *Chem. Age India*, *19*(2), 97 (1968).
186. *Chemical and Process Engineering*, "Project News," *49*(3), 11 (1968).
187. K. S. Chari and Y. Venkatesham, "Phosphatic Fertilizers without the Use of Sulphur," *Chem. Age India*, *19*(4), 265 (1968).
188. *Chemical Age of India*, "Gorakhpur Fertilizers," *19*(6), 417 (1968).
189. *British Chemical Engineering*, "Project News," *13*(6), 763 (1968).
190. *British Chemical Engineering*, "Project News," *13*(8), 1079 (1968).
191. *Indian Chemical Journal*, "Fertilisers by Various Processes," *3*(2), 27 (1968).
192. *British Chemical Engineering*, "Project News," *13*(9), 1213 (1968).
193. *Indian Chemical Journal*, "Fertilizer Corporation of India Ltd. (P & D Division)," *13*(33), 140 (1968).
194. *Chemical Age of India*, "India's First Natural Gas-Based Fertilizer Plant Goes on Stream," *19*(12), 1121 (1968).
195. *Chemical and Process Engineering*, "Project News," *50*(2), 13 (1969).
196. *British Chemical Engineering*, "Project News," *14*(3), 269 (1969).
197. *Chemical Engineering*, "New Plants and Facilities," *76*(7), 141 (1969).
198. *British Chemical Engineering*, "Project News," *14*(15), 613 (1969).

199. *Chemical and Process Engineering*, "Project News—I," *50*(6), 13 (1969).
200. *Chemical Engineering*, "New Plants and Facilities," *76*(23), 120 (1969).
201. N. L. Mukherjee, "Technical Hydrogen Production for Fertilizer Industry in India," *Pet. Hydrocarbons*, *4*(4), 166 (1970).
202. *British Chemical Engineering*, "Project News," *15*(2), 173 (1970).
203. *Chemical Engineering*, "New Plants and Facilities," *77*(7), 123 (1970).
204. *Chemical Age of India*, "New 800 000 p.a. fertilizer complex in Gujarat," *21*(7), 684 (1970).
205. *Chemical Age of India*, "The Fertilizer Industry—Boxscore," *21*(12), 1097 (1970).
206. *Chemical and Process Engineering*, "Project News," *52*(7), 9 (1971).
207. *Chemical and Process Engineering*, "Project News," *52*(11), 9 (1971).
208. *Chemical Age of India*, "Sindri Rationalisation Programme (4)," *23*(1), 38 (1972).
209. *Chemical and Process Engineering*, "New Projects," *53*(3), 9 (1972).
210. A. C. Chatterjee, "Production of Mixed Fertilizer from Distillery Spent-Wash," *Chem. Age India*, *23*(4), 291 (1972).
211. *Chemical and Process Engineering*, "Projects," *53*(6), 9 (1972).
212. H. E. Blake, Jr., W. S. Thomas, K. W. Moser, J. L. Reuss, and H. Dolezal, "Utilization of Waste Fluosilicic Acid—12," *U.S. Bur. Mines, Rep. Invest.*, *7502* (1971).
213. *Chemical and Process Engineering*, "Project News—1," *50*(7), 13 (1969).
214. *British Chemical Engineering*, "Project News," *14*(8), 1045 (1969).
215. *Chemical Age of India*, "The Konkan Chemical Project," *23*(11), 901 (1972).
216. *Chemical and Process Engineering*, "Clean Fuel Gas from Coal by Gasification," *53*(2), 62 (1972).
217. J. H. Prescott, "SNG to Fill Supply Gap?" *Chem. Eng. (N. Y.)*, *78*(21), 90 (1971).
218. G. J. Van den Berg, E. F. Reinmuth, and E. Supp, "OXO Synthesis Gas," *Chem. Process Eng.*, *51*(8), 53 (1970).
219. D. E. Blaser, J. A. Rionda, and A. L. Saxton, "Combine Desulfurizing and Coking," *Hydrocarbon Process.*, *50*(9), 137 (1971).
220. M. G. Sholdrick, "Coal Gasification Warm-up," *Chem. Eng.*, *78*(6), 59 (1971).
221. *Chemical and Process Engineering*, "CPE Processes and Licences—Coal Gasification Process Uses Molten Salt as Catalyst and Heat Transfer Fluid," *58*(12), 13 (1971).
222. P. G. Menon, "Gas Purification—Survey of the Various Processes," *Chem. Age India*, *17*(5), 389 (1966).
223. D. W. Clelland, A. D. Corbett, and R. F. Colts, "Design of a Plant for the Incorporation of Highly Active Wastes into Glass," *Chem. Eng. Prog.*, *65*(94), 89 (1969).
224. *Chemical Process Engineering*, "Project News—1," *50*(8), 13 (1969).
225. C. A. Law, "Heavy Water Production," *Nucl. Eng.*, *13*(145), 510 (1968).
226. *Hydrocarbon Processing*, "HPI Construction Boxscore," *44*(1), 147 (1965).
227. *Hydrocarbon Processing*, "HPI Construction Boxscore," *44*(5), 59 (1965).
228. *Hydrocarbon Processing*, "HPI Construction Boxscore," *44*(9), 56 (1965).
229. *Hydrocarbon Processing*, "1965 Petrochemical Handbook Issue—Processes," *44*(11), 157 (1965).
230. *Hydrocarbon Processing*, "HPI Construction Boxscore," *45*(1), 55 (1966).
231. J. F. Jones, R. T. Eddinger, and L. Seglin, "Multistage, Pyrolysis of Coal," *Chem. Eng. Prog.*, *62*(2), 73 (1966).
232. S. S. Sachdeva, "Economics of Hydrogen Consuming Processes," *Chem. Age India*, *17*(5), 398 (1966).

233. *Hydrocarbon Processing*, "HPI Construction Boxscore," *45*(5), 76 (1966).
234. K. C. Hellwig, S. Felgelman, and S. B. Alpert, "Upgrading Feeds by the H-Oil Process," *Chem. Eng. Prog.*, *62*(8), 71 (1966).
235. *Hydrocarbon Processing*, "HPI Construction Boxscore," *45*(9), 75 (1966).
236. *Hydrocarbon Processing*, "1966 Refining Process Handbook," *45*(9), 173 (1966).
237. R. G. Craig, E. A. White, A. M. Henke, and S. J. Wolek, "HG Hydrocracking Today," *Am. Pet. Inst., Div. Refin., Proc., 364,* (1966).
238. *Hydrocarbon Processing*, "HPI Construction Boxscore," *46*(1), 56 (1967).
239. *Hydrocarbon Processing*, "HPI Construction Boxscore," *46*(5), 57 (1967).
240. *Hydrocarbon Processing*, "HPI Construction Boxscore," *46*(9), 56 (1967).
241. *Hydrocarbon Processing*, "HPI Construction Boxscore," *47*(2), CR 3 (1968).
242. *Hydrocarbon Processing*, "HPI Construction Boxscore," *47*(6, Section 2), CR 3 (1968).
243. *Hydrocarbon Processing*, "HPI Construction Boxscore," *47*(10, Section 2), CR 3 (1968).
244. *Hydrocarbon Processing*, "HPI Construction Boxscore," *48*(2, Section 2), CR 8 (1969).
245. *Hydrocarbon Processing*, "HPI Construction Boxscore," *48*(6, Section 2), CR 3 (1969).
246. J. R. Lambrix, C. S. Morris, and H. J. Rosenfeld, "The Implications of Heavy Fuel Oil Cracking," *Chem. Eng. Prog.*, *65*(11), 65 (1969).
247. *Hydrocarbon Processing*, "HPI Construction Boxscore," *48*(10, Section 2), CR 3 (1969).
248. *Hydrocarbon Processing*, "HPI Construction Boxscore," *49*(2, Section 2), CR 3 (1970).
249. M. S. Michaelian, R. J. Shlegeris, N. J. Haritatos, "Best Economics for Combining Hydrocracking and Reforming," *Hydrocarbon Process.*, *49*(5), 125 (1970).
250. *Hydrocarbon Processing*, "HPI Construction Boxscore," *49*(6, Section 2), CR 3 (1970).
251. *Hydrocarbon Processing*, "HPI Construction Boxscore," *49*(10, Section 2), CR 5 (1970).
252. *Hydrocarbon Processing*, "Worldwide HPI Construction Boxscore," *50*(2, Section 2), CB 5 (1970).
253. *Hydrocarbon Processing*, "NG/SNG Handbook," *50*(4), 93 (1971).
254. *Hydrocarbon Processing*, "Worldwide HPI Construction Boxscore," *50*(6, Section 2), 5 (1971).
255. H. G. Culp and R. R. Tracy, "Expand Natural Gases to Remove N_2," *Hydrocarbon Process.*, *50*(10), 89 (1971).
256. *Hydrocarbon Processing*, "HPI Construction Boxscore," *50*(10, Section 2), 7 (1971).
257. *Hydrocarbon Processing*, "1971 Petrochemical Handbook Issue," *50*(11), 113 (1971).
258. T. Reis, "$C_{5,6}$ Extraction Process Uses Acetonitrile," *Chem. Process Eng.*, *53*(1), 34 (1972).
259. *Hydrocarbon Processing*, "HPI Construction Boxscore," *51*(2, Section 2), 3 (1972).
260. T. Reis, "Low Cost Dehydrogenation of C_5 Streams," *Chem. Process Eng.*, *53*(3), 66 (1972).
261. D. E. Drayer, "How to Estimate Plant Cost–Capacity Relationship," *Pet./Chem. Eng.*, *42*(2), 10 (1970).
262. *Hydrocarbon Processing*, "1972 Refining Process Handbook," *51*(9), 111 (1972).

263. *Hydrocarbon Processing*, "Worldwide HPI Construction Boxscore," *51*(6, Section 2), 3 (1972).

264. C. Hulswitt and J. A. Mraz, "HCl Recovered from Chlorinated Organic Waste," *Chem. Eng.*, *79*(11), 80 (1972).

265. R. B. McBride and D. L. McKinley, "A New Hydrogen Recovery Route," *Chem. Eng. Prog.*, *1*(3), 81 (1965).

266. *Oil and Gas Journal*, "Hydrogen Plants Taking New Stature in Refining Operations," *63*(12), 82 (1965).

267. K. K. Battacharya and S. Patil, "Basic Aspects of Hydrogen Production," *Chem. Age India*, *17*(3), 169 (1966).

268. R. L. Costa and P. G. Grimes, "Electrolysis as a Source of Hydrogen and Oxygen," *Chem. Eng. Prog.*, *63*(4), 56 (1967).

269. R. J. Young, "Development of the Steam Naphtha Reforming Process for the Production of Hydrogen," *Chem. Age India*, *17*(5), 373 (1966).

270. J. Voogd and J. Tielrooy, "Improvements in Making Hydrogen," *Hydrocarbon Process.*, *76*(9), 115 (1967).

271. R. T. Eddinger, J. F. Jones, and F. E. Blanc, "Development of the COED Process," *Chem. Eng. Prog.*, *4*(10), 33 (1968).

272. T. P. Cook and R. N. Tennyson, "Improved Economics in Synthesis Gas Plants," *Chem. Eng. Prog.*, *65*(11), 61 (1969).

273. T. A. Ring, W. L. Mann, and Y. S. Tse, "Innovations in Hydrogen Production," *Chem. Eng., Prog.*, *66*(12), 59 (1970).

274. G. J. Van den Berg, E. Reinmath, and E. Kapp, "Hydrogen from Heavy Residues," *Chem. Process Eng.*, *52*(10), 49 (1971).

275. J. Voogd, "Hydrogen Reforming," *Chem. Age India*, *23*(7), 592 (1972).

276. *Chemical and Process Engineering*, "Hydrogenation of Acetylenics and Diolefines," *53*(7), 41 (1972).

277. H. Morikawa, M. Sagara, and G. Kakemoto, "CDT/CD from C_4 Fractions," *Hydrocarbon Process.*, *51*(8), 102 (1972).

278. *British Chemical Engineering*, "Project News," *13*(5), 615 (1968).

279. *Chemical and Process Engineering*, "CPE Processes and Licences," *52*(10), 11 (1971).

280. *Chemical Age of India*, "New Processes," *19*(11), 1057 (1968).

281. S. Ushio, "Extract Isoprene with DMF," *Chem. Eng. (N. Y.)*, *79*(5), 82 (1972).

282. G. W. Moorehouse, "Construction Performance Index—Engineering and Execution," *Chem. Process Eng.*, *47*(7), 84 (1966).

283. R. L. Mathews and J. F. Adams, "Predicting Manpower Needs in Engineering Departments," *Chem. Eng. (N. Y.)*, *76*(14), 152 (1969).

284. C. L. Amick and G. Gilleard, "Cost and Comfort Factors for Selecting Fluorescent Industrial Lighting," *Chem. Eng.(N. Y.)*, *60*(12), 159 (1965).

285. G. M. Mixon, "Chemical Plant Lighting," *Chem. Eng.(N. Y.)*, *74*(12), 113 (1967).

286. G. P. Cundall and J. S. Humphreys, "Industrial Lighting Design and Its Financial Evaluation," *Illum. Eng. Soc. Trans.*, *33*(2), 27 (1968).

287. F. R. Ainslie, "Central Engineering at U.S. Gypsum," *Plant Eng.*, *19*(2), 131 (1965).

288. J. M. Bourguet, "Economics of Today's Plants," *Hydrocarbon Process.*, *49*(4), 93 (1970).

289. C. H. Jordan, "Liquefied Natural Gas: Natural Gas Processing at Low Temperatures," *Chem. Eng. Prog.*, *68*(9), 53 (1972).

290. D. B. Crawford and G. P. Eschenbrenner, "Liquefied Natural Gas: Heat Transfer Equipment for LNG Projects," *Chem. Eng. Prog.*, *68*(9), 62 (1972).

291. S. Crossland, "Process Liquids to SNG," *Hydrocarbon Process.*, *51*(4), 89 (1972).

292. H. Beuther, R. F. Mansfield, and H. C. Stauffer, "Costs of Lubricating Oil Hydrogenation Processes," *Am. Pet. Inst., Div. Refining Process*, p. 172 (1966).

293. R. Dutriau, "Hydrocracking and Hydrorefining of Lubes IFP Process," *Chem. Age India*, *17*(5), 402 (1966).

294. M. J. Saldanha, "The First Lube Blending Plant in Madras," *Pet. Hydrocarbons*, *5*(1), 3 (1970).

295. P. K. Goel and M. G. Krishna, "Lubricant Production and Utilisation in India in the Seventies," *Pet. Hydrocarbons*, *5*(4), 91 (1971).

296. S. Ushio, "Maleic Anhydride Produced from C_4 Hydrocarbons," *Chem. Eng. (N. Y.)* *78*(21), 107 (1971).

297. N. M. Singh and D. N. Dey, "Feasibility Study for the Manufacture of Electrolytic Manganese Metal in India," *Chem. Age India*, *19*(12), 1170 (1968).

298. *Chemical Age of India*, "Directory of Man-Made Fibre Manufacturers in India," *21*(5), 509 (1970).

299. *British Chemical Engineering*, "BCE Process Scan-Melamine-OSW Process," *14*(10), 1337 (1969).

301. J. Huebler and F. C. Schora, "Coal Hydrogasification," *Chem. Eng. Prog.*, *62*(2), 87 (1966).

302. B. M. Bare and H. W. Lambe, "Economics of Methanol," *Chem. Eng. Prog.*, *64*(5), 110 (1968).

303. J. W. Woolcock, "The New Low Pressure Methanol Process," *Pet. Hydrocarbons*, *3*(3), 110 (1968).

304. *British Chemical Engineering*, "Project News," *13*(12), 1661 (1968).

305. *Chemical and Process Engineering*, "Project News," *49*(12), 11 (1968).

306. B. Hedley, W. Powers, and R. B. Stobaugh, "Methanol: How, Where, Who," *Hydrocarbon Process.*, *49*(6, Section 1), 97 (1970).

307. B. Hedley, W. Powers, and R. B. Stobaugh, "Methanol: How, Where, Who, Future," *Hydrocarbon Process.*, *49*(8), 117 (1970).

308. B. Hedley, W. Powers, and R. B. Stobaugh, "Methanol: How, Where, Who . . . Future," *Hydrocarbon Process.*, *49*(9), 275 (1970).

309. D. Mehta and D. E. Ross, "Optimize ICI Methanol Process," *Hydrocarbon Process.*, *49*(11), 183 (1970).

310. R. J. Kenard, Jr. and N. M. Nimo, "Present Methanol Manufacturing Costs and Economics of Using ICI Process," *Chem. Eng. Prog.*, *66*(98), 47 (1970).

311. *British Chemical Engineering*, "Cost Data File," *16*(2/3), 235 (1971).

312. *Chemical and Process Engineering*, "New Projects," *53*(1), 9 (1972).

313. *British Chemical Engineering*, "Process Scan," *15*(1), 32 (1970).

314. *Chemical Age of India*, "Natural Gas Distribution System in Baroda," *21*(8), 746 (1970).

315. *Chemical and Process Engineering*, "Process Costs—Nitric Acid," *47*(1), 11 (1966).

316. *Chemical and Process Engineering*, "Process Costs—Nitric Acid," *47*(1), 11 (1967).

317. *British Chemical Engineering*, "Project News," *13*(11), 1511 (1968).

318. *British Chemical Engineering*, "Project News," *14*(1), 27 (1969).

319. G. Jorquea, "Nitric vs. Sulphuric Acidulation of Phosphatic Rock," *Chem. Eng. Prog.* *64*(5), 83 (1968).

320. *Chemical and Process Engineering*, "New Projects," *53*(4), 9 (1972).

321. L. Hellmer, "Three New Nitric Acid Processes," *Chem. Eng. Prog.*, *68*(4), 67 (1972).

322. D. Komiyama, T. Ohrui, and Y. Sakakibara, "New Nitric Acid Process," *Hydrocarbon Process.*, *51*(4), 145 (1972).

323. *Chemical Age of India*, "Process for *p*-Nitrobenzoic Acid," *19*(2), 139 (1968).

324. *Chemical Age of India*, "The Norsk Hydronitrophosphate—NPK—Process," *18*(11), 839 (1967).

325. *Chemical Age of India*, "India's First and the World's Biggest Nitrophosphate Plant," *17*(1), 9 (1966).

326. P. Fortescue, R. T. Shantrom, and H. Fenech, "Development of Gas-Cooled First Reactor Concept," *Nucleonics*, *23*(5), 56 (1965).

327. M. E. Friedman, "Utilizacion dela Energia Nucclear," *Rev. Chil. Ing.*, *310*, 8 (1965).

328. I. Wall and H. Fenech, "Application of Dynamic Programming to Fuel Management Optimization," *Nucl. Sci. Eng.*, *22*(3), 285 (1965).

329. H. E. Vann, "Influence of Safety Problems on Development of Economic Nuclear Power," *Nucl. Struct. Eng.*, *2*(5), 475 (1965).

330. G. R. Fanjoy, "Canadian Nuclear Fuel Manufacture," *Eng. J.*, *49*(10), 32 (1966).

331. W. S. Buttler, "Nuclear Power in the CPI," *Chem. Eng. Prog.*, *63*(4), 39 (1967).

332. S. A. Bernsen and R. P. Schmitz, "Process Heat Costs and Nuclear Power," *Chem. Eng. Prog.*, *63*(4), 65 (1967).

333. G. B. Zorzoli, "Role of Heavy Water Advanced Converters in Mixed Fast and Thermal System," *Energ. Nucl.*, *14*(12), 693 (1967).

334. G. L. Decker, W. B. Wilson, and W. B. Biggo, "Nuclear Energy for Industrial Heat and Power," *Chem. Eng. Prog.*, *64*(3), 61 (1968).

335. *Chemical Engineering*, "Chemical Power Plants: Nuclear or Conventional," *75*(25), 66 (1968).

336. P. Haack, "The Reduction of the Specific Installation Cost of Large Power Plants," *Brennst.-Waerme-Kraft*, *21*(6), 299 (1969).

337. B. Leo, "Brief Comparison of Nuclear Power stations (Comparison rapide des centrales nucleaires)," *Energ. Nucl.*, *11*(1), 33 (1969).

338. P. Nathschlaeger and J. Gruemm, "Average Fuel Cycle Costs for Nuclear Power Stations with Variable Load Factor (Mittlere Brennstoffzykluskosten von Kernkraeewerken bei veraendevlichem Lastfaktor)," *At. Strom*, *15*(5/6), 90 (1969).

339. W. Steigelmann, "Outlook for Nuclear Power—Station Capital Costs," *React. Technol.*, *13*(1), 14 (1969).

341. J. P. Starkey and B. A. Bangent, "Review of the U.K. Nuclear Power Programme—2," *Nucl. Energy*, *11*(5), 159 (1970).

342. S. A. Ghalib, "Steam-Generating Heavy Water Reactor," *Nuclear Energy Costs and Economic Development, Proc. Symp. October 20–24 1969, Int. At. Energy Agency, STI/PUB 239*, 89 (1970).

343. *British Chemical Engineering*, "Project News," *15*(1), 35 (1970).

344. *Chemical and Process Engineering*, "Project News," *46*(3), 151 (1965).

345. *Chemical and Process Engineering*, "Project News," *46*(12), 680 (1965).

346. *Chemical and Process Engineering*, "Project News," *48*(6), 124 (1967).

347. *Chemical and Process Engineering*, "Project News," *47*(2), 100 (1966).

348. *Paint Manufacture*, "A Five Million Gallon Paint Plant in Georgia," *38*(8), 25 (1968).

349. *Oil and Gas Journal*, "Molex Becomes New Tool for Recovering Normal Paraffins," *63*(24), 113 (1965).

350. *Chemical Age of India*, "National Organic Chemical Industries Limited Report," *19*(1), 41 (1968).

351. *Petroleum and Hydrocarbons*, "Indian Petrochemicals Corporation Limited," *4*, 175 (1970).

352. *Chemical Engineering Progress*, "PPG Industries Bring Variety," *66*(4), 30 (1970).

353. *Chemical Age of India*, "Nagpal Ambadi," *22*(9), 581 (1971).

354. A. S. Banciu, "Phenol Manufacture," *Chem. Process Eng.*, *48*(1), 31 (1967).

355. *Chemical Age of India*, "Durgapur Chemicals Limited," *19*(5), 329 (1968).

356. *British Chemical Engineering*, "Cost Data File," *14*(6), 851 (1969).

357. S. D. Gupta and S. C. Joshi, "Process for Preparation of a Strictly Hindered Phenol Antioxidant," *Chem. Age India*, *22*(8), 537 (1971).

358. T. W. Segar, "Improved Route to Phosphoric Acid," *Chem. Eng. Prog.*, *62*(2), 123 (1966).

359. J. G. Kronseder, "Phosphoric Acid Plant Design," *Chem. Eng. Prog.*, *63*(10), 52 (1967).

360. J. G. Kronseder, "Effect of Wet Process Phosphoric Acid Technology on Fertilizer Production and Distribution Costs," *Chem. Age India*, *18*(11), 826 (1967).

361. G. S. G. Beveridge and R. G. Hill, "Phosphoric Acid—Process Survey," *Chem. Process Eng.*, *49*(7), 61 (1968).

362. J. G. Kronseder, "Economics of Phosphoric Acid Processes," *Chem. Eng. Prog.*, *64*(9), 97 (1968).

363. *British Chemical Engineering*, "Phosphoric Acid Hemihydrate Process—BCE Process Scan," *14*(1), 24 (1969).

364. *Chemical Age of India*, "New Processes—RABS and R-2 Processes for Superphosphoric Acid," *21*(2), 204 (1970).

365. H. S. Bryant, N. G. Holloway, and A. D. Silber, "Phosphorous Plant Design New Trends," *Ind. Eng. Chem.*, *62*(4), 8 (1970).

366. *Petroleum and Hydrocarbons*, "Process Developments—Dimethyl and Diethyl Phthalates," *4*(2), 71 (1969).

367. R. A. Duckwarth, "Phthalic Anhydride—Process Survey," *Chem. Process Eng.*, *50*(1), 69 (1969).

368. *British Chemical Engineering*, "BCE Process Scan—Phthalic Anhydride Von Heyden Process," *14*(9), 1169 (1969).

369. A. H. Truman, "Batch Distillation of Phthalic Anhydride," *Chem. Eng. Prog.*, *66*(3), 62 (1970).

370. N. E. Ockerbloon, "Xylene and Higher Aromatics, Part 3: Phthalic Anhydride," *Hydrocarbon Process.*, *50*(9), 162 (1972).

371. V. N. Lele, "Pigment Graded Red Oxide of Iron," *Chem. Age India*, *19*(12), 1164 (1968).

372. J. W. Drew and A. F. Ginder, "Pilot Plant Equipment. How to Estimate the Cost," *Chem. Eng. (N. Y.)*, *77*(3), 100 (1970).

373. *SPE Journal*, "1966 Plastics Industry Profit Guide," *22*(10), 17 (1966).

374. J. Walsh, "Market Opportunities in Plastics Industry," *SPE—25th Ann. Tech. Conf.*, *13*, 10 (1967).

375. J. J. Voci and T. D. Jannazzi, "Economics of Spent Liquor Recovery," *Chem. Eng. Prog.*, *61*(5), 110 (1965).

376. N. S. Lea and E. A. Christoferson, "Save Money by Stopping Air Pollution," *Chem. Eng. Prog.*, *61*(11), 89 (1965).

377. J. C. Barber, "Cost of Pollution Control," *Chem. Eng. Prog.*, *64*(9), 78 (1968).

378. K.-D. Werner, "Catalytic Oxidation of Industrial Waste Gases," *Chem. Eng. (N. Y.)*, *75*(23), 179 (1968).

379. R. Ashman, "A Practical Guide to Industrial Dust Control," *Inst. Heat. Vent.*

Eng. J., *38*, 273 (1971).

380. H. Popper, "The CPT's Cost of Meeting Environmental Standards," *Chem. Eng. (N. Y.)*, *78*(19), 106 (1971).

381. C. J. Stairmand, "The Chemical Engineers Contribution to Air Pollution Control," *Chem. Eng. (London)*, *254*, 375 (1971).

382. C. B. Barry, "Reduce Claus Sulfur Emission," *Hydrocarbon Process.*, *51*(4), 102 (1972).

383. *Chemical Engineering*, "New Plants and Facilities, CE Construction Alert," *19*(22), 72 (1972).

384. *Chemical Engineering*, "33rd Inventory of New Processes and Technology," *79*(16), 129 (1972).

385. D. S. Hall and J. W. Davison, "Development of *cis*-4 Polybutadiene Process," *Chem. Eng. Prog.*, *64*(2), 49 (1968).

386. *British Chemical Engineering*, "Process Scan—Polyester Fibre," *14*(2), 153 (1969).

387. P. E. Lesquen, "Low Density Polyethylene Made in Tubular Reactor," *Chem. Eng. (N. Y.)*, *79*(12), 42 (1972).

388. S. K. Sanghvi and S. M. Shah, "Bulk Polymerization," *Pet. Hydrocarbons*, *6*(3), 165 (1971).

389. *Chemical Age of India*, "Propylene Industries Ltd," *19*(11), 961 (1968).

390. *Popular Plastics*, "Polyolefines Industries Ltd," *13*(11), 24 (1968).

391. *Chemical and Process Engineering*, "Project News," *47*(4), 196 (1966).

392. *Chemical and Process Engineering*, "Project News," *50*(3), 13 (1969).

393. *British Chemical Engineering*, "Project News," *14*(7), 919 (1969).

394. D. Lommel, "Granulieranlagen fuer PVC-hart and PVC-weich," *Kunstst. Gummi*, *7*(4), 123 (1968).

395. S. K. Verma and A. N. Shah, "Production and Application's of PVC Film/Sheet," *Pop. Plast.*, *13*(6), 20 (1968).

396. A. N. Shah and S. M. Mundkur, "PVC Heavy Duty Sacks for Packaging," *Pop. Plast.*, *13*(11), 73 (1968).

397. V. N. Lele and J. M. Joshi, "Cost Estimate for a Potassium Carbonate Plant," *Chem. Age India*, *18*(12), 961 (1967).

398. *Chemical Age of India*, "New Processes," *19*(7), 535 (1968).

399. A. M. Squires, "Top Heat Cycle for Clear Power," *Chem. Eng. Prog.*, *62*(10), 74 (1966).

400. *Chemical Age*, "Chemical Age Survey—U.K. Projects," *97*(2516), VII (1967).

401. *Chemical Age*, "Chemical Age Survey—Overseas Projects," *99*(2604), S3 (1969).

402. *Chemical Age*, "Chemical Age Survey—United Kingdom Projects," *99*(2619), S1–S60 (1969).

403. *Chemical Age*, "West German Projects," *99*(2628), S3 (1969).

404. A. H. Mazumdar, "Bridging the Protein Gap—A Review," *Indian Chem. J.*, *6*(1), 117 (1971).

405. *Chemical and Process Engineering*, "New Equipment at KEM TEK Scandinavian Exhibition for the Process Industries—Protein Recovery Process Uses New Flotation Unit," *52*(12), 58 (1971).

406. G. Gavelin, "What Engineer Thinks of Russian Paper Industry," *Pap. Trade J.*, *151*(41), 42 (1967).

407. J. Diaz-Barreiro S. Pimentel, "Pulp and Paper Industry in the Process of Latin American integration (La industria de la celulosa yel papel enel proceso de integracion Latino americano)," *23rd Conference, November 3–7, 1968, Houston*, Paper 5–3.

408. B. P. Chaliha, S. B. Lodh, and M. S. Iyengar, "Pilot Plant Production of

Newsprint from Deinked Old Newspaper," *Chem. Age India*, *19*(1), 25 (1968).

409. G. H. Chidester, "Fiber for Tomorrow's Paper Requirements," *Tappi*, *51*(8), 464 (1968).

410. *Chemical Engineering*, "New Plants and Facilities," *75*(21), 167 (1968).

411. *Chemical Engineering*, "New Plants and Facilities," *77*(23), 141 (1970).

412. *Chemical and Process Engineering*, "Project News," *52*(3), 9 (1971).

413. *Chemical Engineering*, "New Plants and Facilities," *78*(8), 111 (1971).

414. *Chemical and Process Engineering*, "Project News," *52*(4), 9 (1971).

415. *Chemical Engineering*, "31st Inventory of New Processes and Technology," *78*(17), 105 (1971).

416. *Chemical Engineering*, "New Plants and Facilities—CE Construction Alert," Vol. 78, No. 24, 123 (1971).

417. *Chemical Engineering*, "Radiation System Piloted for Pulpwood Pretreatment," *79*(7), 48 (1972).

418. *Chemical Engineering*, "New Plants and Facilities CE Construction Alert," *79*(7), 89 (1972).

419. D. E. Harmer and D. S. Ballantine, "Radiation Processing," *Chem. Eng.* (*N. Y.*), *78*(9), 98 (1971).

420. R. A. Erickson and G. F. Asselin, "Isomerization Means Better Yields," *Chem. Eng. Prog.*, *61*(3), 53 (1965).

421. *Chemical and Process Engineering*, "Project News," *46*(5), 269 (1965).

422. *Chemical and Process Engineering*, "Project News," *46*(7), 374 (1965).

423. *Chemical and Process Engineering*, "Project News," *48*(7), 15 (1967).

424. *Chemical and Process Engineering*, "Project News," *48*(10), 9 (1967).

425. *Chemical and Process Engineering*, "Project News," *49*(2), 11 (1968).

426. *Chemical and Process Engineering*, "Project News," *49*(5), 11 (1968).

427. *Chemical and Process Engineering*, "Project News," *49*(8), 11 (1968).

428. S. Ellison, "New Automated Refinery Takes Shape in Britain," *Pet. Hydrocarbons*, *3*(2), 48 (1968).

429. *Chemical and Process Engineering*, "Project News," *49*(10), 11 (1968).

430. *Chemical and Process Engineering*, "Project News," *49*(11), 11 (1968).

431. *Chemical and Process Engineering*, "Project News," *50*(1), 13 (1969).

432. *Chemical and Process Engineering*, "Project News—1," *50*(5), 13 (1969).

433. *Petroleum and Hydrocarbons*, "India's Ninth Refinery on Stream," *4*(3), 89 (1969).

434. *Chemical and Process Engineering*, "Project News," *50*(11), 13 (1969).

435. *Chemical and Process Engineering*, "Project News—1," *50*(12), 13 (1969).

436. *Chemical and Process Engineering*, "Project News," *51*(1), 11 (1970).

437. *Petroleum and Hydrocarbons*, "Lube India Limited," *4*(4), 123 (1970).

438. *Chemical Age of India*, "Lube India Limited Facts and Figures," *21*(2), 880 (1970).

439. *Chemical and Process Engineering*, "Project News," *51*(2), 13 (1970).

440. *Chemical and Process Engineering*, "Project News—1," *51*(4), 13 (1970).

441. J. J. Bollward and D. F. Hagan, "Data Index System Estimates Costs," *Oil Gas J.*, *68*(29), 59 (1970).

442. J. R. Dosher, "Refinery Economics—Refinery Trends Report," *Chem. Eng.* (*N. Y.*), 77, 108 (1970).

443. *Chemical and Process Engineering*, "Project News," *51*(11), 11 (1970).

444. *Chemical and Process Engineering*, "Project News," *51*(12), 9 (1970).

445. *Chemical and Process Engineering*, "Project News," *52*(1), 9 (1971).

446. *Chemical and Process Engineering*, "Project News," *52*(8), 9 (1971).

447. G. E. Weismantel, "Can the Small Refiners Survive?" *Chem. Eng.* (*N. Y.*), *78*(21), 84 (1971).

448. *Chemical and Process Engineering*, "Project News," *52*(9), 9 (1971).
449. M. J. Stacey, "Influence of Transportation of Refineries in Europe," *Chem. Process Eng.*, *52*(9), 70 (1971).
450. *Chemical Age of India*, "Bongaigaon Refinery and Petrochemical Complex," *23*(1), 40 (1972).
451. D. L. Hendrickson, "Effect of Changing Gas Volatility on Refining Costs," *Chem. Eng. Prog.*, *65*(2), 51 (1972).
452. *Chemical and Process Engineering*, "New Projects," *53*(2), 9 (1972).
453. *Process Engineering*, "Projects," (August 1972).
454. *Process Engineering*, "Projects," (September 1972).
455. *Chemical Age of India*, "A Pioneer New Project to Manufacture Ketone Resins for the First Time in India," *23*(11), 891 (1972).
456. A. K. Bose, "Soda Ash—Its Development and Production in India during the Last Twenty-Five Years—And the Future Trends," *Chem. Age India*, *23*(6), 475 (1972).
457. T. Yamaguchi, "Tekkosha's New Metallic Sodium Process—By Electrolysis of Sodium Amalgam," *Chem. Age India*, *23*(3), 169 (1972).
458. E. Haidegger, I. Peter, I. Gemes, and J. Karolyi, "Production of Sorbitol by Use of Ammonia Synthesis Gas," *Ind. Eng., Chem., Process Des. Dev.*, *7*(1), 107 (1968).
459. A. Kuhn, "Modern Trends in Electrochemical Industry," *Br. Chem. Eng.*, *16*(2/3), (1971).
460. D. A. Lihou, "Thermodynamics of Steam Reforming," *Chem. Process Eng.*, *46*(9), 487 (1965).
461. *Chemical and Process Engineering*, "Structural Steel Work," *47*(11), 86 (1966).
462. S. A. Miller and J. W. Donaldson, "Styrene Manufacture—Process Survey," *Chem. Process Eng.*, *48*(12), 37 (1967).
463. S. A. Bresler and J. D. Ireland, "Substitute Natural Gas: Processes, Equipment, Costs," *Chem. Eng. (N. Y.)*, *79*(23), 94 (1972).
464. D. P. Thornton, D. J. Ward, and R. A. Erickson, "MRG Process for SNG," *Hydrocarbon Process.*, *51*(8), 81 (1972).
465. K. T. Bhandari and S. K. Mukherjee, "Production of Sulfur from Amjore Pyrites," *Chem. Age India*, *17*(6), 433 (1966).
466. S. Katell, "Removing Sulfur Dioxide from Flue Gases," *Chem. Eng. Prog.*, *62*(10), 67 (1966).
467. S. Katell and K. D. Plants, "Here Is What SO_2 Removal Costs," *Hydrocarbon Process.*, *46*(7), 161 (1967).
468. S. F. Galeano and C. A. Harding, "Sulphur Dioxide Removal and Recovery from Pulp Mill Power Plants," *Air Pollut. Control Assoc. J.*, *17*(8), 536 (1967).
469. S. Ludwig, "Antipollution Process Uses Absorbent to Remove SO_2 from Flue Gases," *Chem. Eng. (N. Y.)*, *75*(3), 70 (1968).
470. J. G. Kronseder, "Cost of Reducing Sulphur Dioxide Emissions," *Chem. Eng. Prog.*, *64*(11), 71 (1968).
471. J. E. Nowell, "Making Sulphur Flue Gas," *Chem. Eng. Prog.*, *65*(8), 62 (1969).
472. C. G. Cortelyou, "Commercial Processes for SO_2 Removal," *Chem. Eng. Prog.*, *65*(9), 69 (1969).
473. J. G. Stites, Jr., W. R. Horlacher, J. L. Bachofer, and J. W. Bartman, "Removing SO_2 from Flue Gas," *Chem. Eng. Prog.*, *65*(10), 74 (1969).
474. H. L. Falkonberry and A. V. Slack, "SO_2 Removal by Limestone Injection," *Chem. Eng. Prog.*, *65*(12), 61 (1969).
475. J. J. Humphries, S. B. Zdonik, and E. J. Parsi, *Economic Factors in the Capital and Operating Expenses of the Stone and Webster Ionics SO_2 Removal and Recovery Process*, AICHE 63rd Annual Meeting, Illinois, 1970, Paper 2e.

476. *British Chemical Engineering*, "BCE Review Guide," *16*(1), 15 (1971).
477. *British Chemical Engineering*, "BCE Process Scan—Sulphur CEGB Recovery Process," *14*(8), 1043 (1969).
478. D. P. McDonald, "Recovery of Materials," *Br. Chem. Eng.*, *16*(1), 69 (1971).
479. *Chemical and Process Engineering*, "Process Costs—Sulphuric Acid," *46*(5), 257 (1965).
480. *Chemical and Process Engineering*, "Project News," *46*(7), 374 (1965).
481. *Canadian Chemical Processing*, "How to Bring H_2SO_4 Process up to Date," *49*(12), 61 (1965).
482. P. T. Shannon, A. I. Johnson, C. M. Crowe, T. W. Hoffmans, A. E. Hamielec, and D. R. Woods, "Computer Simulation of Sulphuric Acid Plant," *Chem. Eng. Prog.*, *62*(6), 49 (1966).
483. M. Kishi, F. Kadota, Y. Kaminura, and M. Suehiro, "Optimum Design for Convertor Group of Sulphuric Acid Plant," *Mitsubishi Heavy Ind. Tech. Rev.*, *4*(2), 144 (1967).
485. *Chemical Age of India*, "Hindustan Zinc Limited," *19*(3), 161 (1968).
486. *Chemical and Process Engineering*, "Project News," *49*(4), 11 (1968).
487. *Chemical and Process Engineering*, "Project News," *49*(6), 11 (1968).
488. J. M. Connor, "Economics of Sulphuric Acid Manufacture," *Chem. Eng. Prog.*, *64*(11), 59 (1968).
489. *Chemical and Process Engineering*, "Project News," *49*(9), 11 (1968).
490. T. D. Wheelock and D. R. Boylon, "Sulphuric Acid from Calcium Sulphate," *Chem. Eng. Prog.*, *64*(11), 87 (1968).
491. *British Chemical Engineering*, "Process Scan—Sulphuric Acid and Cement," *14*(4), 408 (1969).
492. *Chemical Process Engineering*, "Project News," *50*(9), 15 (1969).
493. K. V. Nair and J. Narayanamurthy, "Process Development and Pilot Plant Studies on the Recovery of the Waste Sulphuric Acid from a Titanium Dioxide Plant," *Chem. Age India*, *21*(2), 102 (1970).
494. *British Chemical Engineering*, "Project News," *45*(6), 745 (1970).
495. *Chemical and Process Engineering*, "Project News," *51*(6), 11 (1970).
496. *Chemical and Process Engineering*, "Project News," *51*(8), 11 (1970).
497. *Chemical and Process Engineering*, "Project News," *51*(9), 11 (1970).
498. *Chemical and Process Engineering*, "Project News," *51*(10), 11 (1970).
499. *Chemical and Process Engineering*, "CPE Project News," *52*(2), 9 (1971).
500. *Chemical and Process Engineering*, "Project News," *52*(5), 9 (1971).
501. *British Chemical Engineering*, "Project News," *15*(7), 863 (1971).
502. *Chemical and Process Engineering*, "CPE Project News," *52*(10), 9 (1971).
503. J. Lastowiecky, "Sulphur Supply and Price Forecasts up to 1980," *Chem. Age India*, *22*(12), 935 (1971).
504. *Chemical and Process Engineering*, "Project News," *52*(12), 9 (1971).
505. W. A. Graham, "Alkylation Integrates Acid Plant," *Hydrocarbon Process.*, *51*(8), 87 (1972).
506. N. G. Weston, "Sulphur Trioxide Processing," *Chem. Process Eng.*, *48*(3), 61 (1967).
507. *Chemical and Process Engineering*, "Synthesis Gas Manufacture—Costs," *48*(12), 48 (1967).
508. *British Chemical Engineering*, "Project News," *12*(11), 1673 (1967).
509. *Chemical and Process Engineering*, "Coal Future Source of Synthetic Fuels?" *52*(5), 3 (1971).
510. S. Kato, "THF from Dichlorobutane," *Chem. Eng. (N. Y.)*, *70*(3), 50 (1972).
511. P. H. Dundas and M. L. Thorpe, "Titanium Dioxide Production by Plasma

Processing," *Chem. Eng. Prog.*, *66*(10), 66 (1970).

512. A. S. Banciu, "Titanium Dioxide—Process Survey," *Chem. Process Eng.*, *48*(7), 9 (1972).

513. S. D. Gupta and R. K. Srivastava, "Scheme for the Manufacture of Toothpaste in the Small Scale Sector," *Chem. Age India*, *22*(8), 564 (1971).

514. R. Taggiasco, "Simplified Process Reduces Urea's Production Costs," *Chem. Eng. (N. Y.)*, *73*(16), 52 (1966).

515. *Chemical Age of India*, "The Snam Progetti Process for Low Cost Production of Urea," *18*(11), 843 (1967).

516. A. J. Payne and J. A. Canner, "Urea," *Chem. Process Eng.*, *50*(5), 81 (1969).

517. *Chemical Age of India*, "Report on One of Asia's Largest Urea Facilities," *20*(12), F1 (1969).

518. *Chemical Age of India*, "I Visit the World's Largest Urea Plant," *20*(12), F11 (1969).

519. *Chemical Age of India*, "New Process," *20*(9), 820 (1969).

520. *Chemical and Process Engineering*, "Vinyl Acetate via Ethylene," *48*(3), 71 (1967).

521. *British Chemical Engineering*, "Process Scan—Vinyl Acetate, Vapour Phase Process," *15*(4), 604 (1970).

522. R. B. Stobaugh, W. C. Allen, Jr., and V. R. H. Stermberg, "Vinyl Acetate: How, Where, Who—Future," *Hydrocarbon Process.*, *51*(5), 153 (1972).

523. E. F. Edwards and T. Weaver, "New Route to Vinyl Chloride," *Chem. Eng. Prog.*, *61*(1), 21 (1965).

524. P. H. Spitz, "Effects of Plant Size and Changing Technology—Vinyl Chloride Economics," *Chem. Eng. Prog.*, *64*(3), 19 (1968).

525. R. T. Thampy, "Development of Chemical Intermediates and Monomers for the Plastic Industry. Part V—Vinyl Chloride," *Pet. Hydrocarbons*, *3*(1), 149 (1968).

526. *British Chemical Engineering*, "Process Scan," *14*(5), 608 (1969).

527. M. C. Forbes and P. A. Witt, "Estimate Cost of Waste Disposal," *Hydrocarbon Process.*, *44*(8), 153 (1965).

528. J. Cunetta, "New York City Builds 170 Million Treatment Plant," *Water Wastes Eng.*, *4*(9), 80 (1967).

529. B. J. Ramey, "Deep-down Waste Disposal," *Mech. Eng.*, *90*(8), 28 (1968).

530. W. W. Eckenfelder, Jr., and D. L. Ford, "Economics of Waste Water Treatment," *Chem. Eng. (N. Y.)*, *76*(18), 109 (1969).

531. D. Eynon, "Waste Water Treatment and Reuse of Treated Sewage as an Industrial Water Supply," *Chem. Eng. (London)*, *235*, CE 6 (1970).

532. P. C. G. Isaac, "Disposal of Liquid Effluents from the Chemical Industry," *Chem. Eng. (London)*, *239*, CE 165 (1970).

533. G. Windecker and F. C. Engelmann, "Automatic Refuse Disposal Plant," *Chem. Process Eng.*, *52*(6), 95 (1971).

534. G. J. Crits, "Economic Factors in Water Treatment," *Ind. Water Treatment*, *8*(8), 22 (1971).

535. D. K. B. Thistlethwayte, "Economic Aspects of Waste Disposal," *Aust. Chem. Eng.*, *13*(3), 9 (1972).

536. J. C. Cooper and D. G. Hager, "Water Reclamation with Activated Carbon," *Chem. Eng. Prog.*, *62*(10), 85 (1966).

537. W. G. McGlasson, L. J. Thibodeaux, and H. F. Berger, "Potential Uses of Activated Carbon for Wastewater Renovation," *Tappi*, *49*(12), 521 (1966).

538. L. F. Rehm and W. H. Plautz, "Various Costs of Clarification, Filtration and Lime Softening," in *Illinois University College of Engineering, Sanitary Engineering Conference, 8th Proceedings*, 1966, p. 109.

539. L. Koenig, "Cost of Water Treatment by Coagulation, Sedimentation and Rapid Sand Filtration," *Am. Water Works Assoc. J.*, *59*(3), 290 (1967).

540. M. Brooke, "Process Plant Utilities—Water," *Chem. Eng. (N. Y.)*, *77*(27), 135 (1970).

541. J. E. Browning, "New Water—Cleanup Roles for Powdered Activated Carbon," *Chem. Eng. (N. Y.)*, *79*(4), 36 (1972).

542. *Petroleum and Hydrocarbons* "New Processes," *3*(2), 73 (1968).

543. *Chemical Age of India*, "New Processes—White Oils," *23*(7), 610 (1972).

544. D. L. McKay, G. H. Dale, and D. C. Tabler, "*para*-Xylene via Fractional Crystallization," *Chem. Eng. Prog.*, *62*(11), 104 (1966).

545. *Chemical and Process Engineering*, "Advances in Monomer Production," *52*(3), (1971).

546. *Chemical Age of India*, "The Small Scale Sector—Production of Yeasts from Molasses," *23*(1), 168 (1972).

547. *Chemical Age of India*, "New Processes," *19*(8), 683 (1968).

548. H. A. Stewart and J. L. Heck, "Pressure Swing, Adsorption," *Chem. Eng. Prog.*, *65*(9), 78 (1969).

549. F. W. Lohrisch, "How to Find Optimum Agitator Speed," *Hydrocarbon Process.*, *44*(5), 217 (1965).

550. *British Chemical Engineering*, "Cost Data File," *13*(7), 1017 (1968).

551. *Chemical Process Engineering*, "Project News," *50*(11), 17 (1969).

552. *British Chemical Engineering*, "Cost Data File," *15*(9), 1207 (1970).

553. *British Chemical Engineering*, "Cost Data File," *15*(10), 1347 (1970).

554. *British Chemical Engineering*, "Cost Data File," *15*(11), 1467 (1970).

555. *British Chemical Engineering*, "Cost Data File," *16*(3), 408 (1971).

556. R. L. Adams, "High Temperature Cloth Collectors," *Chem. Eng. Prog.*, *62*(4), 66 (1966).

557. J. Nitschke, "Investment Costs and Utilization, Criteria to Determine Optimum Storage Bin Size (Investitionskosten und Nutzungsgrad als Kriterium zur Bestimmung optimaler Lagerfachgrossen")," *Fordern Heben*, *20*(17), 965 (1970).

558. E. Fehr, "Der moderne celbefeuerte Heizkessel," *Schweiz. Arch.*, *31*(4), 126 (1965).

559. J. Van Loosen, "What You Need to Know about Boilers," *Hydrocarbon Process.*, *47*(6), 95 (1968).

560. *Hydrocarbon Processing*, "Centrifugals Cut Ammonia Costs," *45*(5), 179 (1966).

561. J. E. Flood, H. F. Porter, and F. W. Rennie, "Centrifugal Equipment," *Chem. Eng. (N. Y.)*, *73*(12), 190 (1966).

562. F. W. Keith and T. H. Little, "Centrifuges in Water and Waste Treatment," *Chem. Eng. Prog.*, *65*(11), 77 (1969).

563. D. Carlton-Jones and H. B. Schneider, "Tall Chimneys," *Chem. Eng. (N. Y.)*, *75*(22), 166 (1968).

564. R. C. Raynor and E. F. Porter, "Thickeners and Clarifiers," *Chem. Eng. (N. Y.)*, *73*(12), 198 (1966).

565. D. A. Dahlstrom and C. F. Cornell, "Thickening and Clarification," *Chem. Eng. (N. Y.)*, *78*(4), 63 (1971).

566. L. C. Waterman, "Electrical Coalescer," *Chem. Eng. Prog.*, *61*(10), 51 (1965).

567. D. Hallock, "Quick Method for Centrifugal Compressor Estimates," *Hydrocarbon Process.*, *44*(116), (1965).

568. N. Beck, "Purchase of Compressor Installation," *Chem. Process Eng.*, *47*(4), 172 (1966).

569. G. E. Petty and G. E. Handwerk, "Process Refrigeration Compressor—

Selection and Comparative Costs," *ASME Meeting*, September 17–20, 1967, Paper 67.

570. A. Antonelli and F. Fraschetti, "New Developments in Centrifugal Compressors for Synthesis Services," *Chem. Age India*, *18*(11), 822 (1967).

571. *British Chemical Engineering*, "Cost Data File," *13*(1), 117 (1968).

572. D. J. La Cerda, "Better Instrument and Plant Air Systems," *Hydrocarbon Process.*, *47*(6), 107 (1968).

573. *Chemical Process Engineering*, "Project News—2," *50*(10), 17 (1969).

574. *British Chemical Engineering*, "Project News," *14*(12), 1667 (1969).

575. *British Chemical Engineering*, "Cost Data File," *16*(3), 389 (1970).

576. S. A. Bresler, "Guide to Trouble-Free Compressors," *Chem. Eng. (N. Y.)*, *77*(12), 161 (1970).

577. *Chemical and Process Engineering*, "Project News—2," *50*(12), 17 (1969).

578. *British Chemical Engineering*, "Project News," *15*(9), 1121 (1970).

579. D. G. McAllister, "Process Plant Utilities—Air," *Chem. Eng. (N. Y.)*, *77*(27), 138 (1970).

580. J. D. Stafford, Jr., "The High Pressure Centrifugal Compressor Loop," *Chem. Eng. Prog.*, *67*(4), 54 (1971).

581. A. Hussain, "Application of Computers in Chemical Industry," *Chem. Age India*, *19*(3), 217 (1968).

582. E. F. Schagrin, "How Much Do Minicomputer Systems Cost?" *Chem. Eng. (N. Y.)*, *78*(7), 103 (1971).

583. T. M. Stout, "Justifying Process Control Computers: Selection and Cost," *Chem. Eng. (N. Y.)* (Desk Book Issue), *79*(20), 89 (1972).

584. J. A. Lawrence and A. A. Buster, "Guide to Trouble-Free Plant Operation . . . Computer Process Interface," *Chem. Eng. (N. Y.)*, *79*(14), 102 (1972).

585. R. F. Uncles, "What Engineers Should Know about Containers," *Chem. Eng. (N. Y.)*, *73*(14), 113 (1966); *73*(15), 167 (1969); *73*(17), 151 (1966).

586. B. G. Liptak, "Control Panel Costs," *Chem. Eng. (N. Y.)*, *73*(21), 83 (1970).

587. L. M. Shipman, "Optimum Design of Ammonia Quench Converters," *Chem. Eng. Prog.*, *64*(5), 59 (1968).

588. J. Clerk, "How to Compare Costs of Air vs. Water Cooling—CE Cost File," *Chem. Eng. (N. Y.)*, *72*(1), 100 (1965).

589. C. L. Williams, Jr. and R. D. Damron, "Which Costs Cheaper—Water or Air," *Hydrocarbon Process.*, *44*(2), 139 (1965).

590. H. L. Von Cube, "Wirtschaftlich optimale Konstruktionen Von Waermacaustauschern in der Kaeltetechnik," *Kaeltetechnik*, *17*(3), 90 (1965).

591. F. W. Lohrisch, "Heat Exchanger Air Cooled—How Many Tube Rows for Air Cooled Exchangers," *Hydrocarbon Process.*, *45*(6), 137 (1966).

592. R. T. Mathews, "Economic Applications of Air Cooling to Process Industries," *Br. Chem. Eng.*, *13*(10), 1425 (1968).

593. A. C. Kapadia, "Selecting a Cooling System," *Chem. Age India*, *19*(10), 932 (1968).

594. *Chemical and Process Engineering*, "Air Cooler Estimation," *5*(8), 70 (1969).

595. C. M. B. Russel and J. Tiley, "Air Cooler Estimation," *Chem. Process Eng.* (Heat Transfer Survey), *70*, (1969).

596. L. Gazzi and R. Pasero, "Process Cooling Systems—Selection," *Hydrocarbon Process.*, *49*(10), 83 (1970).

597. L. G. Niccoli, R. T. Jaske, and P. A. Witt, "Systems Costs Say Optimise Cooling," *Hydrocarbon Process.*, *49*(10), 97 (1970).

598. W. Kals, "Wet Surface Air Coolers," *Chem. Eng.* (*N. Y.*), *78*(17), 90 (1971).

599. I. Wigham, "Designing Optimum Cooling Systems," *Chem. Eng.* (*N. Y.*), *78*(18), 95 (1971).

600. J. E. Lerner, "Simplified Air Cooler Estimation," *Hydrocarbon Process.*, *51*(1), 93 (1972).

601. R. W. Maze, "Practical Tips on Cooling Tower Sizing," *Hydrocarbon Process.*, *46*(2), 123 (1967).

602. W. J. Pietrucha, "Evaluating Cooling Tower Pumping Schemes," *Combustion*, *38*(11), 33 (1967).

603. P. M. Paige, "Costlier Cooling Towers Require New Approach to Water-Systems Design," *Chem. Eng.* (*N. Y.*), *74*(14), 93 (1967).

604. A. Rabb, "Are Dry Cooling Towers Economical?" *Hydrocarbon Process.*, *47*(2), 122 (1968).

605. A. R. Thompson, "Cooling Towers," *Chem. Eng.* (*N. Y.*), *75*(22), 100 (1968).

606. D. L. Chatfield and D. F. Streeton, "Cost Analysis of Large Evaporative Type Cooling Towers," *Kerntechnic*, *11*(11), 649 (1969).

607. *Chemical Engineering*, "Basic Techniques (Section II of Water Pollution Control)," *77*(9), 63 (1970).

608. J. F. Roesler and R. G. Eilers, "Simulation of Ammonia Stripping from Wastewater," *ASCE J., Sanit. Eng. Div.*, *97*(S.A. 3, Paper 8182), 269 (1971).

609. A. Zanker, "Estimating Cooling Tower Costs from Operating Data," *Chem. Eng.* (*N. Y.*), *79*(13), 118 (1972).

610. A. M. Kunesch, "Mechanical Draught Cooling Towers," *Chem. Eng.* (*N. Y.*), *253*, 337 (1971).

611. E. E. Dybdal, "Subject Index of CE Cost File 1958 to 1966," *Chem. Eng.* (*N. Y.*), *73*(25), 166 (1966).

612. H. Weiss, "Fahrbare Orossbrechanlagren—Entwicklung, Bauformen, Kosten," *Fordern Heben*, *16*(10), 785 (1966).

613. M. M. Kirk, "Cranes, Hoists and Trolleys," *Chem. Eng.* (*N. Y.*), *74*(5), 168 (1967).

614. V. E. Williams, "Cryogenics," *Chem. Eng.* (*N. Y.*), *77*(25), 92 (1970).

615. *Chemical and Process Engineering*, "Cryogenic Plant Construction," *52*(2), 53 (1971).

616. J. E. Powers, "Recent Advances in Crystallization," *Hydrocarbon Process.*, *45*(12), 97 (1966).

617. H. Popper, "Crystalliser Costs for Fertiliser and Fine Chemicals," *Br. Chem. Eng.*, *73*(13), 246 (1966).

618. *Chemical Engineering*, "Crystalliser Costs for Fertilisers and Fine Chemicals," *73*(13), 246 (1966).

619. *Chemical Process Engineering*, "Project News–2," *50*(9), 17 (1969).

620. *British Chemical Engineering*, "Project News," *15*(12), 1521 (1970).

621. R. Billet and L. Raichle, "Optimizing Method for Vacuum Rectification—Part I," *Chem. Eng.* (*N. Y.*), *74*(4), 145 (1967).

622. *British Chemical Engineering*, "Cost Data File," *12*(6), 921 (1967).

623. *British Chemical Engineering*, "Cost Data File," *12*(7), 1119 (1967).

624. *British Chemical Engineering*, "Cost Data File," *12*(9), 1419 (1967).

625. R. Billet, "Cost Optimisation of Towers," *Chem. Eng. Prog.*, *66*(1), 1 (1970).

626. M. V. Winkle and W. G. Todd, "Minimizing Distillation Costs via Graphical Techniques," *Chem. Eng.* (*N. Y.*), *79*(5), 105 (1972).

627. R. F. Sommerville, "New Method Gives Quick, Accurate Estimate of Distillation Costs," *Chem. Eng.* (*N. Y.*), *79*(9), 71 (1972).

628. J. L. Bellis, "Adjustable Speed Drives for Farrel Continuous Mixers," in *IEEE-*

Electrical Engineering Problems in Rubber and Plastics Industries, Annual Conference, April 1–2, 1968, 20th publication, C24-IGa, p. 47.

629. *British Chemical Engineering*, "Cost Data File," *11*(7), 727 (1966).
630. *British Chemical Engineering*, "Cost Data File," *12*(5), 747 (1967).
631. *British Chemical Engineering*, "Cost Data File," *12*(8), 1253 (1967).
632. *British Chemical Engineering*, "Cost Data," *12*(10), 1619 (1967).
633. D. Papee, C. Menier, A. Bellier, and F. Jouanneault, "Optimize Alumina Gas Drying Systems," *Hydrocarbon Process.*, *46*(10), 142 (1967).
634. C. M. Jones, "Fast Drying at One-Tenth Cost," *Ceram. Ind.*, *90*(6), 56 (1968).
635. F. Rueb, "Drying Plants for Shaped Ceramic Components," (Trocknungsanlagen fuer keramische Farmteile), *Keram. Z.*, *21*(2), 98 (1969).
636. *British Chemical Engineering*, "Cost Data File," *14*(4), 537 (1969).
637. *British Chemical Engineering*, "Cost Data File," *14*(8), 1115 (1969).
638. D. Noden, "Industrial Dryers—Selection, Sizing and Costs," *Chem. Process Eng.*, *50*(10), 67 (1969).
639. *British Chemical Engineering*, "Cost Data File," *15*(1), 111 (1970).
640. K. S. Misra, "Design and Performance of Fixed Bed Adsorption Driers," *Chem. Age India*, *20*(9), 808 (1969).
641. D. Noden, "Trend Towards Use of Dispersion Dryers," *Chem. Process Eng.*, *53*(4), 48 (1972).
642. W. E. Clark, "Fluid-Bed Drying," *Chem. Eng. (N. Y.)*, *74*(6), 177 (1967).
643. S. N. Miner, "Freeze Drying," *ASHRAE J.*, *7*(6), 92 (1965).
644. S. S. Kalbarg, "Freeze Drying," *Indian Chem. J.*, *4*(3), 66 (1969).
645. H. B. Nielson and E. H. Rasmussen, *Cost Versus Performance in Spray Drying*, Nichols Engineering and Research Corporation, 150 William Street, New York, 10038.
646. E. M. Cook, "Estimating Spray Drying Costs," *Chem. Eng. Prog.*, *62*(6), 93 (1966).
647. H. B. Nielson and E. H. Rasmussen, "Cost versus Performance in Spray Drying," in *European Federation of Chemical Engineering, 4th Congress, London*, 1966, Paper 4–7, p. 55.
648. *Chemical Engineering*, "Estimating Costs of Process Dryers," *73*(3), 101 (1966).
649. V. A. Turkot, N. C. Aceto, E. F. Schoppet, and J. C. Craig, Jr., "Continuous Vacuum Drying of Whole Milk Foam," *Food Eng.*, *41*(8), 59 (1969); *41*(9), 97 (1969).
650. J. S. Munson, "Dry Mechanical Collectors," *Chem. Eng. (N. Y.)*, *75*(22), 147 (1968).
651. G. D. Sargent, "Dust Collection Equipment," *Chem. Eng. (N. Y.)*, *76*(2), 130 (1969).
652. J. L. Smith and H. A. Smell, "Selecting Dust Collectors," *Chem. Eng. Prog.*, *64*(1), 60 (1968).
653. *British Chemical Engineering*, "Project News," *15*(4), 453 (1970).
654. R. W. Sickles, "Electrostatic Precipitators," *Chem. Eng. (N. Y.)*, *75*(22), 156 (1968).
655. W. R. Nichols, "Shell Construction of Present Day Precipitators," in *TAPPI, 23rd Engineering Conference, Houston*, November 3–7, 1968, Preprint Paper 8–1.
656. *Chemical Engineering*, "Basis Technology (Section IV of Air Pollution Control)," *77*(9), 165 (1970).
657. S. D. Gupta, "Special Report Filtech/71 Olympia London (Sept. 28–Oct. 1)," *Chem. Age India*, *22*(10), 753 (1971).
658. K. M. Guthrie, "Capital Cost Estimating," *Chem. Eng. (N. Y.)*, *76*(6), 114 (1969).

659. G. J. Rao, "Import Substitution," *Pet. Hydrocarbons*, 6(1), 107 (1971).

660. N. Parker, "Equipment and Economics (Part 2 of Agitated Thin Film Evaporators)," *Chem. Eng. (N. Y.)*, 72(19), 179 (1965).

661. P. F. Sullivan and O. Pelegrin, "Thin Film Evaporators—State of Art," *Semicond. Prod. Solid State Technol.*, 8(12), 19 (1965).

662. A. L. Carter and R. R. Karybill, "Low Pressure Evaporation," *Chem. Eng. Prog.*, 62(2), 99 (1966).

663. E. C. Hise, "Simplified Method for Making Preliminary Thermal and Economic Analysis of Vertical-Tube and Multistage-Flash Multieffect Evaporator Plants," *Chem. Eng. Prog. Symp. Ser.*, 64(82), 43 (1968).

664. *Chemical Engineering*, "Evaporator Tackles Wastewater Treatment," 79(6), 68 (1972).

665. E. V. Yashke et al, "Glass Wool Filters for Removal of Sulphuric Acid Mist from Gases," *Khim. Prom.*, 1965(3), 36.

666. A. Martre and M. Endier, "Note on Economics of Filter Fabrication," *Powder Metall.*, 8(16), 234 (1965).

667. M. E. O. 'K. Trowbridge and D. Bradley, "Cost Conscious Selection of Solid–Liquid Separation Equipment," in *European Federation of Chemical Engineering*, 4th Congress, London, 1966, Paper 4–5, p. 35.

668. D. G. Hill, "Evaluation of Costs of Owning and Operating Air Filters," *Filtr. Sep.*, 4(4), 297 (1967).

669. J. G. Mulvany, "Cost and Performance of Filtration and Separation Equipment—Membrane Filtration in Critical Systems," *Filtr. Sep.* 5(1), 53 (1968).

670. D. B. Purchas, "Cost and Performance of Filtration and Separation Equipment—Cost Factors in Selection of Filter Media," *Filtr. Sep.* 5(1), 57 (1968).

671. D. A. Bennel, "Cost and Performance of Filtration and Separation Equipment—Air Filter," *Filtr. Sep.*, 5(2), 150 (1968).

672. J. A. Hooton and C. M. Thomas, "Cost and Performance of Filtration and Separation Equipment—Filter Processes," *Filtr. Sep.*, 5(3), 238 (1968).

673. W. T. Cosly and G. Punch, "Cost and Performance of Filtration and Separation Equipment—Dust Filters and Collectors," *Filtr. Sep.*, 5(3), 252 (1968).

674. G. H. Duffield, "Cost and Performance of Vacuum Filters," *Filtr. Sep.*, 5(4), 347 (1968).

675. G. B. Cherry, A. A. H. Moss, and E. Scott, "Comparison of fixed and Variable Volume Chamber Mechanised Filters for the Isolation of Dyestuffs," *Chem. Eng. (London)*, 237, CE 95 (1970).

678. R. J. Mattson and V. J. Tomsic, "Improved Water Quality," *Chem. Eng. Prog.*, 65(1), 62 (1969).

679. *Chemical Processing*, "Bacteria-Removal Filters Save Plant $13,300," 35(5), 14 (1972).

680. V. Cariati, "Costs: Bare or Finned Exchange Tubes?" *Hydrocarbon Process.*, 47(2), 106 (1968).

681. T. E. Duckworth, "Froth Flotation—Industrial Applications and Economics," *Chem. Process Eng.*, 47(5), 206 (1966).

682. J. F. Blumenfeld, "Engineering and Economic Considerations in Evaluation of Glass Melting Furnaces," *Am. Ceram. Soc. Bull.*, 46(11), 1079 (1967).

683. J. L. De Blieck and A. G. Gossens, "Optimise Olefine Cracking Coils," *Hydrocarbon Process.*, 50(3), 76 (1971).

684. S. Menicatti and L. Cappiell, "Find Optimum Furnace Efficiency," *Hydrocarbon Process.*, 51(9), 226 (1972).

685. J. R. F. Alonso, "Estimating the Costs of Gas Cleaning Plants," *Chem. Eng. (N. Y.)*, *78*(28), 86 (1971).

686. *British Chemical Engineering*, "Project News," *15*(5), 609 (1970).

687. *Chemical and Process Engineering*, "Fertiliser Granulation Made Simple," *52*(9), 5 (1971).

688. A. M. Lloyd, J. A. W. Macintosh, and W. Z. Paduch, "Economics of Process Heating with Organic Heat Transfer Media," *Chem. Process Eng.*, *46*(8), 445 (1965).

689. J. T. Gallagher, "Estimating Heater Costs," *Chem. Eng. (N. Y.)*, *74*(15), 232 (1967).

690. P. V. Wiesenthal and H. W. Cooper, "Guide to Economics of Fired Heater Design," *Chem. Eng. (N. Y.)*, *77*(7), 104 (1970).

691. R. F. Angel, "Economics of Electric Surface Heating," *Chem. Process Eng.*, *Heat Trans. Surv.*, 46 (1971).

692. W. R. Blackburn, "How to Estimate Extra Installation Costs for Multiple Heat Exchanger, Stack Alternatives," *Pet./Chem. Eng.*, *38*(3), 21 (1966).

693. A. Hillard, "Costs of Graphite Heat Exchanger," *Chem. Process Eng.*, *47*(9), 96 (1966).

694. *British Chemical Engineering*, "Cost Date File," *11*(8), 863 (1966).

695. *Chemical and Process Engineering*, "Data Survey—U Tube Heat Exchangers," *48*(3), 98 (1967).

696. R. R. Hood, "Designing Heat Exchangers in Teflon," *Chem. Eng. (N. Y.)*, *74*(11), 181 (1967).

697. *British Chemical Engineering*, "Cost Data File," *13*(3), 403 (1968).

698. V. A. Vedyaev and E. P. Volkov, "Optimizatsiya osnornykh parametrov dlya drukhstupenchatykh rekuparativnykh teploo mennikov," *Teploenergetika*, *1968*(3), 36; *Therm. Eng.*, *1968*(3), 49.

699. *British Chemical Engineering*, "Cost Data File," *13*(4), 551 (1968).

700. *British Chemical Engineering*, "Cost Data File," *13*(6), 851 (1968).

701. R. A. Crane and A. Mayer, "Computer Coded Performing Parametric Studies On Liquid–Metal Fast Breeder Reactors," *ASME*, Paper 68-VA/NE-6,5 (for meeting of December 1–5, 1968).

702. *British Chemical Engineering*, "Cost Data File," *14*(1), 79 (1969).

703. *British Chemical Engineering*, "Cost Data File," *14*(1), 209 (1969).

704. *British Chemical Engineering*, "Project News," *14*(5), 413 (1969).

705. R. Martin, "Fabrication Shop Check List Guides Kellogg," *Pet. Chem. Eng.*, *41*(4), 13 (1969).

706. *British Chemical Engineering*, "Cost Data File," *14*(5), 707 (1969).

707. S. K. Jenssen, "Heat Exchanger Optimisation," *Chem. Eng. Prog.*, *65*(7), 59 (1969).

708. S. N. Shah, "Product Standardization of Steel and Tube Heat Exchangers," *Chem. Age India*, *20*(8), 725 (1969).

709. *Chemical and Process Engineering*, "Project News–2," *50*(8), 15 (1969).

710. *British Chemical Engineering*, "Cost Data File," *14*(9), 1259 (1969).

711. D. Stuhlbarg, "How to Find Optimum Exchanger Size for Forced Circulation," *Hydrocarbon Process.*, *49*(1), 149 (1970).

712. *Chemical and Process Engineering*, "Project News–2," *51*(2), 13 (1970).

713. D. K. Palit, "How to Optimize Heat Exchangers," *Chem. Age India*, *21*(3), 255 (1970).

714. S. Katel and P. R. Jones, "Programmes for the Price Optimum Design of Heat Exchangers," *Br. Chem. Eng.*, *15*(4), 491 (1970).

715. *Chemical and Process Engineering*, "Project News—2," *51*(4), 17 (1970).

716. B. Jamin, "Exchanger Stages Solved by Graph," *Hydrocarbon Process.*, *49*(7), 137 (1970).

717. J. M. Brooke, "Corrosion Inhibitor Economics: How to Calculate," *Hydrocarbon Process.*, *49*(9), 299 (1970).

718. *British Chemical Engineering*, "Cost Data File," *16*(6), 529 (1971).

719. A. R. Tarrel, H. C. Lim, and L. B. Koppel, "Finding the Economically Optimum Heat Exchanger," *Chem. Eng. (N. Y.)*, *78*(22), 79 (1971).

720. D. L. Peters, "Efficient Programming for Cost—Optimised Heat Exchanger Design," *Chem. Eng. (London)*, *259*, 98 (1972).

721. E. A. Vitunac, "Economic Considerations in Hopper Design," *Chem. Eng. Prog.*, *62*(11), 69 (1966).

722. J. G. Hays, "Cost–Performance Summary of Air Pollution Control Equipment Used for Rule 66 Compliance," *SAE*, Paper 670812, (for meeting of October 2–6, 1967).

723. L. Brewer, "Fume Incineration," *Chem. Eng. (N. Y.)*, *75*(22), 160 (1968).

724. P. A. Witt, Jr., "Disposal of Solid Wastes," *Chem. Eng. (N. Y.)*, *78*(22), 62 (1971).

725. J. F. Hornor, "Instrument Development: Whose Responsibility?" *Chem. Eng. (N. Y.)*, *72*(12), 157 (1965).

726. S. E. Roth, "Instrument Maintenance Cost Trends," *Hydrocarbon Process.*, *44*(5), 199 (1965).

727. O. J. Palmer, "Which Air or Electronic Instruments?" *Hydrocarbon Process.*, *44*(5), 241 (1965).

728. A. L. Stott, *Influence of Computers on Cost Instrumentation Systems*, Institution of Electrical Engineers, London, England (Conference Publication No. 43), 1968, p. 148).

729. *Chemical Age of India*, "The Instrument Industry—Need for Planned Development," *20*(11), 885 (1969).

730. F. Church, "Economics of Instrument Selection," *Meas. Control*, *3*(1), 10 (1970).

731. B. G. Liptak, "Cost of Process Instruments," *Chem. Eng. (N. Y.)*, *77*(19), 60 (1970).

732. B. G. Liptak, "Cost of Viscosity, Weight, and Analytical Instruments," *Chem. Eng. (N. Y.)*, *77*(20), 175 (1970).

733. S. D. Bhasin, "Problems and Progress of Instrumentation in Chemical Industries in Developing Countries," *Chem. Age India*, *21*(9), 834 (1970).

734. B. G. Liptak, "Safety Instruments and Control Valves Costs," *Chem. Eng. (N. Y.)*, *77*(24), 94 (1970).

735. *Rubber and Plastics Age*, "Cost of Thermal Insulation," *46*(1), 43 (1965).

736. D. J. Mistrot, "Flexible Ceramic Fiber Blanket Solves Hydro-Cracker Insulation Problem," *Pet./Chem. Eng.*, *37*(11), 54 (1965).

737. W. H. Dodge, "Optimum Insulation Thickness by Computer," *Chem. Eng. (N. Y.)*, *73*(11), 182 (1966).

738. S. Menicatti, "Check Tank Insulation Economics," *Hydrocarbon Process.*, *48*(4), 133 (1969).

739. G. L. Wells, "The Calculation of the Economic Thickness of Insulation," *J. Inst. Fuel*, *44*(369), 609 (1971).

740. F. X. Pollio, R. Kumin, and J. W. Petralia, "Treat Sour Water by Ion Exchange," *Hydrocarbon Process.*, *48*(5), 124 (1969).

741. D. G. Downing, "Calculating Minimum-Cost Ion Exchange Units," *Chem. Eng. (N. Y.)*, *72*(25), 170 (1965).

742. G. B. Remmy, Jr., "Cost Analysis, Bell-Type or Tunnel Kiln," *Ceram. Age,*

86(9), 32 (1970).

743. M. Kirk, "Cost of Mist Eliminators," *Chem. Eng.* (*N. Y.*), *73*(20), 240 (1966).

744. L. L. Scheiner, "Processor's Guide to Mixing, Blending and Compounding Equipment," *Plast. Technol.*, *13*(6), 39 (1965).

745. *British Chemical Engineering*, "Cost Data File," *12*(12), 1915 (1967).

746. N. Harnby, "Unit Costs of Mixing Dry Solids," *Chem. Process Eng.*, *49*(12), 53 (1968).

747. M. J. Griffins, "Air Vortex Mixing of Solids," *Chem. Process Eng.*, *50*(3), 69 (1969).

748. T. R. Goshorn, "Powder Mixing of Elastomers," *SPE-Plastic Powders III, Tech. Papers 50–4* (Meeting of March 12, 1972).

749. *Chemical and Process Engineering*, "Project News," *51*(3), 15 (1970).

750. C. R. Olson and E. S. McKeloy, "How to Specify and Cost Estimate Electric Motors," *Hydrocarbon Process.*, *46*(10), 118 (1967).

751. C. L. Ward, "Estimating Molding Parts Costs," *SPE, Reg. Tech. Conf. Tech. Paper* for meeting (*Connecticut Sec.*) (Meeting of December 3–4, 1969), pp. 11–25.

752. R. Billet and L. Raichle, "Optimizing Method for Vacuum Rectification—Part II," *Chem. Eng.* (*N. Y.*), *74*(5), 149 (1967).

753. R. Billet, "Development and Progress in Design and Performance of Valve Trays," *Br. Chem. Eng.*, *14*(4), 489 (1969).

754. *British Chemical Engineering*, "Economic Evaluation of Tower Packing," *14*(12), 1641 (1969).

755. R. Billet, "Cost Optimization of Towers," *Chem. Eng. Prog.*, *66*(1), 41 (1970).

756. J. S. Eckert, "Selecting the Proper Distillation Column Packing," *Chem. Eng. Prog.*, *66*(3), 39 (1970).

757. J. R. Fair, "Comparing Trays and Packings," *Chem. Eng. Prog.*, *66*(3), 45 (1970).

758. P. C. Wygren, and G. K. S. Connolly, "Selecting Vacuum Fractionation Equipment," *Chem. Eng. Prog.*, *67*(3), 49 (1971).

759. R. Billet, "Gauze-Packed Columns for Vacuum Distillation," *Chem. Eng.* (*N. Y.*), *79*(4), 68 (1972).

766. J. H. Anderson, "Liquefied-Methane Pipeline—Next Gas Transmission Step," *Oil Gas J.*, *63*(6), 74 (1965).

767. F. C. Jelen, "Economic Comparison of Alternatives in Equipment Service Life," *Mater. Prot.*, *4*(3), 23 (1965).

768. G. C. Thomson, "Plastic Piping Systems," *Aust. Corros. Eng.*, *9*(5), 17 (1965).

769. *Oil and Gas Journal*, "Pipeline Installation and Equipment Costs," *63*(27), 77 (1965).

770. D. De Pugh, "Look at Gas Pipeline Construction Costs," *Pipe Line Ind.*, *23*(1), 24 (1965).

771. G. A. Brighton, "Gas Supply—Continental Lead in Use of Plastics," *Surveyor*, *125*(3813), 26 (1965).

772. D. A. Gill, "Water Supply—Plastics—Corrosion. Costs and Handling," *Surveyor*, *125*(3813), 14 (1965).

773. C. A. Miller, "Factor Estimating Refined for Appropriation of Funds," *Am. Assoc. Cost Eng. Bull.*, *7*(3), 92 (1965).

774. E. D. Edmisten, "Reinforced Plastic Piping in Refining Industry," *Mater. Prot.*, *4*(10), 54 (1965).

775. J. H. Mallinson, "Reinforced Plastic Pipe: A User's Experience," *Chem. Eng.* (*N. Y.*), *72*(26), 124 (1965).

776. W. H. Brouwer, "Multiproducts Oil Pipelines—General Review," in *European*

Federation of Chemical Engineering, *4th Congress London*, 1966, Paper 6.7.

777. J. H. Mallison, "Reinforced-Plastic Pipe: Joining and Supporting," *Chem. Eng.* (*N. Y.*), *73*(2), 168 (1966).

778. R. V. Riley, "Economics of Process Piping," *Chem. Process Eng.*, *47*(2), 55 (1966).

779. C. M. Gull, "Welded Steel Tubing Reduces Costs for Skelly," *Pet./Chem. Eng.*, *38*(3), 32 (1966).

780. W. C. Turner, "How to Support Insulated Pipe," *Chem. Eng.* (*N. Y.*), *73*(13), 211 (1966).

781. O. T. Zimmerman, "Optimisation Pipe Size," *Cost Eng.*, *11*(3), 14 (1966).

782. R. Kean, "Plant Layout and Piping Design for Least Cost," *Hydrocarbon Process.*, *45*(10), 119 (1966).

783. T. J. Smith, "Reinforced-Plastic Pipe vs Stainless—Comparing the Actual Costs," *Chem. Eng.*, *14*(1), 110 (1967).

784. J. Starczewski, "Economy Gained by the Use of Low-Fin Tubes," *Br. Chem. Eng.*, *12*(2), 239 (1967).

785. M. Dimentberg, "Better Economics Promise Eventual Use of LNG Lines," *Oil Gas J.*, *65*(38), 97 (1967).

786. J. A. Masek, "Metallic Piping," *Chem. Eng.* (*N. Y.*), *75*(13), 215 (1968).

787. D. A. Boswarth, "Installed Costs of Outside Piping," *Chem. Eng.* (*N. Y.*), *75*(7), 132 (1968).

788. J. E. Cran, "How to Estimate Piping Costs?" *Chem. Eng.* (*London*), *218*, CE 110 (1968).

789. J. J. Bollwark and M. F. Fischetti, "How Esso Cuts Cost of Piping," *Oil Gas J.*, *66*(22), 83 (1968).

790. L. L. Simpson, "Sizing Piping for Process Plants," *Chem. Eng.* (*N. Y.*), *75*(13), 192 (1968).

791. O. Mendel, "Cost Comparisons for Process Piping," *Chem. Eng.* (*N. Y.*), *75*(13), 225 (1968).

792. T. N. Dining, "Factored System for Pricing Piping Installations," *Heating/Piping/Air Cond.*, *40*(6), 112 (1968).

793. J. P. O'Donnell, "11th Annual Study of Pipeline Installation and Equipment Costs," *Oil Gas J.*, *66*(31), 99 (1968).

794. J. W. Gendron, "Economics and Impact of Supersize Petroleum Transportation Facilities. Pipeline and Water," in *Exploration and Economics of Petroleum Industry*, Vol. 7, 1969, pp. 233–247.

795. *British Chemical Engineering*, "Process Pipelines," *14*(2), PPR 1 (1969).

796. *British Chemical Engineering*, "Pipework Estimating—A Review," *14*(2), PPR 16 (1969).

797. E. S. Sokulu, "Estimating Piping Costs from Process Flow Sheets," *Chem. Eng.* (*N. Y.*), *76*(3), 148 (1969).

798. *Chemical and Process Engineering*, "Project News," *50*(5), 15 (1969).

799. *Chemical and Process Engineering*, "Project News—2," *50*(6), 17 (1969).

800. *British Chemical Engineering*, "Cost Data File," *14*(7), 995 (1969).

801. *Chemical and Process Engineering*, "Project News—2," *50*(7), 17 (1969).

802. R. E. Hughes, W. A. Hunt, and W. H. Pearn, "Solids Pipelining—Field of the Future," *Pipe Line Ind.*, *32*(6), 29 (1970).

803. E. J. Wasp, T. L. Thomson, and T. C. Aude, "Process Basis and Economics of Pipeline Systems," *Can. Min. Metall. Bull.*, *63*(704), 1373 (1970).

804. D. J. Stern, "Pipeline Transportation of Solids," *Certif. Eng.*, *44*(6), 119 (1971).

805. E. J. Wasp, T. L. Thomson, and T. C. Aude, "Initial Economic Evaluation of Slurry Pipeline Systems," *ASCE Transp. Eng. J.*, *97*(TE 2, Paper No. 8106), 271 (1971).

806. H. Spitzer, "The Computer Approach to Pipe Detailing," *Chem. Eng. (London)*, *252*, 305 (1971).

807. J. C. Bruce, "Economics of Solid Pipelining," *Pipeline Gas J.*, *197*(8), 44 (1970).

808. *British Chemical Engineering*, "Nomograms for Estimating Pipework Costs," *14*(9), PPR 11 (1969).

809. *British Chemical Engineering*, "Pipes for Corrosive Services," *14*(12), 1657 (1969).

810. G. D. Shann, D. I. Williamson, and R. W. Edwards, "Pipeline Contracting," *I. G. E. J.*, *10*, 223 (1970).

811. J. W. Leimkuhler, "Refinery Piping Inspection Data Computerised," *Oil Gas J.*, p. 130 (1970).

812. W. G. Canham and J. R. Hagerman, "Reduce Piping Connection Costs," *Hydrocarbon Process.*, *49*(5), 131 (1970).

813. J. P. O'Donnel, "Pipeline Installation and Equipment Costs," *Oil Gas J.*, p. 99 (August 3, 1970).

814. W. P. Long and E. D. Montrone, "Selecting the Proper Material—Economics," *Chem. Eng. (N. Y.)*, *77*, 41 (October 12, 1970).

815. B. Rothfarb, H. Frank, D. J. Kleitman, D. M. Resenbaum, and K. Steiglitz, "Optimal Design of Offshore Natural-Gas Pipeline Systems," *Oper. Res.*, *18*(6), 992 (1970).

816. M. J. Bush and G. L. Wells, "Unit Plot Plans for Plant Layout for Pipes and Pipelines," *Br. Chem. Eng.*, *16*(4/5), 325 (1971); *16*(6), 514 (1971).

817. S. P. Marshall and J. L. Brandt, "Installed Costs of Corrosion-Resistant Piping," *Chem. Eng. (N. Y.)*, *78*(19), 68 (1971).

818. S. P. Marshall and J. L. Brandt, "Corrected Tables—Corrosion-Resistant Piping Costs," *Chem. Eng. (N. Y.)*, *78*(22), 111 (1971).

819. J. R. Haug, "Pneumatic Conveying of Bark and Hogged Fuel," *Tappi*, *48*(2), 80A (1965).

820. J. Corker, "Pressure Vessel Costs," *Chem. Process Eng.*, *47*(7), 80 (1966).

821. R. Chadwick, "Trends in Aluminium Fabrication," *Chem. Process Eng.*, *48*(2), 55 (1967).

822. K. M. Guthrie, "Estimating Cost of High-Pressure Equipment," *Chem. Eng. (N. Y.)*, *75*(23), 144 (1968).

823. T. T. Furman and R. F. Cheers, "Optimising, Pressure Vessels," *Br. Chem. Eng.*, *16*(6), 478 (1971).

824. H. H. Jethanandani, "Pressure Vessel Industry in India," *Chem. Age India*, *22*(6), 351 (1971).

825. S. B. Pandya, "A Case Study in Pressure Vessel Design: The Methanol Convertor," *Chem. Age India*, *22*(6), 377 (1971).

826. R. E. Markovitz, "Choosing the Most Economical Vessel Head," *Chem. Eng. (N. Y.)*, *78*(16), 102 (1971).

827. F. D. Clark and S. P. Terni, Jr., "Thickwall Pressure Walls," *Chem. Eng. (N. Y.)*, *79*(7), 112 (1972).

828. *British Chemical Engineering*, "Cost Data File," *12*(11), 1769 (1967).

829. M. M. Kirk, "Chemical Feed Pumps," *Chem. Eng. (N. Y.)*, *72*(5), 112 (1965).

830. C. Thurlow, "Pumps and the Chemical Plant," *Chem. Eng. (N. Y.)*, *72*(11), 117 (1965).

831. J. K. Jacobs, "How to Select and Specify Process Pumps," *Hydrocarbon Process.*, *44*(6), 122 (1965).

832. F. S. Chapman and F. A. Holland, "New Cost Data for Centrifugal Pumps," *Chem. Eng. (N. Y.)*, *73*(15), 200 (1966).

833. D. C. Beletskii, "Progressivnaya tekhnologlya na zavodakh nasostroeniya," *Khim. Neft. Mashinostr.*, *1967*(7), 25; *Chem. Pet. Eng.*, *1967*(7), 554.

834. H. Speich, "Cost/Pressure Factors in Hydraulic Pump Design," *Fluid Power Int*, *32*(378), 38 (1967).

835. *British Chemical Engineering*, "Project News," *15*(3), 307 (1970).

836. S. J. Yaki and R. Carpenter, "Pumping Packing, A Case History," *Hydrocarbon Process.*, *49*(7), 136 (1970).

837. R. G. P. Kusay, "Vacuum Equipment for Chemical Processes," *Br. Chem. Eng.*, *16*(1), 29 (1971).

838. H. Knoll and S. Tinney, "Why Use Vertical In-line Pumps?" *Hydrocarbon Process.*, *50*(5), 131 (1971).

839. R. R. Schlueter and W. E. Ruther, "Inexpensive Pump for Liquid Sodium," *Nucl. Technol.*, *11*(2), 266 (1971).

840. K. M. Guthrie, "Pump and Valve Costs," *Chem. Eng. (N. Y.)*, *78*(23), 151 (1971).

841. W. H. Stindt, "Pump Selection," *Chem. Eng. (N. Y.)*, *78*(23), 43 (1971).

842. J. E. Ellis, "Economics of Multistage Adiabatic Catalytic Gas-Phase Reactors Related to Hydrocarbon Oxidation," *European Federation of Chemical Engineering, 4th Congress*, London, 1966, Paper 4.8, p. 59.

843. J. R. Kittrell and C. E. Watson, "Don't Overdesign Process Equipment," *Chem. Eng. Prog.*, *62*(4), 79 (1966).

844. J. W. Reilly and M. C. Sze, "Hydrogenating Benzene to Cyclohexane," *Chem. Eng. Prog.*, *68*(6), 73 (1967).

845. G. C. Derrick, "Estimating Cost of Jacketed, Agitated and Baffled Reactors," *Chem. Eng. (N. Y.)*, *74*(24), 272 (1967).

846. T. E. Corrigan, W. E. Lewis, and K. N. McKelvey, "What Do Chemical Reactors Cost in Terms of Volume?" *Chem. Eng. (N. Y.)*, *74*(11), 214 (1967).

847. S. C. Naik, "Performances of Fluidised Bed Reactors," *Chem. Age India*, *19*(4), 276 (1968).

848. C. Salana and D. V. Eyre, "Multiple Refrigerants in Natural Gas Liquefaction," *Chem. Eng. Prog.*, *63*(6), 62 (1967).

849. D. F. Ballou, T. A. Lyons, and J. R. Tacquard, Jr., "Design and Cost Estimating of Mechanical Refrigeration Systems," *Hydrocarbon Process.*, *46*(6), 119 (1967).

850. R. W. Zafft, "How to Size and Find the Cost of Absorption Refrigeration," *Hydrocarbon Process.*, *46*(6), 131 (1967).

851. E. Spencer, "Estimating the Size and Cost of Steam Vacuum Refrigeration," *Hydrocarbon Process.*, *46*(6), 136 (1967).

852. R. Kapita and T. G. Gleason, "Wet Scrubbing of Boiler Flue Gas," *Chem. Eng. Prog.*, *64*(1), 74 (1968).

853. *Modern Manufacturing*, "Recent Developments in Air Scrubbers," *126*(6), 194 (1968).

854. N. F. Imperato, "Gas Scrubbers," *Chem. Eng. (N. Y.)*, *75*(22), 152 (1968).

855. D. F. Ball and P. R. Dawson, "Air Pollution from Aluminium Smeltersl" *Chem. Process Eng.*, *52*(6), 49 (1971).

856. F. Broadbent, *The Economics of Solid–Liquid Separations, with Special Reference to Centrifuges*, Thomas Broadbent and Sons Ltd., Central Iron Works, Huddersfield, Yorks, U.K.

857. F. Broadbent, "Economics of Solid–Liquid Separations, with Special Reference to Centrifuges," *European Federation of Chemical Engineering, 4th Congress, London*, 1966, Paper 4.3, p. 17.

858. R. E. Lacey, "Membrane Separation Processes," *Chem. Eng. (N. Y.)*, *79*(19), 56 (1972).

859. J. M. Ryan, R. S. Timmins, and J. F. O'Donnel, "Production Scale Chromatography," *Chem. Eng. Prog.*, *64*(8), 53 (1968).

860. D. H. Follet, "Plastic Main Planted off Reel," *Am. Gas J.*, *192*(1), 38 (1965).

861. J. Clerk, "Storage Tanks," *Chem. Eng.* (*N. Y.*), *72*(3), 104 (1965).

862. R. L. Bell and E. C. Young, "Filament Wound Fiberglass Tanks Cut Costs," *Chem. Eng. Prog.*, *61*(4), 57 (1965).

863. K. D. Zimmerhaus, "LNG Storage Costs Today," *Hydrocarbon Process.*, *44*(6), 163 (1965).

864. L. K. Crawford, "Cost of Water Distribution Pumping and Storage Facilities," *Illinois University College of Engineering, Sanitary Engineering Conference*, 8th, *Proceeding* 1966, p. 137.

865. J. H. Stannard, Jr., "Storage Methods for Liquefied Natural Gas," *Pac. Coast Gas Assoc. Proc.*, *57*, 79 (1966).

866. *Chemical Engineering*, "Corrosion Resistant Storage Tanks," *73*(7), 150 (1966).

867. *British Chemical Engineering*, "Start Chemical Engineering Cost Data File," *11*(5), 340 (1966).

868. *British Chemical Engineering*, "Cost Data File," *11*(6), 513 (1966).

869. *British Chemical Engineering*, "Project News," *13*(4), 465 (1968).

870. *Chemical and Process Engineering*, "Project News," *48*(9), 8 (1967).

871. *British Chemical Engineering*, "Cost Data File," *14*(12), 1731 (1969).

872. D. O. Newling, "Filament-Wound Equipment for the Process Industries," *Chem. Process Eng.*, *51*(2), 81 (1970).

873. S. A. G. Personn, "Oil Storage Ground: Mechanical and Equipment Costs," *Pet. Rev.*, p. 323 (October 1970).

874. J. Altman, "Optimum Design, Costing for Refrigerated Liquefied Gas Storage," *Heat./Piping/Air Cond.*, p. 83 (November 1970).

875. L. D. Epstein, "Cost of Standard Sized Reactors and Storage Tanks," *Chem. Eng.* (*N. Y.*), *78*(24), 169 (1971).

876. F. F. House, "Pipe Tracing and Insulation," *Chem. Eng.*, *75*(13), 243 (1968).

877. J. C. Frank, G. R. Geyer, and H. Kehde, "Styrene–Ethylbenzene Separation with Sieve Trays," *Chem. Eng. Prog.*, *65*(2), 79 (1969).

878. S. A. Bresler, "Prime Movers and Process Energy," *Chem. Eng.* (*N. Y.*), *73*(11), 124 (1966).

879. S. D. Caplow and S. A. Bresler, "Economics of Gas Turbine Drives," *Chem. Eng.* (*N. Y.*), *74*(7), 104 (1967).

880. H. Steen-Johnson, "How to Estimate the Size and Cost of Mechanical Drive Steam Turbines," *Hydrocarbon Process.*, *46*(10), 126 (1967).

881. S. G. Branch, "How to Estimate the Size and Cost of Gas Turbines and Gas Engines," *Hydrocarbon Process.*, *46*(10), 131 (1967).

882. D. J. Shellenberger, "Comparison of Building Cost Indexes," *Am. Assoc. Cost Eng.*, *7*(2), 62 (1965).

883. J. R. Cherry, "Cost of Concrete Industrial Buildings," *Civ. Eng.* (*N. Y.*), *33*(1), 42 (1968).

884. W. G. Knox, "Estimating Cost of Process Buildings via Volumetric Ratios," *Chem. Eng.* (*N. Y.*), *73*(13), 292 (1966).

885. L. R. Jones, "Building Cost Escalation," *Chem. Eng.* (*N. Y.*), *77*(16), 170 (1970).

886. K. Hey, "Oekonomie der Schweisstechnik," *Schweisstechnik* (*Berlin*), *16*(8), 378 (1966).

887. O. Blodgett, "Estimating Welding Costs," *Weld. Eng.*, *51*(7), 45 (1966).

888. K. Hey, "Fertigungskosten bei Schweissarbeiten," *Schweisstechnik* (*Berlin*), *13*(3), 110 (1965).

889. E. Hawkes, "Workshop Fabrication Costs," *Chem. Process Eng.*, *46*(1), 50 (1965).

890. N. J. Luetzow and D. Z. Kern, "How to Design Economically with Titanium,"

Chem. Eng. (*N. Y.*) *73*(6), 173 (1965).

891. E. R. Func, "The Use of Reactive Metals in the CPI," *Chem. Eng. Prog.*, *61*(5), 12 (1965).

892. M. Krishnan and R. T. Tampy, "Plastics as Substitute for Metals," *Chem. Age India*, *17*(1), 45 (1966).

893. R. E. Petsinger and H. W. Marsh, "High-Strength Steels for Lower Cost Tanks," *Chem. Eng.* (*N. Y.*), *73*(10), 182 (1966).

894. *Chemical Age of India*, "Stainless Steel," *18*(1), 5 (1967).

895. E. Ineson, "Stainless and Special Alloy Steels," *Chem. Process Eng.*, *48*(2), 87 (1967).

896. D. A. Dahlstrom and S. S. Davis, "Plastics in Continuous Filtration Equipment," *Chem. Eng. Prog.*, *65*(10), 80 (1969).

897. L. Adams, "Relative Metal Economy of Pressure Vessel Steels," *Chem. Eng.* (*N. Y.*), *76*(27), 150 (1969).

898. G. Sorell, "Designing for Corrosion Resistance—Process Control," *Chem. Eng.* (*N. Y.*), *77*(22), 83 (1970).

899. M. L. Hosfelt and N. L. Owen, "Heavy Vessel Erection," *Hydrocarbon Process.*, *47*(7), 109 (1969).

900. *British Chemical Engineering*, "Economising in Flanged Joints," *16*(6), 501 (1971).

901. E. Fuchs, "Essential Welding Knowledge for the Project Engineer," *Br. Chem. Eng.*, *16*(10), 905 (1971).

902. F. F. Charlton, "Economics of Corrosion Control by Protective Coatings," *Corros. Technol.*, *12*(1), 11 (1965).

903. W. Gass, "Berechnung des Verbrauchs von Anstrichmitteln," *Werkstatt Betr.*, *99*(1), 39 (1966).

904. J. Rodgers, "Understanding Economic Aspects of Coatings Is Vital to Engineers," *Mater. Prot.*, *9*(5), 26 (1970).

905. B. Wook, "Painting Costs," *AACE Bull.*, *12*(5), 138 (1970).

906. K. B. Tator, "Engineered Painting Pays Off," *Chem. Eng.* (*N. Y.*), *78*(29), 84 (1971).

907. R. E. Culver, "Power Cable Cost," *Chem. Process Eng.*, *47*(9), 92 (1966).

908. M. R. Bauman, "Water-Fire Extinguisher," *Mater. Des. Eng.*, *61*(5), 118 (1965).

909. E. J. Burgin, "Design of Metering and Regulating Stations," *W. Va. Univ., Eng. Exp. Str., Tech. Bull.*, *73*, 320 (1965).

910. J. M. English, "Discount Function Comparing Economic Alternatives," *J. Ind. Eng.*, *16*(2), 115 (1965).

911. J. M. Carson, "Critical Path Method Gives Better Project Control," *Br. Chem. Eng.*, *10*(4), 248 (1965).

912. A. F. Popalisky, "Standard Absorption and Direct Costing Techniques," *Chem. Eng.* (*N. Y.*), *72*(9), 164 (1965).

913. R. B. Norden, "CE Cost Indexes: A Sharp Rise Since 1965," *Chem. Eng.* (*N. Y.*), *76*(10), 134 (1965).

914. W. R. Hirschmann and J. R. Branweiler, "Continuous Discounting for Realistic Investment Analysis," *Chem. Eng.* (*N. Y.*), *72*(15), 210 (1965).

915. D. H. Ferguson, "Manufacturing Plant and Facilities—Cost and Control," *Chem. Process Eng.*, *46*(7), 376 (1965).

916. M. L. Anderson, J. Eschrich, and R. C. Goodman, "Economic Analysis of R&D Projects," *Chem. Eng. Prog.*, *61*(7), 106 (1965).

917. D. A. F. Wite, "Capital Cost Estimating in Chemical Industry," *Chem. Process Eng.*, *46*(9), 508 (1965).

918. J. W. Hackney, "Analysis of Estimating Accuracy," *Am. Assoc. Cost Eng., Bull.*,

7(3), 105 (1965).

919. F. C. de Paula, "Commitment Accounting for Contract Profit Control," *Chem. Process Eng.*, *46*(9), 517 (1965).

920. *Chemical Engineering*, "How to Avoid Common Pitfalls in Making Cost Estimates," *72*(22), 159 (1965).

921. J. A. Dunster, "Cost Control in Civil Engineering, Contracting," *Chem. Process Eng.*, *46*(11), 627 (1965).

922. D. Booth, "Value Engineering," *Chem. Process Eng.*, *46*(9), 630 (1965).

923. A. Hart, "Evaluation of Capital Projects," *AIChE Institution of Chemical Engineer*, 7, *Management Oriented Topics*, London, June 13–17, 1965, Paper 7.8, pp. 50–55.

924. D. W. Jones, "The Work of the Project Engineers," *Chem. Process Eng.*, *47*(1), 42 (1966).

925. D. Stuhlbarg, "Economic Justification for Equipment," *Chem. Eng. (N. Y.)*, *73*(2), 145 (1966).

926. J. D. Wilson, "Want to Predict Cost Escalation?" *Hydrocarbon Process.*, *45*(1), 159 (1966).

927. G. Teplitzky, "Guide to Clearer Cost Estimates," *Chem. Eng. (N. Y.)*, *73*(5), 130 (1966).

928. W. J. Urban and F. A. Holland, "How to Determine Optimum Plant Size," *Chem. Eng. (N. Y.)*, *73*(7), 193 (1966).

929. W. J. Davis, "Purchasing Procedures for Engineers," *Chem. Eng. (N. Y.)*, *73*(7), 109 (1966).

930. N. O. Machnig, "Berechnungsmethode zur Kostenoptimierung von Apparaten and Apparadegruppen," *Chem.-Ing.-Tech.*, *38*(3), 246 (1966).

931. C. H. Chilton, "Plant Cost Index Points up Inflation," *Chem. Eng. (N. Y.)*, *73*(9), 184 (1966).

932. L. Essayan, "Rapid Calculation of Plant Costs," *Br. Chem. Eng.*, *11*(6), 499 (1966).

933. W. W. Twaddle and J. B. Malloy, "Evaluating and Sizing New Chemical Plants," *Chem. Eng. Prog.*, *62*(7), 90 (1966).

934. P. R. Walton, "Cost Estimating at the Research Level," *Chem. Eng. (N. Y.)*, *73*(17), 172 (1966).

935. H. A. Quigley, "Economics of Multiple Units," *Chem. Eng. (N. Y.)*, *73*(18), 97 (1966).

936. R. Billet and L. Raickle, "Verf ahrenstechisehe Bewertung und Kostenoptimierung von Stoffauscheinrichtungen," *Chem.-Ing.-Tech.*, *38*(8), 825 (1966).

937. W. L. Sullivan, "EDP and Ingenuity Help Control Engineering Construction Costs," *Pet./Chem. Eng.*, *38*(9), 46 (1966).

938. C. P. Dillon, "Calculation of Net Present Value and Equivalent Uniform annual Cost," *Mater. Prot.*, *5*(6), 47 (1966).

939. J. H. Lutz, "Estimating Project-Completion Costs," *Chem. Eng. (N. Y.)*, *74*(3), 164 (1967).

940. M. Sundaram, "Chemical Plant Design Principles. Part V," *Chem. Age India*, *18*(2), 103 (1967).

941. S. Black, "Reduce Equipment Costs by Correct Design," *Br. Chem. Eng.*, *12*(2), 233 (1967).

942. D. P. Herron, "Investment Evaluation," *Chem. Eng. (N. Y.)*, *74*(3), 124 (1967).

943. R. A. Volkin, "Economic Piping of Parallel Equipment," *Chem. Eng. (N. Y.)*, *74*(7), 148 (1967).

944. W. M. Thomas, "Do Piping Codes Reduce Costs?" *Hydrocarbon Process.*, *46*(5), 152 (1967).

945. J. T. Gallagher, "Rapid Estimating of Engineering Costs," *Chem. Eng.* (*N. Y.*), *74*(13), 250 (1967).

946. H. Buntzel, Jr., "The Rising Cost of New Plants," *Hydrocarbon Process.*, *46*(6), 182 (1967).

947. R. P. Sturgis, "For Big Savings—Control Costs While Defining Scope," *Chem. Eng.* (*N. Y.*), *74*(17), 188 (1967).

948. J. Hudig, "The Multipurpose Process Unit," *Chem. Eng. Prog.*, *63*(9), 79 (1967).

951. N. Harnby, "Cost of Overdesign in Chemical Industry," *Chem. Process Eng.*, *48*(12), 61 (1967).

952. J. E. Haselbarth, "Updated Investment Costs for 60 Types of Chemical Plants," *Chem. Eng.* (*N. Y.*), *74*(25), 214 (1967).

953. J. T. Gallagher, "Rapid Estimation of Plant Costs," *Chem. Eng.* (*N. Y.*), *74*(26), 89 (1967).

954. G. L. Street and T. E. Corrigan, "Make Quick Evaluation Estimates," *Hydrocarbon Process.*, *46*(12), 147 (1967).

955. S. Katell and W. C. Morel, "Bibliography of Investment and Operating Costs for Chemical and Petroleum Plants, January–December 1966," *U.S. Bur. Mines, Inf. Cir. 8246*, (1967).

956. A. Foord, "Investment Appraisal for Chemical Engineer. Part 3," *Chem. Process Eng.*, *49*(2), 87 (1968).

957. *British Chemical Engineering*, "Cost Cutting at the Design Stage," *13*(3), 391 (1968).

958. R. H. Clay, "Transportation of Chemical Plant," *Chem. Process Eng.*, *49*(3), 63 (1968).

959. C. E. Carrol, "Power and Other Utilities for Small Plants," *Chem. Eng. Prog.*, *64*(3), 69 (1968).

960. R. I. Reul, "Which Investment Appraisal Technique Should You Use?" *Chem. Eng.* (*N. Y.*), *75*(9), 212 (1968).

961. J. T. Gallagher, "Analyzing Field Construction Costs," *Chem. Eng.* (*N. Y.*), *75*(11), 182 (1968).

962. T. C. Ponder, "Keep Them Within Budget (Under Estimate Construction Cost)," *Hydrocarbon Process.*, *47*(7), 107 (1968).

963. F. E. Galopin, "Electrical Equipment and Material Erection," *Hydrocarbon Process.*, *47*(7), 112 (1968).

964. M. N. Krishnamurthi, "Ready Method for Cost Estimating Process Vessels," *Chem. Age India*, *19*(8), 559 (1968).

965. E. F. Hensley and A. A. MacPhail, "Minimizing Investment in Small Scale Plants," *Chem. Eng. Prog.*, *64*(9), 89 (1968).

967. W. A. Krause, "Cost Control: A Contractors View-point," *Chem. Eng. Prog.*, *64*(12), 15 (1968).

968. M. D. Vijayaraghavan, "Project Cost Estimation in India: An Appraisal," *Indian Chem. J.*, *3*(6), 64 (1968).

969. K. M. Guthrie, "Rapid Calculation Charts," *Chem. Eng.* (*N. Y.*), *76*(1), 138 (1969).

970. *British Chemical Engineering*, "Modern Trends in Chemical Plant Cost Estimation," *14*(2), 205 (1969).

971. R. J. Johnson, "Cost of Overseas Plants," *Chem. Eng.* (*N. Y.*), *76*(5), 146 (1969).

972. K. M. Guthrie, "Field Labour Predictions for Conceptual Projects," *Chem. Eng.* (*N. Y.*), *76*(7), 170 (1969).

973. K. M. Guthrie, "Cost," *Chem. Eng.* (*N. Y.*), *76*(8), 201 (1969).

974. G. E. Mapstone, "Find Exponents for Cost Estimates," *Hydrocarbon Process.*, *48*(5), 165 (1969).

975. J. T. Gallagher, "Efficient Estimation of Worldwide Plant Costs," *Chem. Eng.* (*N. Y.*), *76*(12), 196 (1969).

976. P. H. Dundas and M. L. Thorpe, "Economics and Technology of Chemical Processing with Electric Field Plasmas," *Chem. Eng.* (*N. Y.*), *76*(14), 123 (1969).

977. J. L. Maher and R. W. Coggins, "How to Estimate Size Cost of Producing Equipment—1," *World Oil*, *169*(2), 29 (1969).

978. T. G. Papavero, "Piping Efficiency Program (PEP)," *ASTME*, Pet-7, Paper 69 (for meeting of September 21–25, 1969).

979. *Chemical and Process Engineering*, "Chemical Engineering Contractors—2," *50*(12), 74 (1969).

979a. G. E. Mapstone, "Comparison of Amortisation Rates and Capitalised Cost," *Chem. Process Eng.*, *50*(4), 141 (1969).

980. The Institution of Chemical Engineers, *A Guide to Capital Cost Estimation*, London, 1969.

981. J. Matley, "Keys to Successful Plant Startups," *Chem. Eng.* (*N. Y.*), *76*(19), 110 (1969).

982. R. P. Feldman, "Economics of Plant Startups," *Chem. Eng.* (*N. Y.*), *76*(24), 87 (1969).

983. V. F. Moskvin and E. P. Schukin, "Effect of Power Factors on Location in the Chemical Industry," *Khim. Prom.*, *1969*(3), 56.

984. G. Viehweger, "The Forecast of Investment Costs for Chemical Installations with Process-Oriented Ancillary Costs," *Chem. Technol.*, *21*(11), 713 (1969).

985. *Indian Chemical Journal*, "Direction of Development and Choice of Technology," *4*(7), 14 (1970).

986. H. Popper and G. E. Weismantal, "Costs and Productivity in the Inflationary 1970's," *Chem. Eng.*, *77*, 132 (1970).

987. A. Syverson, "An Economic Experiment in Process Design and Economic Evaluations—Parametric Studies Provide a New Dimension," *Chem. Age India*, *21*(4), 366 (1970).

988. J. D. Chase, "Cost vs Capacity: New Way to Use Exponents," *Chem. Eng.* (*N. Y.*), *77*(7), 113 (1970).

989. J. N. King, "Planning of Capital Expenditure in Chemical Industry," *Chem. Ind.*, *1970*, 582 (May 2).

990. J. W. Wallington, "Project Engineering Management," *Chem. Process Eng.*, *51*(5), 71 (1970).

991. E. A. Stallworthy, "Modern Trends in Chemical Cost Estimating—The Viewpoint of a Large Chemical Manufacturing Company," *Chem. Eng.* (*London*), *239*, CE 182 (1970).

992. E. R. Hill, "Modern Trends in Chemical Cost Estimating—The Contractor's Viewpoint." *Chem. Eng.* (*London*), *239*, CE 172 (1970).

993. M. Kneale, "Modern Trends in Chemical Cost Estimating—A Medium Sized Chemical Company's Approach to Project Cost Estimating," *Chem. Eng.* (*London*), *239*, CE 176 (1970).

994. R. Jarold and J. De Devitis, "Minimising Construction Costs," *Chem. Eng. Prog.*, *66*(8), 17 (1970).

995. *Chemical and Process Engineering*, "Revision of CPE Productivity and Cost Indices," *51*(8), 59 (1970).

996. F. Wesser, "Manufacturing of Chemical Plants and Equipment," *Chem. Age India*, *21*(9), 789 (1970).

997. H. J. Abrams, "Economics Investment Criteria. Part 1, Choice When Budget Constraints Are Absent," *Chem. Eng.* (*London*), *241*, CE 252 (1970).

998. G. Sachs, "Economic and Technical Factors in Chemical Plant Layout," *Chem.*

Eng (*London*). *242*, CE 304 (1970).

999. Y. R. Loonkar and J. D. Robinson, "Minimization of Capital Investment for Batch Processes. Calculation of Optimum Equipment Sizes," *Ind. Eng. Chem., Process Des. Dev., 9*(4), 625 (1970).

1000. *Chemical Engineering*, "Process Plant Utilities—Reports," *77*, 130 (1970).

1001. J. Schmidt, "Estimating the Costs of Process Engineering Projects," *Verfahrenstechnik, 4*(2), 58 (1970).

1002. B. F. Cindadar, J. K. Carley, and A. T. Schooley, "Miniplant Design and Use," *Chem. Eng.* (*N. Y.*), *78*(2), 62 (1971).

1003. J. D. Buchler and G. J. Figge, "Operating vs. Capital Costs: Evaluating Trading off Benefits," *Chem. Eng.* (*N. Y.*), *78*, 96 (1971).

1004. D. B. Webster and R. Reed, Jr., "A Materials Handling Systems Selection Model," *AIIE Trans., 3*(1), 13 (1971).

1005. S. Martinez, "Equipment Buying Decisions," *Chem. Eng.* (*N. Y.*), *78*(8), 146 (1971).

1006. J. C. Khurana, "Location of Refining Capacity-Planning Process," *Pet. Hydrocarbons, 6*(1), 23 (1971).

1007. E. O. Ohsol, "Estimating Marketing Costs," *Chem. Eng.* (*N. Y.*), *78*, 116 (May 3, 1971).

1008. G. Enyedy, "Cost Data for Major Equipment," *Chem. Eng. Prog., 67*(5), 73 (1971).

1009. G. T. Wilson, "Capital Investment for Chemical Plant," *Br. Chem. Eng., 16*(10), 931 (1971).

1010. R. Rautenbach and K. W. Witzel, "Rationalisation of Chemical Plant Designing," *Chem.-Ing.-Tech., 43*(7), 413 (1971).

1011. I. Leibson and C. A. Trischman, Jr., "Should You Make or Buy Your Major Raw Materials?" *Chem. Eng.* (*N. Y.*), *79*(4), 76 (1972).

1012. S. A. Bresler and M. T. Kuo, "Cost Estimating by Computer," *Chem. Eng.* (*N. Y.*), *79*(12), 84 (1972).

1013. S. A. Bresler and M. T. Kuo, "More Programmes for Cost Estimating by Computer," *Chem. Eng.* (*N. Y.*), *79*(14), 130 (1972).

1014. V. F. Cappello, "Simplifying Scaleup Cost Estimation," *Chem. Eng.* (*N. Y.*), *79*(17), 99 (1972).

Index

Chemical Plants

Chemical Processes

Chemical Equipment

General

D. VENKATESWARLU

K. D. CHANDRASEKARAN

Cost, Start-up for New Plants

Introduction

A completely constructed chemical plant, even one that is well-designed and had a good research and development effort before the engineering phase, is not an automatically operable plant. The more novel and complex the operation, with new types of equipment, advanced instrumentation, special materials of construction, and necessary pollution abatement facilities, the more likely it is that the raw materials will not flow through the process to form the final product exactly as designed without problems. Even if this were not so, an extensive planning and training effort would be needed to assure that a capable and informed group of people will be in full control of the entire process at all times. The time, manpower, and materials needed to bring a plant up to design conditions result in costs beyond those needed solely to construct the plant and to operate it routinely after it is built. These are plant start-up costs, which are often defined as those plant-associated costs incurred after construction in excess of "standard" production costs needed to bring the unit to budgeted levels of production rate, quality, and operating costs. Other costs related to the product but not to the plant, such as those for sales and marketing, are not considered to be plant start-up costs.

Although the definition of start-up costs is clear and concise, the actual accounting for start-up costs is not straightforward. Problems arise because of the need to separate costs in an ongoing operation between those activities that are ultimately making useful product and those that are not. One does not know immediately if, say, maintenance on a pump, after start-up has been underway a while, is a routine cost or is due to some special process problem which is just surfacing. A tank car of product could be rejected by the customer weeks after the usual accounting procedures had credited it as a normal operating transaction due to erratic initial analysis or quality control, resulting in reprocessing costs chargeable to start-up. Also, "standard" costs may change during the course of a start-up due to changes in such things as raw material costs, for instance. Unless start-up costs are determined by a poststart-up financial audit, i.e., total costs less credit for product made and in-process inventories, they cannot be determined exactly, and particularly not on a day-to-day basis. For these reasons, start-up costs are often allocated based on the best judgment of excess current costs, with any major corrections being made later as they are determined.

Start-up Cost Estimating

Start-up cost estimates, together with those for engineering, construction, and operating costs, are needed before funds for a new project can be appropriated.

Estimates of start-up cost and time have been made by formula [1]. This is an empirical expression based on a limited number of variables (technology newness, equipment newness, labor quality and quantity, and interplant dependency) and proportional to the fixed capital cost or the construction time. In practice, there are probably over a hundred variables in a plant start-up, any of which could be significant, and many of them are not linear. Everything else being equal, start-up costs will decrease as a percent of fixed capital with increasing plant size, since some items such as technical start-up manpower remain relatively constant. Depending on plant location, travel and living expenses for the start-up group may vary from nothing for a new plant in a complex with a large pool of technical manpower to a significant cost for a plant at a remote location where all manpower has to be brought in. Use of a formula to determine start-up costs should be limited to areas where one has extensive knowledge and information.

Where there is previous experience with similar processes, a consensus or "Delphi" method [2] for estimating start-up costs may be used. It will be as good as the experienced judgment of the people involved. There is an element of psychological motivation if the people who make the estimate also do the start-up work.

A person with some start-up experience who knows the company's strengths and weaknesses, however, can probably make a better estimate of both costs and time by going through a detailed analysis of the start-up based on expected occurrences for the entire period [3]. While this may seem extremely complicated and almost impossible for a project where unpredictable events are almost certain to occur, the cost items that enter into such an estimate are finite and can easily be tabulated on a single sheet of paper. These items are shown in Table 1. They can be time- and cost-estimated for each of the major parts of the start-up described below. The final cost estimate is the sum of all the sections. Estimates for some areas, e.g., manual preparation and operator training, can be made relatively accurately. Estimates for others, e.g., operating the plant to bring it up to capacity, are best made by realistic-to-generous allowances for unpredictable factors.

The costs in a start-up at a specific time involve some or all of the following items (see Table 1).

1–2. *Salaries*: The start-up coordinator's and shift engineers' salaries, overhead, and overtime pay, usually at straight time.

3–4. *Wages*: Plant operators' and laboratory technicians' wages, overhead, and overtime, at $1\frac{1}{2}$ times hourly rate. These people may be brought in from other parts of the company for start-up assistance. Regular plant operators' charges during their training period are considered part of start-up costs.

5–9. *Travel*: The cost of travel for employees to an away-from-home location. For long start-ups, families may be temporarily moved and their travel costs are included. Living costs for employees, as for food, motels, car rental, and incidentals, can be estimated based on the approximate daily rate for each times the frequency.

TABLE 1 Estimated Start-up Costs

Item	Calculations	Cost ($)
1. Coordinator	Salary ( ) × (1 + overhead) ( ) × mo. ( )	
Overtime	Salary ( ) × (fraction of mo./mo.) ( ) × mo. ( )	
2. Engineers	No. ( ) × (salary) ( ) × (1 + overhead) ( ) × mo. ( )	
Overtime	No. ( ) × (salary) ( ) × (fraction of mo./mo.) ( ) × mo. ( )	
3. Operators	No. ( ) × (wages) ( ) × (1 + overhead) ( ) × mo. ( )	
Overtime	No. ( ) × (wages) ( ) × (fraction of mo./mo.) ( ) × mo. ( ) × 1.5	
4. Technicians	No. ( ) × (wages) ( ) × (1 + overhead) ( ) × mo. ( )	
Overtime	No. ( ) × (wages) ( ) × (fraction of mo./mo.) ( ) × mo. ( ) × 1.5	
5. Air travel	Flights/mo. ( ) × (cost/flight) ( ) × mo. ( )	
Technical and operating allowance	Flights/mo. ( ) × travel pay ( ) × mo. ( )	
Family travel	No. ( ) × cost/round trip ( )	
6. Motels/apartments	No. ( ) × days/mo. ( ) × cost/day ( ) × mo. ( )	
7. Food	No. ( ) × days/mo. ( ) × cost/day ( ) × mo. ( )	
8. Car rental	No. ( ) × days/mo. ( ) × cost/day ( ) × mo. ( )	
9. Incidentals	No. ( ) × cost/mo. ( ) × mo. ( )	
10. Maintenance and materials	No. ( ) × mo. wages ( ) × factor ( ) × mo. ( )	
11. Outside assistance	No. ( ) × cost/man-mo. ( ) × mo. ( )	
12. Lab analyses	No./mo. ( ) × cost/analysis ( ) × mo. ( )	
13. Operating materials	Cost/mo. ( ) × months ( )	
14. New fixed capital		
15. Bad batch losses	Cost/day ( ) × days material lost ( )	
16. Special materials		
17. Additional overhead	Cost/mo. ( ) × no. months ( )	
Total		

10. *Maintenance* : Estimated as a combination of wages, overheads, and materials. Because the degree of problems to occur in a plant start-up is not known ahead of time, these costs are difficult to estimate accurately. Some estimate of the man-months for the various crafts is made first, such as for instrument men, pipefitters, and machinists. They will be used around-the-clock at times, and an estimate of their average costs should include an allowance for premium pay and overtime. A simplified method for estimating costs is to first estimate total man-months required, multiply this by average monthly wage rate, and finally multiply by a figure to cover the costs of overhead and materials. This multiplier will generally be 3 or higher.

It should be realized that the amount of mechanical failures can be minimized by initially selecting reliable equipment manufacturers and thoroughly inspecting and testing the equipment in their shops. Deviation from this policy adds to the amount of plant maintenance required on start-up.

11. *Outside Assistance* : From other company groups (research and development, materials group, engineering, etc.) or consultants outside the company and vendors' technical personnel. This can be cost estimated by first determining the nominal amount of assisting manpower expected and multiplying by an average monthly rate, which would include travel and living expenses. Extra technical assistance from vendors' service personnel can run as high as several hundred dollars per day plus expenses.

12. *Lab Analyses* : For special samples, estimated by using the average lab cost per analysis times an educated guess at the number of samples to be analyzed in this manner.

13. *Operating Materials* : Such as safety equipment, special log sheets, sample bottles, and analytical reagents; taken at some nominal figure.

14. *New Fixed Capital* : Such as extra pumps, tanks, and pipelines needed during start-up. This will depend on company experience with this type of plant, but a value around 1% of the fixed capital of the project may be in the right order of magnitude.

15. *Bad Batches* : And other losses such as spills of the chemicals being handled in the process. The same is true for disposal costs of unusable production material or excessive residues. A nominal figure may be obtained by multiplying days (or amount) of production expected to be lost in this manner by the standard manufacturing cost per day (or per amount).

16. *Special Materials* : Process chemicals, etc., may be considered start-up costs when outside the scope of the original operating concepts. Generally, normal raw materials end up in the salable product and would be an operating cost rather than a start-up cost.

17. *Additional Overhead* : Charges for relocation, associated with a new plant manager, his technical and nontechnical operating group, as well as the office force, guards, etc., can be considered start-up costs if not covered by any other budget. This is more likely to occur in a pronounced way at a grass roots plant site. At an older plant, most services are already provided and the production group usually is more readily assembled from the existing organization.

Start-up costs may be broken into many categories. For purposes of estimating, planning, and control, they are best broken into a few significant categories that can be identified distinctly even though they may overlap in actual time. Seven such major activities are used here to define a start-up for cost estimating purposes. Others could be added if one wishes to enlarge the importance of any special function within a category.

Once a company defines its start-up procedure by written or verbal policy, and a person is put in direct charge of start-up operations, the seven activities are:

1. Planning and scheduling
2. Preparation of operating manuals
3. Checkout of plant equipment
4. Training of operators and others
5. Operation of the plant
6. Follow-up
7. Preparation of a technical audit

In using a limited number of key categories, it must be remembered that a number of important secondary activities are going on. For instance, a plant safety inspection may take place during the plant checkout period. Each major area is discussed below after a review of prestart-up considerations.

Since the product made during the start-up period is generally credited to the production account, not all costs should go against start-up. One simplified system for obtaining a value for start-up charges directly from the field is to assign the costs for raw materials, the operator crew after production starts, and the regular production staff to the production account while putting the remaining charges, including maintenance manpower and materials, on the start-up account. The cost of the raw materials and production manpower will approximate the standard costs during the period while rates are brought up to design values. The costs of operator training, technical start-up manpower, maintenance, and nonroutine items will approximate start-up costs.

For a very rough guideline, start-up costs have occurred in the range of 5 to 10% of the plant direct fixed capital cost for established processes, 10 to 15% for relatively new processes, and from 15 to 20% for radically new processes [4]. However, this can vary considerably with the size and complexity of the plant, interrelationship with other plants, etc. [1, 3].

Prestart-up Considerations

The manner in which a company handles plant start-ups is usually determined at the executive or top-management level. Start-up organizations can range from a highly staffed force to as little as one engineer working with the existing production group on a small system.

For large plants the start-up function may be under the control of an overall project manager who also directs engineering project management and the support activities of various staff groups. In addition, an executive Planning Committee, consisting of the vice-presidents of Engineering, Purchasing, Finance, and the Production Division plus any other important staff groups, may be formed to review project status and focus on major problem areas. This group would meet periodically to make sure that the resources of the company are available when they are needed and to expedite critical activities such as major purchase orders with a minimum of time delay and red tape.

A separate accounting cost center should be set up to receive all charges for the start-up operation for both internal control and possible tax purposes. The more this can be broken down by time and activity, the more readily the costs of start-ups can be analyzed and a firmer basis laid for estimates of future start-ups. The Purchasing Department must establish contacts with suppliers for the needed materials at budgeted rates. The availability of alternative suppliers should be explored early. Auxiliary services, such as an adequate laboratory for analysis of process streams, must be planned and provided for well in advance of start-up.

Start-ups in foreign countries often involve special problems. Pressures may exist to purchase as much as possible within the country where the plant will be built, even though this may not always fit the original requirements of the plant. Equipment supplied by a foreign affiliate of a United States company may be different from a similar item made in this country.

Regardless of where the start-up decisions originate, the most capable people in the company should be assigned to start up any major new plant or expansion. The costs of start-up together with lost profits can be exceedingly large, justifying the use of the best talent available. The assignment of the start-up coordinator for the field operations should be made based on several considerations. He must not only have a good technical background to know all technical aspects of the project, he should have some practical experience with plant practices, some background in plant design and equipment selection, understand the basic principles of construction practices, and have a knowledge of the project's economics, including the sales requirements and customer needs. He should have administrative and organizing ability, a personality to get along with others, and drive and creative vision to complete the job. He will generally be a chemical engineer, as will most of his shift engineers. While the nontechnical operating personnel and plant service personnel may be under budget control of the Production Department, decisions on operation of the *process* must be in the hands of the assigned start-up coordinator and technical members of his group, and they must have sufficient authority to carry them out.

Trends must be remembered. We have been in a period when plants are growing in size and designed for minimum capital expenditures, often single-line with little standby equipment and flexibility, highly dependent on instrument control and increasingly tending toward some type of computerized operation, and frequently interdependent on another large plant for supplies or production. In addition, occupational and environmental considerations allow little room for losses that cannot be contained. These factors can make future start-ups more difficult than some of a similar nature in the past.

Start-up Planning and Scheduling

In contrast to the next six sections, in which all or most of the start-up group is actively involved with the plant or its data, this section is limited mostly to the start-up coordinator and the preparations he must make for the start-up. During this time, which can be a period of months, he may be learning about similar processes at other locations, carrying on discussions with research and development, process design, project engineering, and construction personnel about their respective responsibilities and plans, assisting in decisions on personnel to be assigned to the start-up group. Liaison and coordination with plant people and related groups (sales, purchasing, accounting, industrial relations) are necessary to assure a proper supply of raw materials, operating supplies, operators, lab technicians, and maintenance people.

Well in advance of start-up, and before the plant is completely designed, there should be a hazards review of the engineering flow sheets by a committee of experienced company personnel highly knowledgeable in plant operations. This may occur before the start-up coordinator is assigned, in which case he should review their safety recommendations with regard to the plant being built. If he is assigned early enough, the start-up coordinator should be on this committee. In theory, a start-up coordinator should be assigned to follow activities from the time funds are appropriated for construction of the new facility. In practice, this is likely to occur only on projects where the coordinator has some project responsibility.

During this period, plans should be drawn for the personnel to be involved in the start-up. A technical group of start-up engineers should be scheduled to be brought in before completion of construction. Usually this would be four shift engineers, sufficient to cover each shift when the plant is ready to run. A variation requiring less personnel and more overtime involves three engineers on 12-h shifts. This is often used where many of the start-up people must come from other locations, with a full week off to return home routinely scheduled. These people are preferably obtained from within the company and should have an ability to analyze problems involving mechanical equipment. A group of specialists for calculations, evaluation of data, computer work, or bench studies may also be planned for if this type of work can be seen as necessary to the effectiveness of the start-up team. On any medium-to-large project, clerical assistance on accounting records, supplies, etc. is very worthwhile. Prior to start-up, the technical personnel should have an introductory course in instrumentation, preferably with actual "hands-on" demonstration equipment. Special reviews of plant safety requirements, analytical test procedures, and computer operation may be in order.

Liaison should be established with the plant production management group as soon as it forms. These people will generally have budgetary and long-range responsibilities relative to the nontechnical operating personnel. They will also take over full control of the plant once start-up is over. While the start-up group has control over the process during start-up, regular plant supervisory personnel have the right to order a shutdown of the process for reasons of safety, pollution, emergencies, and general plant welfare.

The start-up coordinator should meet well in advance of start-up with the maintenance superintendent to plan expected maintenance manpower requirements. At an established plant site these needs will subtract from existing plant manpower, and additional people may have to be provided. At a new plant, maintenance workers will have to be permanently hired or obtained through outside contractors. If more than one start-up is occurring at a time at a given plant, the projected number of men and time required for each craft should be tabulated over the period of the start-ups so that the full plant maintenance load can be appropriately manned. For a large project, a maintenance supervisor may be assigned to the start-up full time. Spare parts should be ordered early and a maintenance schedule, including routine maintenance, inspections, and testing, should be established.

In many cases an analytical group will exist at the location of the new process. The head of this group should be contacted early to arrange for new analytical procedures. If a new group is needed, a laboratory system for general and specific analytical tests must be provided. Special analytical equipment should be ordered early because delivery time is often lengthy.

The start-up coordinator should develop a schedule of events or goals in order to plan work assignments to achieve them. All people on the start-up staff should know what their job functions are to be and what specific things are their responsibility. A chart or graph can be prepared which shows the relative sequence of events in engineering, construction, and start-up, and the expected degree of completion of each particular phase with time [4, 6]. This serves as a basis for scheduling of personnel, need for raw materials, etc.

Office space must be provided for members of the start-up group as they begin direct work on the project. In general, space should be provided as close as possible to the new plant, often in an adjacent building or a trailer.

Preparation of Operating Manuals

One of the first things a start-up coordinator and his staff can do directly for start-up, before construction is even near completion, is prepare manuals for plant personnel. Primary sources of information will be research and development reports and the engineering design manual. Other sources include vendors' information on equipment, previous plant manuals on similar processes, and discussions with personnel who are or have been associated with the project. As in any written communication, plant manuals must consider the people for whom they are written. One procedure is to first write a "master manual" covering all details describing the plant and how to operate it. This may consist of a dozen or more major sections.

Typically they are:

Introduction (product made, how much, where used, etc.)
Process description
Operating procedures (detailed, including normal start-ups and shutdowns)
Emergency procedures
Control samples and limits
Instrument settings and control action (including alarms)
Engineering flow sheets
Safety
Analytical procedures
Equipment description

Pumps and pump curves
Tank capacity charts
Rupture disk and safety valve index
Utilities
Flow sheet and drawing index

Sections can be removed from the master manual and made into smaller manuals which are used for specific groups; e.g., the operators and maintenance workers.

Since the operating instructions are the first direct assembly of all operating information, it is also a good time to prepare a filing system. This information cannot only be used by technical personnel for reference during the start-up of the process, but for training operators, maintenance workers, analytical people, etc. The filing system should have sufficient project reports, flow sheets, equipment descriptions, and vendor operating details so that some explanatory information is immediately available on any major problem area. However, it should not be necessary to duplicate the entire set of engineering file drawings, for instance, since in most cases this additional detail can be obtained relatively quickly if needed. If the plant has special procedures, such as for an emergency, these should not only be on file but they should be reviewed periodically with members of the start-up group. It is desirable to have a "contingency" plan for possible problem areas before actual process start-up. This would include handling off-grade intermediates or product, ways of checking materials of construction for high corrosion rates, and a list of possible replacement alternatives.

Checkout of Plant Equipment

The coordinator and his technical staff will be involved in seeing that the plant is in proper condition for start-up. Equipment and piping must be systematically checked out to be leak-free and devoid of obstructions or contaminants, such as water in an anhydrous process. This is a detailed operation where experience and patience are needed. As much as possible, everything must be put into readiness to take process fluids without chemical or mechanical problems.

A list of inspection items should be prepared ahead of time so that this work can progress efficiently. This list [4, 7, 9] should include such areas as:

Inspect insides of all tanks, vessels, and closed equipment
Flush or otherwise make sure all lines are clear (use screens in lines to protect
    pumps from solids, etc.)
Arrange for special treatment of lines (e.g., pickling steel lines carrying HCl,
    degreasing of lines for oxygen use)
Pressure (vacuum) check all lines and equipment for leaks (flanges should not
    be insulated at this point)
Identification of piping (install pipe markers, etc.)

Check mechanical operation of motors, pumps, agitators, conveyors, etc.
Precheck all instrument systems from panelboard to control unit
Check for safety hazards, particularly those left by construction
Check safety equipment (relief valves, rupture disks, etc.)
Check interlocks, alarms, shutdown devices
Steam tracing should be operable, especially in winter
Check availability of all utilities to all equipment needing them
Make "dry" runs, hot tests, and, if possible, closed loop dynamic instrument
    checks

In the case of instrumentation it is advisable to have the project instrument engineer or the field instrument engineer assist the start-up group. He in turn may be assisted by the company's instrument mechanics or by an outside instrument subcontractor. In a highly instrumented plant the time needed to check out instruments and interlock circuits may run into thousands of manhours, based on $1\frac{1}{2}$ to 3 h per instrument and 2 to 3 h per interlock circuit [9].

Plant checkout will proceed in a more orderly manner if a schedule of items or areas to be inspected is made out ahead of time. Certain sections of the plant will be completed by construction and released to the start-up group ahead of others, and proper planning will take this into account. As the inspections progress, a preacceptance checklist should be prepared, indicating deviations from design and corrections to be made, and should be issued to the interested parties periodically. During the period of plant checkout, maintenance personnel should be available from the contractor or the plant.

A formal procedure for transfer of equipment from construction may be used in which members from construction and from start-up each sign a statement indicating acceptance of the equipment and noting any exceptions which must be corrected shortly thereafter. Another form, used for equipment which will require maintenance, will provide a record of equipment information, initial maintenance work, and test data (see Fig. 1). Copies of all documents, including manufacturers test certificates, should be sent to the start-up coordinator as well as to the local engineering and maintenance groups.

As construction of the plant nears completion, with all major pieces of equipment installed and tied together to a large extent, a field review by a plant safety group is needed. Many important nonprocess items must be checked at this time; e.g., safety showers, eyewash fountains, and hand rails. In addition, government O.S.H.A. requirements for personnel in the operating area must be satisfied.

Training Operators and Others

The people who are to run the plant, the operators and foremen, must have a clear and detailed understanding of how the plant is to be operated. Using the

Date_____

Property No_____

Original P.O. No._____ Mfr._____ Pump Ser. No. _____

Dept._____ Bldg._____ Process _____

Type_____ Mat'l. of Constr._____

Mat'l. Pumped_____ Concn._____ Sp. Gr._____ Temp._____

Suction Dia._____ Disch. Dia._____ Suction Screen _____

Pump RPM_____ No Belts_____ Belt No._____

Motor HP_____ RPM_____ Rotation_____ Full Load Amps_____

Starter No._____ Location_____ Breaker Size_____

Overload Cutout_____ Amps. Fuse_____ Amps_____

Design Flow GPM_____ at_____ psi or ft. head.

Impeller Size_____ Rotation_____ By_____ Date_____

Pump Lubricant_____ By_____ Date_____

Motor Lubricant_____ By_____ Date_____

Packing_____ By_____ Date_____

Mech. Seal-Coolant/Lubricant_____ By_____ Date_____

Coupling/Belt Alignment_____ By_____ Date_____

Coupling/Belt Guard_____ By_____ Date_____

Flex Connections: Suction_____ Dischg._____

Relief Valve Setting_____ psi_____ By_____ Date_____

Run-in Test:_____ Test Liquid_____

Packing/Seal Leakage_____

Pressures, psi: System Wide Open: Suction_____ Disch._____ Amps_____

Dead End: Suction_____ Disch._____ Amps_____

Design flow_____ Suction_____ Disch._____ Amps_____

Flow Rates: Max_____ gpm @_____ psi (Test Liquid)

Max_____ gpm @_____ psi Process Liquid

Bearing Temp._____ Packing/Seal Temp._____ after 1 hour

Test Liquid Drained By_____ Date_____

Tested By:_____ Date_____

Witnessed By:_____ Date_____

Accepted for Start-up By:_____ Date_____

FIG. 1. Pump checkout report.

operating manual prepared earlier by the start-up group, training sessions should be given in suitable surroundings by technical start-up members. Initial instructions are best given in a lecture room using simplified flow sheets. Specialists can be called in to lecture on such subjects as instrumentation, compressors, and analytical instruments. Visits to specific parts of the plant should supplement lectures after initial instructions. Technicians, who may be needed to run frequent chemical analyses during the first few weeks at start-up, should be given sufficient training to do their job properly and also understand how their results relate to plant operation. Maintenance personnel should be trained adequately to be immediately effective on special equipment as well as on routine repairs.

When inexperienced operators are involved, special instructions and textbooks can be used to provide their introduction to chemical and mechanical principles. The more complex the operation, the more useful will be such items as three-dimensional scale models of the plant, panelboard mock-ups, and use of computer-based or other types of process control system simulation. In some cases, equipment manufacturers may supply additional training on specific items. At locations where the operators are allowed to do maintenance work, further training in instruments, pumps, and specific process equipment like compressors and filters is desirable. All personnel must know how to use safety equipment, the elements of fire protection, and the need for good housekeeping. Plant safety should be explicit in the operating manual in such things as methods for taking samples and what to do when chemicals are spilled on the body. When it can be arranged, use of the operators to assist in "dry runs," runs with water or other nonprocess liquids, are very useful for obtaining a feel for the way the process will be controlled.

In systems with several complex pieces of equipment to be started up in special order, a sequence diagram or critical path type of drawing showing the relative timing for starting up sequential and parallel operations should be provided by the technical people to show the importance of each item in the system to the overall process [8]. The sequence for shutting the process down is equally important.

Operate Plant

The technical responsibility for bringing the plant into operation and taking it up to full capacity is that of the start-up coordinator. His technical crew will generally consist of one or more engineers per shift together with whatever special assistance is needed for such areas as chemical analysis and instrumentation. Supervision of the operators and other employees in nontechnical matters will generally reside with the normal plant production supervisory staff, but directives of the start-up crew in process matters must be quickly implemented.

Plant maintenance will often be scheduled around the clock, particularly instrument mechanics, electricians, pipe fitters, and machinists. The amount of skilled maintenance people needed in the first weeks or months of a start-up can be very high, after which such services will drop to more normal levels.

The heavy emphasis on maintenance in a start-up is based on historical experience [4, 5] which has shown that equipment-related difficulties cause over half and possibly as high as 95% of the problems. Chemical or process problems occur much less frequently on a properly developed process. However, when they do occur, failures in the process can be exceedingly difficult to solve, sometimes requiring a return to laboratory or pilot-plant studies. Troubleshooting to locate the cause of a problem is an area where the capabilities and experience of the technical crew become evident. Mechanical problems are usually spotted by close observation of the components making up a piece of equipment, while process problems often require a study of localized material balances and special analyses to pinpoint the cause. Properly filled out log sheets for the various operating areas, usually on an hourly basis with comments on unusual occurrences, form the basis of plant data collection. Special problem areas may require an additional tabulation of operating data. Analysis of these data to solve persistent problems is a primary job for the technical crew. Computers have a direct value in troubleshooting due to their data logging capabilities and by making material balances and special calculations.

Final tune-up of instrument loops by instrument mechanics or engineers cannot generally be done until the process is running. In plants designed to operate with closed-loop computer control, initial operation with manual control and off-line computer assistance is often preferred. Trying to start up both the computer control system and the plant can at times severely hamper the start-up operation.

Feed rates are usually brought up to design conditions in steps, such as half rate and then full rate, so that bottlenecks or problem areas can be spotted and corrected without major mishaps or pileups of undesired material. Operators should be trained on-the-job and allowed to run the unit as soon as possible. It should always be remembered that making production is not as vital as maintaining safe working conditions. The operating crew should not hesitate to use an emergency shutdown procedure whenever conditions reach a point where there is a risk to life or property, as from fire or explosion.

Laboratory chemical analyses are generally used to monitor all streams, from raw materials through intermediates and waste streams to final product, to see if they are within original planned specifications or requirements. Other samples must be taken and analyzed when problems occur in the plant that affect the composition, quality, or handling capabilities of process streams. The analytical laboratory is normally under the control of the plant production group, and routine analyses may be considered part of their costs during start-up. However, special or unusual analyses would be considered legitimate start-up costs. Extra assistance in the form of an analytical chemist or regular chemist may be necessary where chemical problems, or related areas such as corrosion, are likely to occur. The capability of detecting trace quantities of impurities in the feeds or process streams, which can be severely detrimental as they accumulate in the system, must be readily available.

The start-up group must provide brief and concise verbal and written reports to all interested parties on the status of the start-up. Issuance of a short one-page daily report, giving the status of all major steps or pieces of equipment, production rates and yields, and describing any problem areas, will provide the minimum information needed by research and development, engineering, and production personnel, as well as groups concerned with outside committments such as sales and marketing. This is especially true when a plant is at a location remote from management and the groups that developed the process. Daily meetings, or every other day if problems are infrequent, should be held by the start-up coordinator, members of his staff who have developed important information, and production personnel, plus any other groups which may be directly involved in solving plant problems, such as Engineering or Development. As repair items and equipment change items increase, preparation of a checklist with the status and follow-up responsibility of all items should be issued to all involved groups, including purchasing.

In general, the process is under the control of the start-up group until design capacity or some steady high percentage of it is reached, the desired yields are achieved, and the quality of the product is assured. Demonstration periods of from 5 days to a month are often used. The final agreement on acceptance of the plant by production should take place with a single page "Acceptance Form" (see Fig. 2) which lists the agreed upon rates and quality, notes any exceptions to full acceptance by production, and is signed by responsible representatives of both groups. The exceptions must be corrected shortly thereafter.

If the process in the plant has been licensed, guarantees on the process must be demonstrated by tests performed by the licensor or the engineering firm involved. These tests must be well defined and must be carried out under the supervision of highly capable technical personnel. The methods for making measurements, calculating yields, and determining production rates must be agreed upon in advance. Contractural obligations will usually impose a penalty against the final fee or royalty payment based on the ability to meet prespecified requirements.

DATE _____
PROJECT _____

TRANSFER TO OPERATIONS _____
The Start-up Coordinator transfers to Production, and Production accepts the _____
_____ system located at _____
_____

This transfer is subject to the following exceptions:

The Start-up Coordinator and Production agree that the _____
_____ has been demonstrated by actual plant operation as being capable of producing _____ . It is recognized that the above rate is contingent on compliance with the operating procedures described in the revised **Start-up Manual.** Furthermore, it is agreed that the operating personnel and supervision have been trained in and are capable of following the operating instructions.

Date _____

Released By: Accepted By:

_____   _____
Start-up Coordinator   Production Representative

FIG. 2

Follow-up

After the plant has reached its design goals, the start-up technical group is dissolved as a team and the production staff is given complete responsibility. However, some work remains before all start-up people return to their regular company jobs.

The start-up coordinator with some technical manpower may be expected to stay on the site or be called in for assistance on minor problems which often occur after turnover to production. These include the "exceptions" noted on the Acceptance Form, and a variety of problems which reoccur in different ways with the elimination of technical shift coverage, where problems were identified more quickly and appropriate action taken. In addition, engineering drawings should be updated, operating manuals should be revised, and any changes in analytical procedures should be written up for permanent use. Time for one or two technical men, and possibly some maintenance assistance, should be allowed for in the start-up budget. Final data for a technical audit, discussed in the next section, should be obtained at this time.

Technical Audit

The final task of members of the start-up group involves documenting the information obtained during start-up. A technical audit, or summary report, should be prepared which will pass this valuable information on to all interested personnel in the company, such as those in production technical groups and engineering design personnel. This is of particular value if this is the first plant of its type to make the product. Items to be covered in such a report are:

1. *Material Balance* on the system and a comparison with flow-sheet values. This includes determination of raw material usage factors, conversions, and yields. This information is useful for evaluating the chemical efficiency of the process.

   Information on process losses, such as vent losses and sewer losses, should be included as a means of ascertaining the ability of the plant to stay within pollution limits with the available equipment. An environmental group may be assigned to provide these data.

2. *Design Information* on the process in general and on specific equipment (e.g., distillation column plate efficiencies, heat transfer coefficients, pressure drops, filtration rates). Also a description of operating problems which occurred and how they were overcome. Specialty areas, such as instrumentation, should be placed in separate sections. This information is of use in evaluating the current design as well as for future designs of similar equipment.

3. *Maximum Rates of Operation* of the various major pieces of process

TABLE 2   Start-up Cost Summary ($): Plant X at Location Y, Process Start-up Date: June 1977

Month: Activity: Cost Area	0–2.0 Planning and Scheduling	2.0–3.5 Manual Preparation	3.5–5.0 Checkout Plant Equipment	5–5.5 Train Operators and Technicians	5.5–7.5 Operate Plant	7.5–8.0 Follow-up	8.0–9.5 Technical Audit	Totals
Engineer salaries and overhead	8,800	15,600	24,600	8,200	38,900	4,200	12,600	112,900
Operators, technicians, wages and overhead	0	0	4,700	7,800	0	0	0	12,500
Travel and living expenses	1,600	1,600	12,900	5,100	14,900	3,200	0	39,300
Maintenance wages, overhead, and materials	0	0	8,600	2,900	41,600	2,400	0	55,500
Outside assistance	1,000	1,000	1,500	2,000	9,000	0	2,000	16,500
Operating materials, lab analysis, and miscellaneous	0	0	2,500	500	5,000	1,000	500	9,500
Fixed capital	0	0	0	0	25,000	0	0	25,000
Bad batch allowance	0	0	0	0	3,000	0	0	3,000
Special materials	0	0	0	0	1,000	0	0	1,000
Additional plant overhead	0	0	7,000	6,000	0	0	0	13,000
Totals	11,400	18,200	61,800	32,500	138,400	10,800	15,100	288,200
							Say	290,000

equipment. This information is very important for identifying bottlenecks to be overcome for future plant expansions and in new plant designs.

4. *Recommendations* on methods for improving plant operation in the future in the existing plant or in new ones. Typical items to be covered are technical changes, operating methods, and safety.

This technical review report should be written and issued as soon after the end of plant work as possible while details of the operation are still fresh in everyone's mind. Copies should be sent to all groups involved in the project, and not just filed away.

Summary

Although start-up costs are difficult and in many cases probably impossible to estimate accurately, this can best be done by making a written month-by-month, or period-by-period, tabulation of expected occurrences and placing the best realistic costs obtainable on them. This not only provides an estimate of overall costs for this part of the project, but a time framework is set up which gives an early indication of approximate requirements for funds and manpower. An example of a typical start-up cost summary table based on methods described here is shown in Table 2 for a relatively small unit and using the simplified accounting procedure. In practice, the start-up will not follow these plans exactly. Completion of construction may be delayed, manpower often is assigned later than desired, the separate periods of activity will overlap, and the weather can cause unexpected problems, particularly in winter. But if realistic costs are assigned for all the steps in the start-up, the pluses and minuses along the way can result in an estimate that will be close to the actual final cost.

While everything in a project is important, a few considerations are critical to a successful start-up. These are proper responsibilities of company management and, with sufficient time and authority, of the start-up coordinator. They are (1) proper plant design, (2) a qualified and adequately staffed start-up organization, and (3) provision for sufficient technical assistance to solve the most difficult problems when they occur.

Bibliography

Brown, R. Y., and Brewer, C. C., "Startup and Acceptance," *Ind. Eng. Chem.*, *48*(1), 68A–71A (1956).

Bruner, J. R., "Utilizing Contractor's Services for Efficient Plant Startup," *AIChE Symp. Ser.*, *70*(142), 91–95 (1974).

Clark, M. E., DeForest, E. M., and Steckley, L. R., "Aches and Pains of Plant Startup," *Chem. Eng. Prog.*, *67*(12), 25–28 (1971).

Hansen, C. A., "Startup and Shutdown Procedures," *Chem. Eng.*, 66(17), 154–160 (August 24, 1959).

Parsons, R. H., "Guidelines for Plant Startup," *Chem. Eng. Prog.*, 67(12), 29–31 (1971).

Severa, J. E., "Startup of a Crude/Vacuum Distillation Unit," *Chem. Eng. Prog.*, 69(8), 85–88 (1973).

Swain, R. T., and Hopper, B. J., "Contractor Problems during Startup," *Chem. Eng. Prog.*, 67(12), 32–35 (1971).

Troyan, J. E., "Troubleshooting New Processes," *Chem. Eng.*, 67(23), 223–226 (November 14, 1960).

Troyan, J. E., "Elements of Operating Manuals," *Chem. Eng.*, 68(5), 134–138 (March 6, 1961).

Troyan, J. E., "Troubleshooting New Equipment," *Chem. Eng.*, 68(6), 147–150 (March 20, 1961).

Troyan, J. E., "Pumps, Compressors and Agitators," *Chem. Eng.*, 68(9), 91–94 (May 1, 1961).

References

1. R. P. Feldman, "Economics of Plant Startups," *Chem. Eng.*, 76(24), 87–90 (November 3, 1969).
2. T. G. Ray, "Confidence Limits on Startup Costs," in *1975 Transactions of the American Association of Cost Engineers*, pp. 316–318.
3. G. C. Derrick and W. L. Sutor, "Estimation of Industrial Chemical Plant Startup Costs," in *1975 Transactions of the American Association of Cost Engineers*, pp. 308–315.
4. J. Matley, "Keys to Successful Plant Startups," *Chem. Eng.*, 76(19), 110–130 (September 8, 1969).
5. R. Landau (ed.), *The Chemical Plant*, Reinhold, New York, 1966, pp. 270–289.
6. J. E. Troyan, "How to Prepare for Plant Startups in the Chemical Industries," *Chem. Eng.*, 67(18), 107–126 (September 5, 1960).
7. J. P. Eames, M. H. Sturgis, and C. F. Weeks, "The Contractor's Role in Plant Startup," *Chem. Eng. Prog.*, 55(8), 47–50 (1959).
8. G. T. Ryan, "Managing the Project Startup," *Chem. Eng. Prog.*, 68(12), 65–71 (1972).
9. M. Gans, "The A to Z of Plant Startup," *Chem. Eng.*, 83(6), 72–82 (March 15, 1976).

WALTER L. SUTOR

Cottonseed

Introduction

Cotton fibers have been used for textiles for as long as 5000 years. Excavations in Pakistan, Peru, and Mexico have uncovered cotton cloth dated from 4000 to 7000 years old. Cotton growing in the United States developed after the American Revolution and increased rapidly with the invention of the cotton gin and the development of varieties which were suitable to soil and climate conditions.

The southern part of the United States is now the world's greatest cotton producing area. The 1976 cotton planting was estimated at 11,300,000 acres, which produced about 4,800,000 tons of seed. About 5% of the seed is used for planting seed with the balance used by the cottonseed processing industry. India, China, Russia, Brazil, and Egypt also produce significant quantities of cottonseed, and the world production is about 12,000,000 tons.

Cottonseed is the second major oilseed crop in the United States, having been surpassed by soybean production in 1945. The seed is a by-product of the production of cotton. The average yield per acre is 500 lb of fiber and 850 lb of seed. The value of the fiber is nearly five times that of the seed, so the demand for seed has little effect on the planting of cotton.

The cottonseed crushing industry processes the seed to produce approximately 184 lb of short cotton linters, 480 lb of hulls, 340 lb of crude oil, and 947 lb of 41% protein meal per ton of seed. These products may be considered as raw materials for the manufacture of various end products.

The linters are made into low grade fabric and felt, upholstery and mattress padding, and chemical cellulose. The hulls can be used in the production of furfural but have a higher value as a roughage feed for cattle. Cottonseed oil is refined and processed to produce salad and cooking oils, shortening, margarine, and dairy cream substitutes. The high protein meal is blended with other materials to produce stock feeds. Cottonseed flour was produced for a number of years in Fort Worth, Texas and used in the baking industry.

Composition of Cottonseed

Over 90% of the cotton grown in this country is of the so-called upland variety, *Gossypium hirsutum*, and the seeds do not vary widely in type or composition. As received from the gin the seeds contain:

Linters	8–12%
Hulls	30–35%
Kernels	50–55%
Moisture	10–15%

The linters are short cotton fibers not removed by ginning and are about 90% pure cellulose.

Table 1 contains typical analyses of cottonseed hulls and kernels. The hulls contain 20 to 25% of pentosan cellulose which may be converted to the sugar xylose or to furfural by hydrolysis with dilute acid.

Cottonseed protein is lower in lysine and methionine than soybean protein. Most of the protein can be extracted by dilute alkali at a pH of 9 to 10 or by a dilute salt solution. It is also fairly soluble in dilute hydrochloric acid. The minimum solubility is at a pH of 4.8, the isoelectric point. The amino acid composition of cottonseed protein is shown in Table 2.

The water-soluble sugar fraction is composed of approximately 81% raffinose, 12% sucrose, 6% stachyose, and small amounts of fructose and glucose. Raffinose is a trisaccharide which may be hydrolized to give the monosaccharides, fructose, glucose, and galactose.

Gossypol and related color substances are contained in small spherical cells or glands distributed throughout the kernel. The glands are visible to the eye and can be separated by a flotation process or air classification from pulverized oil-free meats.

Gossypol is an oil-soluble complex poly phenolic compound which is highly toxic to swine and nonruminant animals. During the processing of the kernel the cell walls are destroyed by moisture and shearing action and the gossypol either dissolves in the oil or combines with certain amino acids to form a "bound" inert compound.

Glandless, gossypol-free cottonseed has been developed, and the production of this variety is steadily increasing.

Phytin is a salt of inositol hexaphosphoric acid combined with alkali and alkaline earth metals. It becomes soluble as phytic acid in dilute hydrochloric acid and is precipitated by neutralization.

Bailey [2] classifies cottonseed oil in the oleic-linoleic group of oils along with peanut, olive, corn, and other oils. Its average properties and fatty acid analysis are given in Table 3.

TABLE 1   Average Analysis (in %) of Cottonseed Hulls and Kernels [1]

	Hulls	Kernels[a]	
		Glandless	Glanded
Crude fiber	48.5	1.7	1.6
Nitrogen-free extract	35	—	
Protein	3.6	38.9	39.3
Oil	0.9	39.7	37.8
Sugars	—	7.6	7.4
Gossypol	—	0.02	1.2
Phosphorus	—	1.1	0.8
Moisture	12	—	

[a]Dry weight basis

TABLE 2 Approximate Amino Acid Analysis of Cottonseed Protein [1]

Amino Acids	Quantity (g/16 g N)
Lysine	4.4
Histidine	2.6
Ammonia	2.0
Arginine	11.7
Tryptophan	1.2
Cystine	2.4
Aspartic acid	9.1
Threonine	3.2
Serine	4.3
Glutamic acid	20.5
Proline	3.8
Glycine	4.2
Alanine	3.9
Valine	4.5
Methionine	1.4
Isoleucine	3.1
Leucine	6.0
Tyrosine	3.3
Phenylalanine	5.5

Cottonseed oil is somewhat higher in palmitic and linoleic acids than other members of its group. The crude oil contains 2% or more of nonglyceride substances other than free fatty acids. These have been identified as raffinose, pentosans, phytosterols, phytosteroline, peptones, xanthophyll, chlorophyll, gossypol, and mucilaginous substances [5].

TABLE 3 Composition and Properties of Refined Cottonseed Oil [3]

Iodine value	106–113
Saponification value	190–198
Refractive index	1.470
Unsaponifiable matter	1.5%
Fatty acids (%):	
Myristic	1.4
Palmitic	23.4
Stearic	1.1
Arachidic	1.3
Tetradecenoic	0.1
Hexadecenoic	2.0
Oleic	22.9
Linoleic	47.8

Standards and Trading Rules

The cottonseed processing industry of the United States has established standards of quality for cottonseed and its products [4]. Official methods of analysis and calculations enable the oil mill processor to estimate the quantity and quality of products that may be expected from a particular batch of seed.

Prime quality seed that by analysis contains not more than 1.0% foreign matter, 12% moisture, or 1.8% free fatty acids in the oil is given an index of 100.

For seed below prime quality the index is reduced by 0.4 unit for each 0.1% in excess of 1.8% free fatty acid, 0.1 unit for each 0.1% in excess of 1.0% foreign matter, and 0.1 unit for each 0.1% of moisture in excess of 12.0%.

A quantity index is also given. In the case where the value of linters is not considered and the oil content is more than 16.5%, the index is 4 times the percentage of oil plus 6 times the percentage of ammonia plus 5.

For cottonseed containing less than 16.5% oil, the index shall equal 6 times the percentage of oil plus 6 times the percentage of ammonia minus 28.

The seed grade is the quality index times the quantity index. The market price is adjusted by premiums or penalties for variations in seed grade.

Quality standards are also recommended for the primary seed products; that is, the linters, hulls, meal, and oil.

Processing of Cottonseed

The sequential operations of an oil mill include (1) handling and storage, (2) seed cleaning, (3) delinting, (4) hulling and hull separation, (5) meats preparation, and (6) oil expression or extraction.

Some oil mills have facilities for refining the crude oil, and mills using solvent extraction may caustic refine the oil while it is disolved in the solvent. Most cottonseed oil is refined as soon as possible to limit color reversion caused by gossypol. Gossypol acts as an antioxidant but it degrades to a red colored compound which is difficult to remove from the oil. The color reversion does not occur if the gossypol level is below 0.007% [5].

The processing of cottonseed oil into finished products is done at the vegetable oil refinery.

Storage and Handling

Cottonseed is a seasonal crop which is received during a short period of time and must be stored to permit a processing period of up to 11 months.

If the seed temperature and moisture content are high, rapid deterioration occurs due to germination, spontaneous heating, increased mold growth, and insect damage. The result is increased free fatty acids, darkened oil, and lower nutritive value of the meal.

When seed is to be stored at 80°F or above, the maximum safe moisture content is 11 to 12%. High moisture seed is usually processed as quickly as possible.

Nearly all seed storage facilities have aerating systems to help control seed temperature by drawing air through the pile. Some reduction of moisture occurs if the air humidity is low and evaporative cooling occurs. Equilibrium moisture contents of whole cottonseed, meats, hulls, and cake are shown in Table 4.

Storage Facilities

Cottonseed is stored in seed tanks or silos, in varying design wooden structures, in all-metal "Muskogee" type seed houses, or in dry climates may be stored in uncovered piles.

The "Muskogee" house has 12-ft high sidewalls with a roof-pitch of 45°. The seeds have a slightly lower angle of repose which allows room for an air space and a loading belt conveyor above the pile. Storage capacity may be estimated on a density of 25 lb/ft³ although the range may vary from 18 to 32 lb/ft³ depending on the amount of lint and the degree of packing.

Aeration is accomplished by drawing air down through the seed pile to collecting ducts located in the floor. Down-flow of air prevents condensation of moisture at the surface in cool weather.

Design data for aeration systems are scarce. However, Harris [7] has developed a computer program for determining air flow patterns for piles of different shapes, varying permeability, and different duct arrangements. Permeability was defined by a modified form of the Darcy equation $q = K(\Delta p/\Delta x)$ or K (permeability) $= (\Delta x/\Delta p)$, where q is the flow in cubic feet per square foot of cross section, Δx is the length of section in feet, and Δp is the

TABLE 4 Equilibrium Moisture Contents of Whole Cottonseed, Meats, Hulls, and Oil-Free Meats at Various Relative Humidities [6]

Relative Humidity (%)	Moisture Content (wet basis)			
	Whole Seed (%)	Meats (%)	Hulls (%)	Oil-Free Meats (%)
31	6.03	5.13	7.67	8.32
43	7.23	5.92	9.60	9.33
62	9.25	7.73	11.85	12.11
71.2	10.27	8.89	12.62	13.72
81.1	13.21	11.72	15.31	18.29
93.0	22.19	21.40	22.35	32.80

pressure drop in inches of water. It is assumed that the viscosity of air does not vary appreciably with temperature and the flow is laminar.

Permeability tests made in a 67-ft deep pile of cottonseed gave a range of K from 70 at the top to 5.5 at the bottom.

Since the lowest permeability and highest air flow is adjacent to the collecting duct, the entrance area must be large enough to prevent excessive pressure drop. Smith [8] recommends a maximum flow of 10 to 20 ft/min into the duct.

Cleaning

Seed cleaning is important to prevent damage to processing equipment and to avoid contamination of the product, lint, hulls, or meal.

Trash which occurs in mechanically picked cotton may be carried into the seed or the seed may be contaminated with soil scrapings after temporary storage on the ground.

Common impurities are sand and small rocks, cotton burrs, leaves and stems of the plant, wood seeds, burrs, and stray metal.

Cleaning is done after storage since some contamination may occur in storage at the mill. The separation methods include screening, air separation, and magnetic separation. A typical cleaner uses shaking screens to separate oversize and undersize impurities followed by air aspiration to remove light particles. A permanent magnet separator is usually included in the seed flow path.

Cylindrical reel screens containing beaters are also used. They are useful for cleaning sticky seed having excessive residual linters.

Some cleaning may be done with screen-bottom conveyor sections. They may be used as a supplement to other cleaners and are also used between the delinter and huller to separate abrasive sand as well as small black and immature seed.

Practically all screens used in cleaning consist of perforated metal. Wire screens have some tendency to clog and are more difficult to clean. The perforations are usually circular or slotted, and they may be "flanged" or "indented" to assist in separation of sticks. The open area of perforated screens ranges from 25 to 45%. The cleaner capacity is limited by the rate of separation of oversize (boll deck) from the seed where the seed must pass through the screen. Tonnages are reported in the range of 0.5 to 2.0 tons/day per square foot of screen area [9].

Delinting

Removal of most of the lint fibers from cottonseed facilitates the hulling operation and separation of hulls from kernels.

Delinting is performed in a manner similar to that of the ginning process for removal of staple cotton. A standard delinter has 176 fine toothed circular saws, $12\frac{1}{2}$ to $11\frac{1}{2}$ in. in diameter with 330 teeth which project through a slotted plate.

The seed are fed into a roll box which covers the projecting saws and the lint is cut or pulled off. A rotating brush blows the linters from the saws into an air stream which carries the lint to a lint beater where hull pepper is removed. The linters are collected on a rotating cylindrical screen.

The bank of saws is easily removable so that it can be transferred to a saw filing machine when the saws become dull. This is required after 12 to 24 h of operation.

Depending on market demand, one or two lint cuts may be made. The first cut will have longer fibers and will be of better quality than the second cut, and the latter will have more hull particles.

Two other methods of delinting are in commercial use. In one the seed is rubbed against a concave surface of abrasive stone in a closed cylinder. The linters are transferred to a dust separator by an air stream as the seed travels through the cylinder.

In the other method the linters are rendered brittle by treatment with a small amount of sulfuric acid or hydrochloric acid vapor. They are then easily removed by abrasion. This process is used to delint planting seed.

The delinting process is omitted in some mills because of low demand for the fibers.

Decortication or Hulling

The older type of huller, known as the Bauer Mill, is a ribbed disk mill. The seed are fed at the center of the stationary disk and are sheared by the ribs of the disks. The disk faces may be replaced when worn. However, the disk mill has been mostly displaced by the bar huller. It has a greater capacity, cuts the seed better, and has a lower fire hazard.

The bar huller consists of a cylinder studded with parallel steel knives which rotate adjacent to an adjustable set of stationary knives.

The cracked seed pass through a hull separator where the kernels pass through a shaking screen, the hulls are removed by suction near the end of the screen, and the uncracked seeds pass over the end of the screen and are recycled to the huller. Additional meat particles are separated in a hull beater which is a cylindrical screen containing an internal mixer. The amount of hull particles entering the meats stream is usually adjusted to regulate the percentage of protein in the meal product.

Oil Separation

Cottonseed meats contain about 30% oil which is separated by four methods; hydraulic pressing, expeller or screw pressing, direct solvent extraction, and prepressing plus solvent extraction.

When the industry developed in about 1850, hydraulic pressing was the best method for separating the oil. This method, with improved mechanics and

TABLE 5 Residual Oil Left in Meal by Different Extraction Methods (Basis: 1.0 ton of seed)

	Cottonseed (lb)	Soybeans (lb)
Oil-free meal	947	1600
Total oil	372	400
Residual oil in meal:		
Hydraulic press, 5.5%	52.1	93
Expeller, 3.5%	33.1	66.7
Direct solvent extraction, 1.0%	9.5	16.2
Prepress and extraction, 0.6%	5.7	—

conditions, is still in use. However, after World War II labor costs and prices increased, transportation became easier, and competition for seed became greater. The small mills have been gradually forced out and the large mills converted to the more efficient screw pressing or solvent extraction processes.

Table 5 is a comparison of typical extraction results obtained by each of the four methods of oil recovery.

Meats Preparation

It is customary to flake cottonseed meats in a vertical, "five high" set of rolls. In a typical set the top four rolls are 14 in. in diameter by 48 in. long and the bottom roll is 16 in. in diameter. With a peripheral speed of about 630 ft/min, the mill has a rated capacity of 80 tons of cottonseed per 24 h. The seed is fed uniformly between the two top rolls and makes four passes before being discharged.

The optimum seed moisture is 12 to 14% and the flakes are rolled to 0.005 to 0.010 in. The effect of rolling is to rupture the oil cells and gossypol glands. Thin flakes are desirable for hydraulic pressing and solvent extraction of the oil.

Cooking

Proper cooking of cottonseed flakes is necessary for maximum oil yield by either hydraulic or expeller presses. Cooking has the following objectives:

1. To complete rupturing of oil cells
2. To coagulate protein which facilitates the separation of oil
3. To raise the temperature to reduce oil viscosity during pressing
4. To sterilize the material and destroy enzyme action

5. To detoxify gossypol by causing it to combine with protein or carbohydrate
6. To adjust moisture content
7. To fix some phosphatides in the cake

Several types of cookers are in use but the most common is the vertical stacked cooker. This consists of 4 to 6 steam-jacketed vessels placed on top of each other. Each contains a scraper-stirrer mounted on a vertical shaft which extends through the entire stack.

Each vessel has an opening in the bottom and a gate for regulating the flow of material. The gates may be automatically controlled for continuous flow or they may be used for multiple batch operation where the lower vessel is emptied before receiving material from above.

In a typical operation the moisture is maintained at about 14% in the upper vessels and lowered to 10% when the material is discharged. Residence time is about 45 min and the final temperature is 220°F. Cooking takes place at atmospheric pressure although pressure cookers have been used.

Precooking followed by rerolling may also be done before direct solvent extraction. This helps bind gossypol and produces firm flakes for better solvent percolation.

Hydraulic Pressing

The hydraulic press normally contains about 15 rectangular boxes. These are manually loaded with prepared cakes wrapped in press cloth. The loaded press contains about 350 lb of flakes and cooked meats.

The preformed cakes are made by withdrawing a measured amount of flakes from the cooker by a hydraulically operated former buggy. This is deposited on a press cloth in the former bed. Press cloths were formerly made of hair but are now made of nylon. The cloth is folded over the flakes and slight hydraulic pressure preforms the cake so it can be transferred to the press box.

When the press is filled with cakes, hydraulic pressure is applied at a controlled rate until a maximum pressure of 2000 lb/in.^{2} of area is applied. The rate of application is important to obtain a maximum yield of oil. The time of pressing ranges from 30 to 60 min depending on the desired balance between capacity and oil yield.

After pressing, the cakes are removed by hand and the press cloths are stripped off for re-use. The edges of the cakes are trimmed since they are wet with oil. They are then stored or ground into meal.

Expeller or Screw Pressing

Screw presses have higher power requirements than hydraulic presses but they can reduce the residual oil in the cake to around 3.5%. They also have the advantages of continuous flow of material and low labor requirements. However, the pressed oil must be filtered since some solids pass through the drainage slots with the oil.

Continuous pressing is accomplished by the use of a heavy screw with decreasing pitch which forces the material into a slotted cylinder at high pressure. The cylinder is built up of heavy bars separated by spacers which produce slot spacings from 0.005 to 0.025 in. depending on the nature of the material to be processed. A spring-loaded back-pressure valve controls the pressure in the cylinder.

Modern expellers operate at higher speeds and have higher capacities than the older presses and have equal efficiencies. Screw press capacities range from 5 to 460 ton/24 h with horsepower requirements ranging from 1.0 to 4.0 hp/ton of feed capacity [10].

Solvent Extraction

A number of different types of extractors have been developed for solvent extraction of vegetable oils. These are classified into two types: percolation and immersion.

Percolation Extractors

In the percolation types the prepared seed flakes are charged into baskets or chambers with perforated bottoms. The solvent is sprayed over the bed of flakes and drips into collecting basins.

The baskets are conveyed on chains from the charging to discharge points and are dumped by inverting the baskets. An example is the Bollman or Hansa-Muhle extractor in which the baskets are conveyed in a vertical cycle in a large vapor-tight enclosure. Each basket is charged with about 800 lb of flakes as it starts down, and half-concentrated miscella is sprayed in to drip through the baskets in parallel flow on the downward leg. Fresh solvent is sprayed in near the top of the upward leg and drips in counterflow through the ascending baskets.

Horizontal chain and basket extractors, such as the Lurgi, are now preferred over the vertical type. They are more convenient to operate and better control of solvent flow is possible.

Another percolation extractor is the Dravo Rotocell. This is constructed of a number of compartments which move in a circular path around a vertical axis. The bottoms of the compartments are perforated doors which drop open at the discharge point.

The French extractor is similar in configuration except that the compartments are stationary while the charging and discharge mechanisms rotate.

A successful continuous horizontal pan filter extractor was developed for cottonseed by the Southern Regional Research Laboratory of the U.S.

Department of Agriculture. For this extractor the cottonseed flakes are prepared and slurried with concentrated miscella in a mixing conveyor which discharges onto the filter. A vacuum is used to assist flow through the filter, and the filtrate flows into a compartmented collection chamber. A number of pumps are used to spray the liquid back over the bed to obtain countercurrent extraction.

Immersion Extractors

In immersion extractors the solids move down through an upward flowing solvent. Then they are elevated above the solvent level for drainage and discharged into a desolventizer. The Bonatto is an early successful extractor of this type. It consists of a cylindrical tower which contains fixed trays and rotating scrapers. The trays have pie-shaped ports which are staggered so that the solids follow a circular path on each plate before dropping to the next. The scraper speed is adjusted to keep the tower nearly full of solids for maximum retention time. The solids are discharged from a conical bottom into an elevator which is usually of the drag type.

Solvent is introduced either in the elevator or tower bottom and flows upward to discharge through a screen at the top of the tower.

Presently the Direx process uses a combination of percolation and immersion for high oil content materials [11]. The prepared seed is first extracted in a horizontal basket extractor to lower the oil to 10 to 15%. The wet material is then reflaked in a solvent-tight flaker and transferred to the immersion extractor for final extraction.

Recovery of Oil, Meal, and Solvent

It is desirable to reduce the solvent content of the extracted material (marc) as much as possible before it enters the desolventizer. With gravity drainage the minimum retention is about 50% by weight of solvent.

The older but still used desolventizers are called Schneckens. They consist of vertical banks of steam-jacketed ribbon or screw conveyors. Vapor take-offs are located on each section, and direct steam is added to the final sections for improved stripping of the solvent.

The desolventizer-toaster has been developed for large capacity requirements. However, toasting is not needed for cottonseed but is used for soybeans. The machine resembles the stack cooker previously described. It consists of circular steam-heated trays located vertically in a cylindrical shell. Rotating sweep arms attached to a vertical shaft operate with close tolerance to the trays to improve heat transfer. Direct steam is also admitted through the sweep arms to increase solvent stripping, and the moisture is increased to over 20% to assist cooking.

Moisture is then reduced by indirect heating in follow-up units or by hot air drying, and the meal is cooled on cold trays or by means of louvered cylindrical coolers. Pneumatic conveyors may also be used for cooling.

Oil Recovery

The miscella is usually concentrated in a single-stage rising-film evaporator. In large installations heat is saved by operating the evaporator under reduced pressure and using the desolventizer vapors as a source of heat.

Evaporator concentrate contains about 10% solvent which is removed by single- or double-stage steam stripping. The oil is flash dried under vacuum and then cooled.

Minimum temperature and time are used in the oil recovery process to reduce the fixation of a red color due to the presence of gossypol. Miscella refining is also used to remove the gossypol before the oil is heated in the stripping operation. This is discussed in the section entitled "Refining."

Solvent Recovery

Solvent vapor from the desolventizers is scrubbed with a spray of hot water to remove entrained solids before it enters the condensers.

During the desolventizing, evaporation, and stripping operations, some air may leak into the system. This noncondensible gas is purged through a vent recovery system. Solvent may be separated from the vent gas by a refrigerated condenser followed by charcoal adsorption or mineral oil absorption.

Solvents

The standard solvent used for extracting vegetable oils is known as commercial normal hexane. It is a petroleum fraction with a boiling range of 145 to 155°F and is essentially free of sulfur and unsaturated compounds. The solvent is used for the following reasons.

1. It is fairly specific for the oil and extracts a minumum of nontriglyceride materials.
2. The choice of boiling range gives enough volatility for evaporation from the marc and miscella yet the vapor is readily condensed at atmospheric pressure and normal cooling water temperatures.
3. It is chemically inactive and does not impart taste or odor to the oil or meal.

The disadvantage of hexane is its fire and explosion hazard. The industry has well recognized this hazard and the incidence of accidents has been very low.

Trichlorethylene was used in several early soybean extraction plants because of its low flammability. It was found that traces of it left in the oil acted as a poison to the hydrogenation catalyst and that traces left in the meal were extremely toxic to cattle.

Ethanol and isopropanol [12] have also been examined as solvents for cottonseed oil. They have the ability to extract gossypol as well as oil which produces a higher quality meal. However, the added material extracted goes into the oil and increases the difficulty of refining. Other disadvantages are the reduced miscibility of oil and alcohol when water is present and the higher heat requirements for solvent recovery.

Refining

Crude cottonseed oil contains nonglyceride substances which include free fatty acids, phosphatides, gums, gossypol, and other pigments. These must be removed to make the oil suitable for human consumption.

Refining is accomplished by treatment with strong alkali followed by bleaching with clay and finally vacuum deodorization. The term refining is usually applied only to the alkali treatment.

In smaller operations, batch refining is used. A predetermined amount and concentration of sodium hydroxide is mixed vigorously with the oil until the reaction is complete. The insoluble soapstock formed is coagulated by heating with slow stirring and is allowed to settle. The clear oil is then decanted.

During refining the alkali reacts with the free fatty acids and part of the phosphatides to form soap. Meal particles and color bodies are also removed by absorption in the soap phase. Gossypol also combines with strong alkali and is removed.

The optimum adjustment of the variables of alkali quantity, concentration, temperature, and mixing time results in the lowest loss of neutral oil and the best reduction of color during refining. The treatment depends on the free fatty acid and gossypol contents of the crude oil. A sodium hydroxide concentration of $18°$ Be is customary and from 0.15 to 0.20% excess over that required to neutralize the fatty acids is used.

Continuous Refining

In this method, correctly proportioned streams of crude oil and caustic solution are fed into a high-speed mixer for rapid mixing. The mixture is heated in an exchanger to keep the soapstock fluid and it is then separated in a centrifuge. Usually the oil is washed once or twice with water to remove traces of soap and alkali.

Continuous refining has the advantage of less saponification of neutral oil and better separation of oil from the foots.

Soda Ash–Caustic Soda Methods

When the crude oil has a high free fatty acid content, a single-stage caustic treatment gives excessive loss of neutral oil. A two-stage treatment may be used. An excess of 15 to 20° Bé soda ash solution is used to neutralize the fatty acids in the first treatment since the sodium carbonate will not saponify neutral oil. After mixing and heating, the mixture is sprayed into a vacuum chamber to remove carbon dioxide and lower the moisture content. This gives a cleaner separation of the foots in the centrifuge.

The second stage treatment is with a small percentage of 20° Bé caustic. This strong caustic is better for removing color substances. Oil from the first treatment is cooled to about 100°F, mixed with the caustic, and reheated to 150°F before centrifuging.

Miscella Refining

In solvent extraction of cottonseed, considerable gossypol and related pigments are extracted with the oil. During the solvent stripping the oil is heated and the color is "fixed." It is then more difficult to remove the red color during the refining and bleaching operations.

In some plants miscella refining is used to reduce the color problem and lower the refining loss. The low density and viscosity of the hexane–oil solution aids in the separation of oil from the foots.

In this operation part of the hexane is removed before refining to give a 40 to 50% oil solution. Either batch or continuous refining may be used. In either case the soapstock is separated by centrifuging and the miscella is filtered and stripped of hexane to produce a neutral oil.

The crude oil–hexane miscella may be preconditioned with chemicals to assist in the removal of gums [14]. Then it is treated with 18 to 26° Bé caustic in 0.3 to 0.6% excess over the theoretical amount required to neutralize free fatty acids. The mixture is strongly stirred, heated to 140 to 150°F to melt the soapstock, and then cooled to about 120°F for centrifuging.

The refined miscella, after filtering, may be chilled to separate solid fats (winterizing) or precisely hydrogenated in a continuous process before removal of the hexane [13].

Bleaching

Refined oil contains more color than is desired for certain end products. The pigments include carotene, xanthophyll, chlorophyll, gossypol, and other substances. These are reduced and removed by treatment with acid-activated bleaching clay and activated carbon.

The residual color in the bleached oil is measured by matching a $5\frac{1}{4}$-in. column of the oil against the arbitrary standard red and yellow Lovibond glasses.

The treatment also removes certain desirable as well as undesirable substances. The natural antioxidents are removed as well as some odor and flavor substances.

Bleaching is done in either batch or continuous equipment and is conducted under vacuum to eliminate moisture and dissolved air.

From 0.5 to 2.0% of activated clay is added and mixed while the oil is heated from 185 to 195°F. Contact time is 15 to 30 min.

The oil is cooled before contact with air, and the clay is removed in filter presses.

Winterizing

Liquid cottonseed oil products may become cloudy and precipitate solids when stored at cool temperatures. The solids are glycerides of saturated fatty acids.

These are removed by chilling the oil and holding it until crystal growth is complete, and then filtering out the crystals. The addition of seed crystals will speed up the crystalization process. Winterization reduces the content of saturated fats. Approximately 30% of the oil is separated as filter cake with an iodine value of 90 to 95, and the winterized oil has a value of 110 to 115.

Hydrogenation

Hydrogenated vegetable oils have largely supplanted the solid animal fats for shortening and cooking. The first margarine used a mixture of beef tallow, coconut oil, and other refined vegetable oils. Since hydrogenation was introduced early in the twentieth century, it has had a great effect on the edible oil industry.

Hydrogenation is not carried to completion since this would produce too hard a fat. It is desired to have a mixture of triglycerides of a range of melting points to give a product of proper consistency with a low tendency to form large crystals.

Hydrogen reacts with the unsaturated double bonds of the triglycerides, and this raises the melting point and increases stability toward oxidation. Selective hydrogenation refers to the reaction of fatty acids having the greatest number of double bonds reacting first. With cottonseed oil the linoleic acid is converted to oleic acid before the oleic is converted to stearic acid. Selectivity also refers to the suppression of formation of higher melting isooleic acid from oleic acid.

Bailey [14] discusses in detail the effect of variables on hydrogenation. These variables include temperature, hydrogen pressure, agitation, catalyst concentration, and the nature of the catalyst. The catalyst used is finely divided

TABLE 6   Typical Conditions for Margarine and Shortening [15]

		Shortening Sample Number	
Conditions and Analysis	Margarine	1	2
Hydrogenation conditions:			
Temperature, °F	275	218	275
Pressure, lb/in.^{2}gauge	5	50	50
Catalyst, % nickel	0.08	0.07	0.10
Hydrogenation time, min.	230	175	63
Iodine value at completion	67	63.8	62.8
Fatty acid composition of finished product:			
Saturated, %	26.5	32.7	28.8
Isooleic, %	20.8	10.2	14.0
Oleic, %	48.2	50.7	55.4
Linoleic, %	4.5	6.4	1.8
Stability by Swift method, h		75	200

nickel which is prepared by reduction of such compounds as nickel sulfate, carbonate, or formate.

Hydrogenation is almost entirely carried out as a batch operation in a pressure tank equipped with heating or cooling coils and in some cases an agitator. Agitation may also be accomplished by action of the hydrogen gas.

The progress of the reaction is monitored by the refractive index of the oil, which correlates with the degree of unsaturation, or the iodine number, which is also a measurement of unsaturation. Typical conditions for production of hydrogenated shortening and margarine oil are given by Bailey in Table 6.

Deodorizing

Deodorization removes undesirable odors or flavors either occurring naturally in the oil or developed during bleaching and hydrogenation. It is done by either batch or continuous operation, and the flavor and odor bodies, free fatty acids, and small amounts of other nonoil materials are removed by high vacuum steam distillation.

In the batch process a charge of up to 60,000 lb is treated by applying maximum vacuum, injecting superheated steam through spargers, and heating the oil to about 450°F. The treatment may take from 2 to 5 h. However, excessive time will cause the loss of natural antioxidents such as tocopherols.

A problem with batch operation is the reduction of stripping efficiency by the high hydrostatic head and the tendency of steam to form large bubbles at the sparger surface.

A number of continuous deodorizers are in use. These are designed as stripping towers and have the advantage of lower effective hydrostatic pressure and counterflow contact with the steam.

After deodorizing, the oil is cooled and filtered.

References

1. J. T. Lawhon et al., "Evaluation of the Food Use Potential of Sixteen Varieties of Cottonseed," *J. Am. Oil Chem. Soc.*, *54*, 75–80 (1977).
2. A. E. Bailey, *Industrial Oil and Fat Praducts*, Interscience, New York, 1945, pp. 154–155.
3. T. P. Hilditch and L. J. Madison, *Soc. Chem. Ind.*, *59*, 162–168 (1940).
4. *Rules Governing Transactions between Members, Annual*, National Cottonseed Products Association, Memphis, Tennessee.
5. L. R. Watkins, Anderson Clayton Co., Private Communication.
6. M. L. Karon, *J. Am. Oil Chem. Soc.*, *24*, 56–58 (1947).
7. W. B. Harris, "Cottonseed Aeration," Ph.D. Thesis, Colorado State University, 1973.
8. A. Smith, *Oil Mill Gazet.*, *61*(6), 16–18 (1956).
9. S. P. Clark, *Cleaning of Cottonseed*, Texas Engineering Experiment Station, 1976.
10. L. H. Tindale, "Current Equipment for Mechanical Oil Extraction," *J. Am. Oil Chem. Soc.*, *52*, 656–659A (1975).
11. E. D. Milligan, "Survey of Current Solvent Extraction Equipment," *J. Am. Oil Chem. Soc.*, *53*, 286–290 (1976)
12. W. D. Harris et al., "Isopropanol as a Solvent for Extraction of Cottonseed Oil," *J. Am. Oil Chem. Soc.*, *24*, 370 (1947).
13. G. C. Cavanagh, "Miscella Refining," *J. Am. Oil Chem. Soc.*, *53*, 361 (1976).
14. Ref. 2, p. 558.

W. D. HARRIS

Coumarin-Indene Resins (see Resins)

Coumarone-Indene Polymers (see Polymers)